Persistent Phosphors

Woodhead Publishing Series in Electronic and Optical Materials

Persistent Phosphors

From Fundamentals to Applications

Jianrong Qiu
Zhejiang University, Hangzhou, China

Yang Li
Guangdong University of Technology, Guangzhou, China

Yongchao Jia
Universite Catholique de Louvain, Louvain-la-neuve, Belgium

ELSEVIER
WP
WOODHEAD PUBLISHING
An imprint of Elsevier

Woodhead Publishing is an imprint of Elsevier
The Officers' Mess Business Centre, Royston Road, Duxford, CB22 4QH, United Kingdom
50 Hampshire Street, 5th Floor, Cambridge, MA 02139, United States
The Boulevard, Langford Lane, Kidlington, OX5 1GB, United Kingdom

Library of Congress Cataloging-in-Publication Data
A catalog record for this book is available from the Library of Congress

British Library Cataloguing-in-Publication Data
A catalogue record for this book is available from the British Library

ISBN: 978-0-12-818637-4 (print)
ISBN: 978-0-12-818772-2 (online)

For information on all Woodhead publications
visit our website at https://www.elsevier.com/books-and-journals

Publisher: Matthew Deans
Acquisitions Editor: Kayla Dos Santos
Editorial Project Manager: John Leonard
Production Project Manager: Surya Narayanan Jayachandran
Cover Designer: Mark Rogers

Typeset by SPi Global, India

Contents

Contributors

Mohamed A. Ali College of Optical Science and Engineering, Zhejiang University, Hangzhou, China

Xiaofeng Liu School of Materials Science and Engineering, Zhejiang University, Hangzhou, China

Ibrahim Morad Zhejiang University, Hangzhou, China; Suez University, Suez, Egypt

Preface

This book is intended as an introductory level book for graduate students, professional scientists, and engineers interested in persistent phosphors. We hope that the readers have knowledge about luminescence, defect, energy transfer, etc.

Chapter 1 describes the basic concepts related with persistent phosphorescence. We pay special attention to the similarities and differences of fluorescence, phosphorescence, and persistent phosphorescence; defects and traps related with the charge carrier release and storage in persistent phosphors; the luminescent centers including rare earth ions and transition metal ions; and the effects of codoping of ions on the persistent phosphorescence. In addition, the development history of persistent phosphors is described.

In Chapter 2, we introduce various synthesis methods for the control of size, morphology, and homogeneity; the design of functionality; and the optimization of afterglow performance. We focus our discussion on the synthetic methods for persistent phosphorescence nanoparticles, which can be classified into two types: (1) wet-chemical approach, including sol–gel, combustion synthesis, hydro(solvo)thermal, coprecipitation, microwave synthesis, and template methods, and (2) physical process, such as laser ablation and electron beam irradiation processes. The methods and moieties for the functionalization of the surface of the persistent phosphors and synthetic methods for persistent luminescent glasses are also introduced.

In Chapter 3, we review the mechanisms of persistent phosphorescence. A comprehensive discussion is given on the characteristics of traps, including the trap type, trap concentration and distribution, and detrapping probability of captured charge carriers.

In Chapter 4, we introduce experimental techniques together with theoretical approaches based on the first principles and semiempirical methods. Special attention has been paid to the methodology for the determination the nature of luminescent center and trap at atomic scale and their energy levels in the bandgap of the persistent phosphors.

In Chapter 5, various persistent phosphors are introduced, which are activated with luminescent centers of rare earth (RE) ions, transition metal (TM) ions, main group elements, and native crystalline defect. The influence of different hosts on excitation band, emission band, and decay time of persistent phosphors are discussed.

In Chapter 6, we give a detailed description of the defect implanting approaches, modulating principles, and performance tunability through defect engineering.

In Chapter 7, we introduce the applications of persistent phosphors as safety indicators, road traffic signs, luminous decoration, luminous fibers, luminous watches, alternating current light-emitting diode (AC-LED), temperature sensors, photovoltaic materials, and photocatalysts.

Yang Li and Jianrong Qiu wrote Chapters 1, 2, 3, and 6; Yongchao Jia wrote Chapter 4; Mohamed Ali and Jianrong Qiu wrote Chapter 5; Ibrahim Morad and Xiaofeng Liu wrote Chapter 7; and Jianrong Qiu, Xiaofeng Liu, and Yongchao Jia revised and checked the whole manuscript.

I encountered persistent phosphor $SrAl_2O_4:Eu^{2+}$, Dy^{3+} in Kyoto when I was a researcher in Hirao Active Glass Project, Japan Science and Technology Agency in 1997, and was astonished by the performance of the phosphor. I then started to study persistent phosphorescent glasses, glass ceramics, and phosphors. I also found a proverb "It glows only because of defects" during my research. It is really a mysterious field, and many things are still unclear and are waiting to be explored. I hope this book can give readers helpful guidance into this field.

Jianrong Qiu
July 1, 2020
Hangzhou, China

Introduction

1

1.1 Definitions of optical phenomena

1.1.1 Fluorescence

This book deals with the persistent phosphor, which shows a unique luminescence phenomenon. Usually the luminescence is classified into fluorescence, phosphorescence, and persistent phosphorescence [1]. When the luminescent materials are irradiated by high-energy particles, such as photons, protons, or electrons, they emit photons with energies lower than that of the excitation photons, resulting in the occurrence of fluorescence. Fluorescence is the inversed process of absorption where the excited electrons return to their ground state along with a spontaneous emission of photons. Its decay lifetime can be down to 10 ns. Therefore the luminescence almost stops immediately after removal of the excitation [2].

Depending on the type of excitation source, fluorescence can be divided into several categories, such as photoluminescence excited by electromagnetic radiation (usually ultraviolet light), cathodoluminescence excited by high-energy electron beams, and electroluminescence by electric field stimulation. Notably, thermoluminescence is a reluminescence process characterized by the release of energy accumulated upon heating, regardless of the excitation source.

1.1.2 Phosphorescence

Unlike fluorescence, phosphorescence is a slow process that is associated with a "forbidden" process (with respect to the selection rules) in which the optical centers relax from the excited state to the ground state. In phosphorescence a typical process is that the electron that absorbs the photon (energy) undergoes an unusual intersystem crossing into an energy state of higher spin multiplicity, usually a triplet state. As a result the excited electron can be trapped in the triplet state, from which the return to the lower energy singlet state is not allowed by selection rules. These transitions, although "forbidden," will still occur in quantum mechanics but are kinetically unfavored and thus progress at significantly long timescales. Therefore phosphorescent material does not immediately reemit the radiation that it absorbs, but reemits at a much lower intensity and has a longer emission lifetime in the range of about 10^{-6} to 1 s. Thus the difference between fluorescence and phosphorescence is the transition nature, which leads to the different decay lifetimes. Any type of phosphorescence can show its decay process in the form of luminescence [3].

Persistent Phosphors. https://doi.org/10.1016/B978-0-12-818637-4.00001-X

1.1.3 Persistent phosphorescence

Persistent phosphorescence is an interesting optical phenomenon, whereby a material is excited with high-energy radiation (typically ultraviolet light, x-rays and beta rays, etc.) and the resulting emission decays very slowly for an appreciable time—from seconds to many hours after the excitation source is removed. It is also called long persistent luminescence or long-lasting phosphorescence. Generally the slow liberation of trapped charge carriers by thermal excitation results in the persistent phosphorescence. Therefore it is worthy to note that the mechanisms of persistent phosphorescence and phosphorescence are totally different [4].

The intensity and lifetime of the persistent phosphorescence are the two important parameters to evaluate the persistent phosphorescence properties of materials [5]. The lifetime of persistent phosphorescence is defined as the time length between the stopping of the excitation and the dropped of the luminescence intensity to 0.32 mcd/m^2, which is 100 times higher than the sensitivity limit of human eyes in the dark. Commercial persistent phosphors have very high intensity of persistent phosphorescence and long persistent duration. Up to the present a myriad of long persistent materials have been developed, and the number of new materials is growing year by year. For instance, blue persistent phosphor $CaAl_2O_4$:Eu^{2+} and Nd^{3+}; green persistent phosphor $SrAl_2O_4$:Eu^{2+} and Dy^{3+}; red persistent phosphor Y_2O_2S:Eu^{3+}, Mg^{2+}, Ti^{4+}; and near-infrared persistent phosphor $ZnGa_2O_4$:Cr^{3+} have been developed in recent years [6–9].

1.2 Defects

Defect usually means imperfection and should be avoided during material design, but defect plays indispensable roles in persistent phosphorescence. In an ideal and perfect crystal, all atoms are located strictly in regular, periodic lattice sites. However, the arrangement of atoms is not perfect in real crystals, and some atoms are not always located at their ideal sites in a three-dimensional crystal structure. The regions in the crystal structure that are different from the ideal lattice structure are considered as the structural defect of the crystal. Defects in the real crystals include thermal defect, point defect, vacancy, interstitial atom, and substitution defects [10].

1.2.1 Thermal defects

The generation and movement of vacancy and interstitial atoms are mainly dependent on the thermal fluctuation of atoms. Thermal defects, also known as intrinsic defects, are generated by thermal vibration of atoms in the crystal lattice when the temperature is higher than the absolute temperature. Some atoms, which have a high thermal energy, move away from their equilibrium position, resulting in the formation of defects. There are two types of intrinsic defects: Frenkel and Schottky defects [11, 12].

Frenkel defect: A atom jumps from the ordinary lattice site to the interstitial position due to thermal fluctuations, creating a vacancy and an interstitial atom at the same time. The features of Frenkel defect are that vacancy and interstitial atoms are generated simultaneously with the same numbers, such that the crystal volume and density remain unchanged, as shown in Fig. 1.1A [13].

Schottky defect: An atom at its ordinary lattice site, due to large thermal vibration, squeezes out the nearest neighbor atom and occupies the nearest neighbor lattice site. In the earlier process a vacancy is formed in the original position of the atom, and this process results in the formation of Schottky defects. Similar as a row of elastic balls, the most superficial atoms eventually shift to a new position. The characteristics of Schottky defect are that new atomic layer is added to the crystal surface, and only vacancy defect is found in the bulk of the crystal; therefore the crystal expands, and the density decreases, as shown in Fig. 1.1B [13].

1.2.2 Point defects

The process of persistent phosphorescence occurs through a sequence of electronic transitions between the energy bands of the host lattice and energy states of dopants, as well as the trap in intrinsic defects. The emission wavelength of activators strongly depends on the energy levels of the emitters, as well as the influence of surrounding crystal field provided by the host crystal. The region where the crystal structure deviates from the ideal lattice structure is usually called the structural defect in the crystal [14].

Intrinsic defects, or native defects, relate only with the atoms in a perfect crystal. The simplest example of a native defect is a missing atom, or vacancy. Extrinsic defects, in contrast, refer to impurity atoms that are foreign to the crystal. Dopants are usually described as impurity atoms that are introduced intentionally. A defect complex is a combination of two or more impurities or native defects (e.g., the boron–hydrogen complex in silicon and VN center).

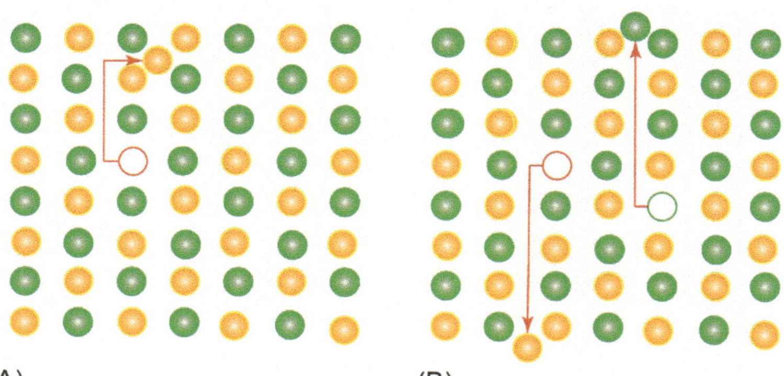

Fig. 1.1 The formation process of (A) Frenkel defect and (B) Schottky defect. Reproduced with permission from K.W. Böer, Handbook of the Physics of Thin-Film Solar Cells. Springer Science & Business, 2014, Copyright Springer.

Three different types of native defects are of importance in most inorganic crystals. They are lattice vacancy, interstitial atom, and substitution defects [12].

Vacancy. Atomic vacancies are usually present in real crystals, as shown in Fig. 1.2A. This kind of defect is known as a vacancy. The vacancy defect is the simplest, most common defect in real crystals. This defect plays an important role in the electrical properties, optical properties, and high-temperature dynamic process of materials.

Interstice atom. Foreign impurity atoms with a size smaller than the constituent atoms of the crystals are likely to enter into the interstitial site of the lattice, resulting in the formation of interstitial impurities, as shown in Fig. 1.2B.

Substitution defect. Foreign dopant atoms replace the native atoms and occupy their lattice sites, resulting in the formation of substitutional defect in which the foreign atom may have a different size or oxidation state, as shown in Fig. 1.2C and D.

Now we take the typical III–V semiconductor GaAs as an example to illustrate these point defects. A gallium vacancy is denoted as V_{Ga} in GaAs. In the GaAs crystal, there is a new kind of defect, the antisite (e.g., a gallium atom residing on an arsenic site, Ga_{As}, or an arsenic atom sitting on a gallium site, As_{Ga}). Deviations from exact stoichiometry lead to the formation of large concentrations of interstitials and antisites

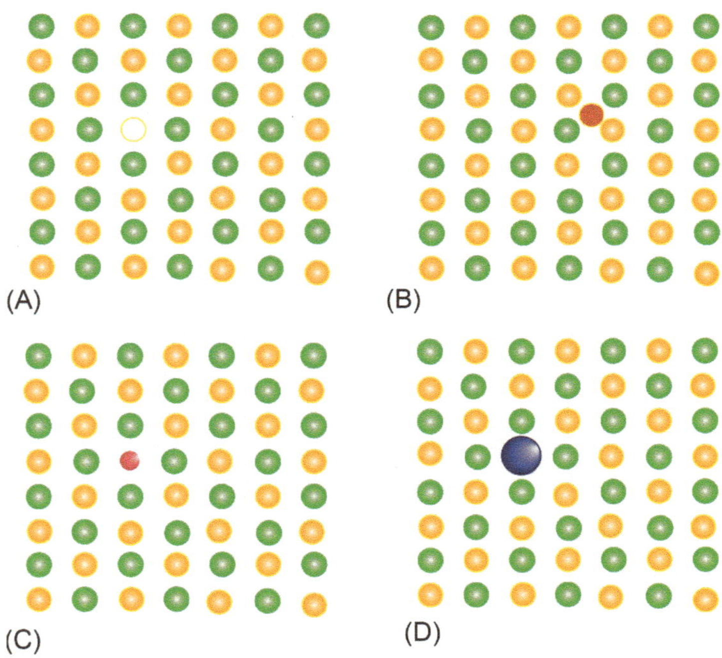

Fig. 1.2 Point defects: (A) vacancy, (B) interstitial atom, (C) small substitutional atom, and (D) large substitutional atom (all of these defects disrupt the perfect arrangement of the surrounding atoms).
Reproduced with permission from K.W. Böer, Handbook of the Physics of Thin-Film Solar Cells. Springer Science & Business, 2014, Copyright 2013, Springer.

of the excess constituent. Moreover, impurities can enter into different lattice sites depending on their size, oxidation state, and electronegativity. In GaAs, for example, silicon can substitute for gallium or arsenic sites, which are denoted as GaAs/Si_{Ga} or GaAs/Si_{As}, respectively. A site that is not substitutional is called interstitial and is identified by the subscript i. Interstitial copper in GaAs, for example, is denoted as Cu_i [15].

1.3 Traps

Generally speaking, traps, which originate from the aforementioned lattice defects, are responsible for the storage and release of charge carriers [8, 14]. It plays a dominant role in all the persistent luminescence process. Traps are characterized by their depth inside the bandgap, which is determined by the energy difference between the energy level of the trap and the conduction band (in the case of electron traps) or the valence band (in the case of hole traps). Generally, shallow traps, whose depth is lower than around 0.4 eV [16], are fully emptied at low temperatures and do not actively take part in persistent luminescence at room temperature. On the other hand, very deep traps, whose depths are around 2 eV or larger, require more energy to be emptied at room temperature. A trap depth of around 0.6–0.7 eV is often considered to be an ideal trap for persistent luminescence [17].

1.3.1 Trap depths

The replacement of atoms in a persistent phosphor with substitutional impurities, interstitial impurities, intrinsic defects, complexes between point defects, and dislocations can bring new energy states inside the bandgap of the host matrix. The creation of these energy states, also called energy levels, is the major reason why persistent phosphors have gained enormous scientific and technical importance.

In persistent phosphors, it is generally accepted that after excitation, charge carriers could be captured for a long duration by the "traps," which create energy levels inside the bandgap. The captured charge carriers are stored in the traps, until the thermal excitation is strong enough to allow the slow liberation of trapped charge carriers. The released electrons can then return to the activators and produce luminescence [18].

Generally, though not universally, it is accepted that the traps are associated with the lattice defects mentioned earlier, that is, intrinsic defects, intentionally introduced defects, or both. The main trap levels are shallow traps and deep traps. The trap depth determines the release rate of the electrons captured in traps. Shallow traps are easily emptied at room temperature with a fast release rate. Deep traps with the slower release rate, on the other hand, are difficult to evacuate at room temperature in short time. A fraction of electrons are even stored in deep traps under the room temperature. A high release rate is beneficial to obtain a strong phosphorescent intensity but will quickly exhaust the stored energy. In contrast a slow release rate leads to a weaker phosphorescent intensity but with a slower decay rate.

Actually the deep trap energy states are located far from the bandgap edges. Such impurities do not generate a hydrogenic ground state because the carriers are bound strongly at a short distance from the impurity core. Due to this localization the averaging of the ground state wave function over a large volume does not occur. For these reasons, deep traps strongly depend on the particular details of the potential near the impurity core. There exists no simple model that accurately describes deep levels. Impurities that generate deep levels in the bandgap typically do not fit well into the crystal, generating significant local strains. The localized levels can trap carriers, impeding conduction. Historically, deep levels and shallow levels were always defined according to their proximity to the band edges. More recently, however, localized states have been discovered that happen to lie close to the conduction or valence band, and deep states can act as dopants. It is therefore useful to define a defect as localized or delocalized. In literature, "deep" generally refers to a defect wave function that is localized in real space.

1.3.2 Trap examples

The deep traps control the position of the Fermi level of semiinsulating persistent phosphors. The center of trap energy band has a number of unusual properties that are related, in part, to the change of its relative positions. This process is now known to occur for a large number of deep-level centers in various persistent phosphors.

In extreme cases the lattice relaxation can be strong enough to allow a center to bind two electrons and assume a lower total energy than binding one electron, despite the fact that two equally charged carriers repel each other. Such defects are called negative-U, where U refers to the Coulomb repulsion term in the Anderson model. One example of a negative-U center is the oxygen vacancy (V_O) in ZnO, which is a deep trap. An illustration of this vacancy is shown in Fig. 1.3. When an oxygen atom is removed from the lattice, two electrons are left in the vacancy. These electrons are not part of a Zn—O bond, so they are dangling bonds. This is the neutral charge state, V_O^0. If the Fermi level is sufficiently low, then the V_O defect will give up two electrons, becoming V_O^{2+}. The V_O^+ is not thermodynamically stable; its energy is always higher than V_O or V_O^{2+} for the whole Fermi level in the bandgap region [19].

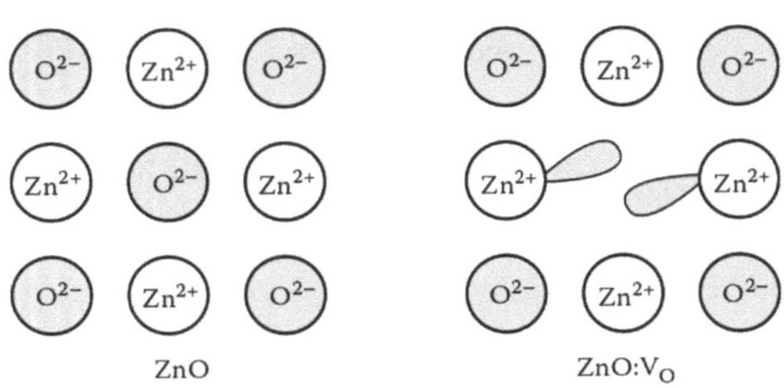

Fig. 1.3 Ionic picture of an oxygen vacancy in ZnO. The removal of an oxygen atom leaves two dangling bonds.
Reproduced with permission from J. Pelleg, Defects in Materials. Diffusion in Ceramics. Springer, Cham, 2016, Copyright 2016, Springer.

In persistent phosphors, anion vacancies are generally donors, while cation vacancies are acceptors. These native defects are important as compensating centers. The formation of vacancies introduces acceptor levels into the bandgap, enabling electrons to reduce their energy by several eV. This reduction in energy encourages the formation of vacancies. The formation energy of electrically active intrinsic defects depends on the position of the Fermi level [20].

Vacancies also form complexes with hydrogen impurities. The hydrogen atoms can be usually captured by dangling bonds, which is a process known as passivation. By providing an electron to form a saturated bond, hydrogen passivation removes the energy level associated with the dangling bond from the bandgap. Hydrogen also forms complexes with cation vacancies in compound semiconductors. Nevertheless, hydrogen can also occupy anion vacancies and act as shallow donors, which has been confirmed in several persistent phosphors.

Interstitials are usually generated through electron irradiation at cryogenic temperatures. At room temperature, as in the case of vacancies, interstitials tend to diffuse rapidly until they reach another defect or the crystal surface. In addition to the high-symmetric interstitial sites, a split interstitial in which two atoms share a single lattice site is known as an interstitialcy. Intrinsic point defects can agglomerate into larger, extended defects. When interstitials condense, forming an extra section of a lattice plane, we obtain an extrinsic stacking fault, which are surrounded by dislocations. Analogously the condensation of vacancies results in the formation of intrinsic stacking faults [21].

1.3.3 Isoelectronic traps

Isoelectronic traps have the same number of valence electrons as the atoms they replace. If there is no significant lattice strain, the isoelectronic trap is electrically neutral. This kind of traps is usually found in the persistent phosphors with equivalent substitution during the doping design. Such as Cr^{3+} ions can substitute the octahedral lattice of Ga in Cr^{3+}-doping $ZnGa_2O_4$ materials, generating Cr^{3+}-Ga^{3+} trap cluster [4]. Actually, not so many isoelectronic traps are discovered in persistent phosphors, and now the mainly researches focus at semiconductor and metal materials. For example, in silicon, carbon is an isoelectronic impurity if it replaces Si sites. Because of its small size, it produces considerable local strain. One way to relax this strain is to have a pair of carbon atoms occupy one silicon site, similar to the case of split interstitial. Although carbon by itself does not produce any energy levels in the gap, it participates in many defect reactions that occur at high temperatures. The X center, for example, is a defect pair in silicon with a group-III acceptor on one substitutional site and a carbon on a neighboring substitutional site. The acceptor level is slightly shallower than that of the isolated group-III acceptor.

1.3.4 Doping traps

Ion doping has been a preferred strategy in the creation of traps in persistent phosphors, and it gives great freedom in the choice of the trap species and depth. In contrast to free diffusion, ion doping is a process far from equilibrium. The ions are stopped by various interactions, some of which lead to the displacement of host material atoms.

The dopant concentration profile is dependent on thermal equilibrium for a particular dopant/solid combination. Ion doping is a nonequilibrium process, the dopant atoms being driven into the solid by kinetic energy. As doped, the atoms are not necessarily on their substitutional sites. In practice, shallow implants are used to produce a well-defined Gaussian doping profile [22].

Dopants must occupy the correct substitutional lattice position to become electrically active carriers. Heating, or thermal activation, is the principal process used to move impurities into their substitutional sites. In all discussions of dopant activation, it should be remembered that the solid solubility of any particular impurity could not be exceeded for equilibrium conditions. If the doping leads to concentrations larger than the solubility limit, then precipitation and cluster formation will occur during thermal annealing [23].

1.4 Luminescent centers

1.4.1 Coordination geometry of isolated ions in solids

Luminescent centers, such as transition metal (TM) and lanthanide (Ln) ions, can emit light by d-d and f-f transitions and interband f-d transitions. Since the d electrons of TM and Ln ions are very sensitive to the surrounding environment, the energy levels of these ions can vary significantly when they are doped in different hosts. In contrast, due to the shielding effect of the outer s and p orbitals, the energy of 4f states of Ln ions are almost independent of the host material. However, the coordination environment can still control the emission of Ln ions by the constraints of the selection rules [24].

The effect of the host on the emitter's energy levels can be described by crystal field theory (CFT). CFT is a classic model of the electronic structure commonly used to estimate ligand-stabilized transition metal complexes. In CFT a metal ion is regarded as a "free ion," and a ligand is considered a point charge. It is assumed that the orbitals of the metal ion have negligible overlap with that of the ligand. Consequently the luminescence of a complex, such as a TM complex, mainly depends on the character of energy levels according to the following factors [25, 26]:

(1) the nature of transition metal ions, in particular the number of d electrons;
(2) the spatial arrangement and distance of ligands around the central metal ion;
(3) the nature of the ligand: the stronger the ligand field strength, the greater the energy difference between the high-energy state and the low energy state.

Fig. 1.4 presents spatial orientations of the five degenerate d orbitals. If TM/Ln ions are placed at the center of a spherically symmetric negative electric field, the repulsive action of the d orbital is the same; therefore the d orbital energy is elevated with respect to the energy in the "free atom" state but without splitting. On the other hand, in real crystals, the metal ions are placed in a nonspherical negative electric field, and the d orbitals become nondegenerate. The splitting of energy level depends on the local geometry where the metal ions are located, such as octahedron, tetrahedron, and planar square [27].

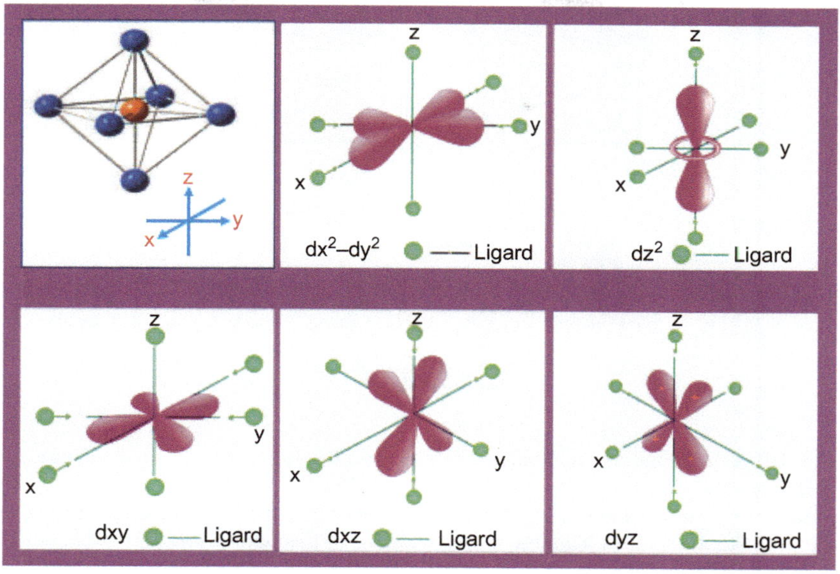

Fig. 1.4 Partial orientations of d orbitals in octahedronal coordination environment. Reproduced with permission from L.E. Orgel, Spectra of transition-metal complexes. J. Chem. Phys. 1955, 23, 1004–1014, Copyright 1955, American Institute of Physics.

Octahedron: Fig. 1.4 shows a reprehensive example about the d orbitals in the octahedral coordination environment that six ligands are distributed along the positive and negative directions of the x, y, and z axes. Due to the anisotropic spatial distribution of the electric field associated with the crystal field, the degree of energy increase is different in each direction. In this figure the coordinating ligand with the dz^2, dx^2-y^2 orbital's head meets, and the collision force is relatively close, resulting in larger increase in orbital energy compared with other orbitals. In comparison the coordination ligands with dxy, dxz, and dyz are farther from the side interaction distance; thus the orbital energy is lower [28].

Tetrahedron: In the tetrahedron coordination the five *d* orbits do not collide in the head-on fashion. The heads of dz^2 and dx^2-y^2 orbitals both point to the face center, while the d_{xy}, d_{xz}, and d_{yz} orbitals all point to the middle of each edge of the cube. As shown in Fig. 1.5, the face to the vertices is the face of the right triangle, and the point from the edge to the vertex is the right side of the right triangle. The latter is closer to the ligand than the former; therefore the former has a smaller repulsive force than the latter [29].

Fig. 1.6 presents the splitting of d orbitals and their relative ligand field strength (Δ) values in octahedral and tetrahedral complexes. Quantitative calculations have been developed to determine the energy levels of a d^n system in which their ligand fields strength vary from weak to strong. Based on Tanabe-Sugano diagram, the energy levels of the TM ions in different host can be calculated [30]. In this diagram the energies of d orbitals are plotted against Δ. The energy unit is the Racah parameter B.

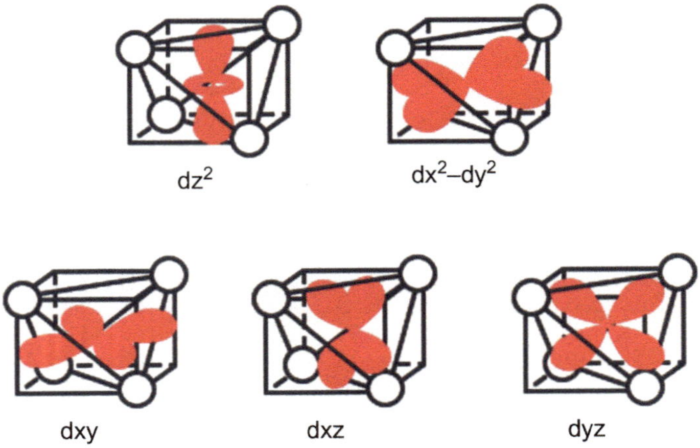

Fig. 1.5 Different partial orientations of d orbitals overlapping with the ligands in tetrahedron environment.
Reproduced with permission from L.E. Orgel, Spectra of transition-metal complexes. J. Chem. Phys. 23 (1955), 1004–1014, Copyright 1955, American Institute of Physics.

The Tanabe-Sugano diagram can be used for complexes with different B values in an octahedral coordination.

Two parameters B and C are needed to describe the energy levels of the d electron system with known d electron numbers. In the Tanabe-Sugano diagram, each graph the ratio of C/B only affects the relative energy of the excited states with different spin multiplicities and ground states. It is worth noting that the Tanabe-Sugano diagram is calculated for the O_h symmetric ligand field. If the C/B ratio varies with the number of d electrons, the same graph can be used for the tetrahedral ligand field (d^{10-n} equivalent to d^n). The use of this energy-splitting map allows for prediction of the optical properties of TM-based coordination compounds [31].

1.4.2 Lanthanide ions

Trivalent lanthanide ions feature a Xe core electronic configuration with n 4f electrons, in which n varies from 0 (for La(III)) to 14 (for Lu(III)). Lanthanide ions are characterized by their rich $4f^n$ levels, which endows Ln-doped materials with intriguing optical properties that are attractive for applications in diverse fields, such as lasers and phosphors. For free (gaseous) Ln^{3+} ions, the energy level of each electron configuration split into a series of energy levels due to the interaction of the electron orbital magnetic moment with the magnetic moment [1, 2]. These states can be described by the multistate $^{2S+1}L_J$ consisting of the $(2S + 1)$ $(2L + 1)$ state under the Russell-Saunders coupling scheme, where S and L represent the total electron spin and orbital angular momentum, respectively.

The values of $L = 0, 1, 2, 3, 4, 5, 6, 7$, and 8 are used to be designated by the letters S, P, D, F, G, H, I, K, and L in spectroscopic term. Due to the coupling of the orbital

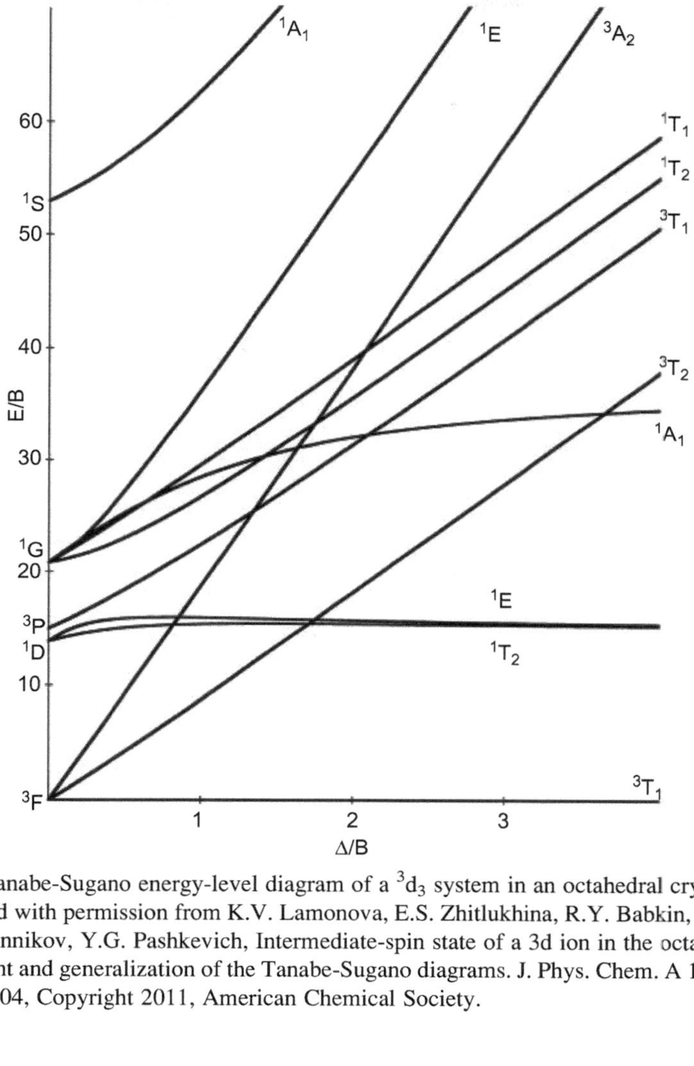

Fig. 1.6 Tanabe-Sugano energy-level diagram of a $^3d_3$ system in an octahedral crystal field. Reproduced with permission from K.V. Lamonova, E.S. Zhitlukhina, R.Y. Babkin, S.M. Orel, S.G. Ovchinnikov, Y.G. Pashkevich, Intermediate-spin state of a 3d ion in the octahedral environment and generalization of the Tanabe-Sugano diagrams. J. Phys. Chem. A 115 (2011), 13596–13604, Copyright 2011, American Chemical Society.

angular momentum and the spin angular momentum in quantum mechanics, these multiplets ^{2S+1}L are usually composed of total 2 J + 1 possible states, where J is the total (i.e., vector sum of S and L) angular momenta.

For lanthanide ions, their 4f electron shell is not fully filled; meanwhile, electrons in 4f, 5d, and 6 s configurations have similar energies, making their energy levels extremely complicated. Lower energy $4f^n$, $4f^{n-1}6\,s^1$, and $4f^{n-1}6p^1$ group state produces a number of energy levels. Nevertheless, in lanthanide ions' electronic configurations, $4f^n$ is the lowest energy configuration and is therefore the most important in the study of spectral properties. Lanthanide ions have an extremely complex line-shaped spectrum due to its unfilled 4f electron shell and that 4f electrons are shielded by 5 s and 5p electrons. For example, Pr^{3+} $4f^2$ and Nd^{3+} $4f^3$ configuration have 13 and 41 J energy levels, respectively, as shown in Fig. 1.7 [31].

The absorption spectrum of lanthanide ions can be differentiated into three types: interlevel f-f and f-d transitions and charge transfer absorption. The intensity of

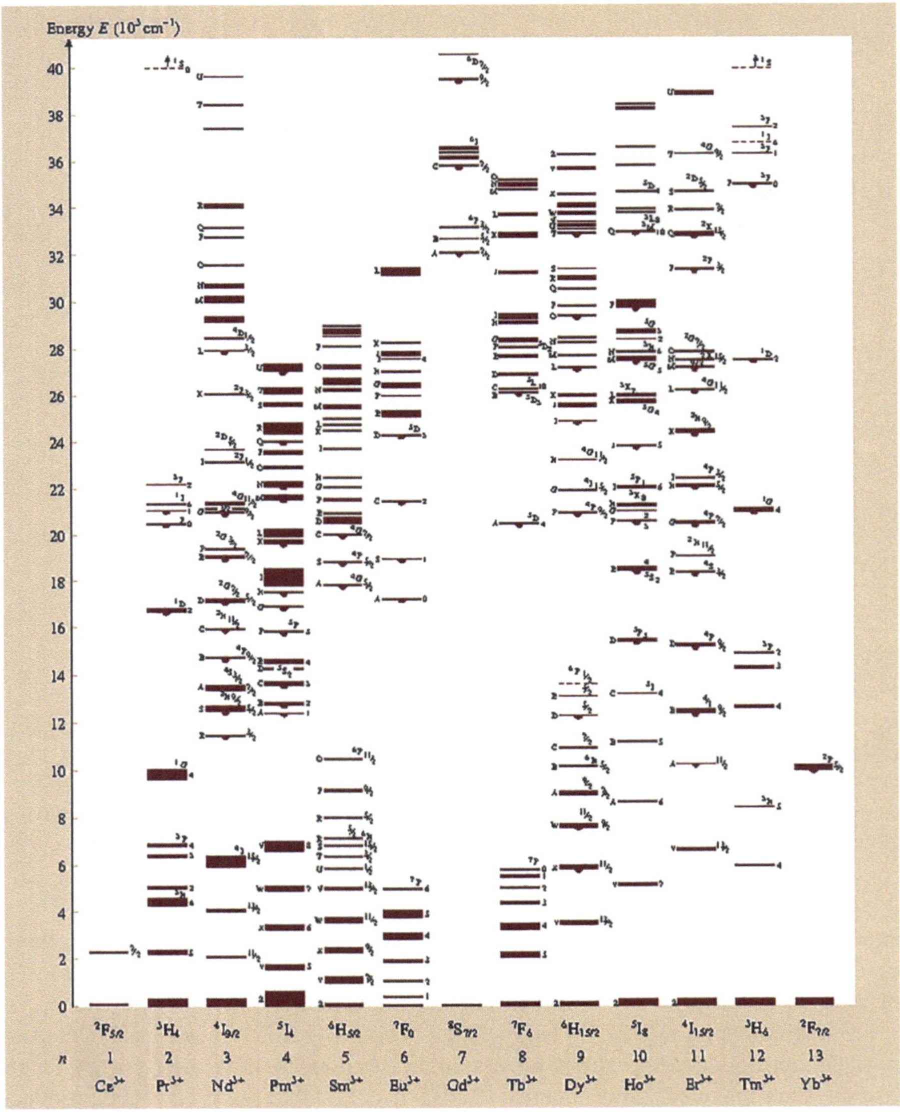

Fig. 1.7 Energy level diagram of trivalent lanthanide ions.
From P.S. Peijzel, A. Meijerink, R.T. Wegh, M.F. Reid, G.W. Burdick, A complete $4f^n$ energy level diagram for all trivalent lanthanide ions, J. Solid State Chem. 178 (2) (2005) 448–453.

absorption spectrum mainly depends on the transition strength and transition type. The transition strength is determined by the possibility of transition between energy states, that is, the oscillator strength (*f*), which can be expressed as follows [32]:

$$f = 4.32 \times 10^{-9} \int \varepsilon_i(\sigma) d\sigma \tag{1.1}$$

where oscillator strength (*f*) is a dimensionless quantity; ε_i is the absorption rate at energy (cm^{-1}), that is, the molar extinction coefficient; and σ is the energy. f is equal to the integral of ε_i on energy σ. Generally the larger the molar extinction coefficient is, the larger the vibration strength (*f*) of the band will be; meanwhile the stronger the transition intensity is, the stronger the absorption band will be. The optical transitions can be divided into the following types [33]: electric dipole transition, magnetic dipole transition, and electric quadrupole transition. The maximum oscillator strength (*f*) of electric dipole transition is typically 10^{-5}–10^{-6}, while the intensities of the other two transitions are weak. The f values of magnetic dipole and electric quadrupole transition are in the range of 10^{-8} and 10^{-11}. The f value of electric quadrupole transition is so small that the corresponding absorption is hardly observed. Therefore the observable transition is caused by the interaction of photon with electric and magnetic dipoles, that is, $f_{total} = f_e + f_m$ [34]. In other words the intensity of the absorption band is determined by the transition of the electric and magnetic dipoles. If electric dipole transition is not allowed by selection rules, magnetic dipole transition and electric quadrupole transition become important.

Irrespective of the transition types, selection rules are valid in all the materials. The selection rules can be summarized as follows (Table 1.1) [2]:

ΔJ is the difference between the J values of the two spectral branches, ΔL is the difference between the L values of the two spectral branches, and ΔS is the difference between the S values of the two spectral branches. The parity selection rule can be always broken, and some restrictions are lifted due to the influence of the external field and the internal field.

The spin selection rule requires that $\Delta S = 0$. The validity of the rule depends on the degree of strength of the spin-orbit coupling of the transitions. Since rare earth element has a strong spin-orbit interaction, some transitions of the rare earth ions do not strictly obey the spin selection rule, that is, $\Delta S \neq 0$ can also occur for some transitions.

The parity selection rules are generally strict, but they are often lifted for f-f transitions of lanthanide ions. Since the f-f transition of lanthanide ions is a transition

Table 1.1 Summary of selection rules of lanthanide ions.

Transition type	**ΔJ**	**ΔL**	**Parity transition**	**ΔS**
Electric dipole	$0, \pm 1, J_1 + J_2 \geqq 1$	$0, \pm 1$	Yes	0
Magnetic dipole	$0, \pm 1, J_1 + J_2 \geqq 1$	$0, \pm 1$	No	0
Electric quadrupole	$0, \pm 1, \pm, J_1 + J_2 \geqq 2$	$0, \pm 1, \pm 2$	No	0

within the $4f^n$ configuration, the parity of the initial state and the final state is the same, so the electric dipole is strictly forbidden, and the magnetic dipole is allowed. However, due to the symmetry of the metal ions' coordination environment and the vibrational movement of the crystal lattice in the complex and the crystal, different configurations of the opposite parity are mixed into the f^n configuration, resulting in the partial reduction of the parity selection rule. In this case the f-f transitions of lanthanide ions are usually known as the induced electric dipole or forced electric dipole transition. Their intensity is weaker than the allowed *f-d* electric dipole, but it is still 1–2 orders of magnitude stronger than that of the magnetic dipole transitions of the f^n configurations. [35]

Typically, trivalent lanthanide-doped complexes exhibit line-shaped emission because the $4f^n$ subshells are shielded by $5s^2$ and $5p^6$ subshells. Therefore the most important Hamiltonian for determining the 4f energy level is the atomic number, and the positional pattern of the spectral term of each lanthanide ion is mainly determined by the spin-orbit interaction of electrons. Therefore their luminescence properties are similar to those of free ions. However, when lanthanide ions are introduced into the crystal lattice, the crystal field can distort the partially filled 4f electron. The multiplets then split and shifted due to the interaction of the localized 4f electrons with the electric field of the crystal. The number of energy levels is determined by the J-term, and the splitting depends on the crystal field strength. Hereinafter, we mainly introduce electronic configurations and spectroscopic characteristics of lanthanide ions [36, 37].

(a) Gd^{3+} ($4f^7$): The 4f layer of Gd^{3+} is half-filled, producing a very stable $^8S_{7/2}$ ground state. The first excited state is higher than 32,000 cm^{-1}. Therefore the emission of Gd^{3+} is located in the ultraviolet region, and the $^8S_{7/2}$ ground state does not split in a crystal field. The low-temperature emission spectrum presents only one emission line, originating from the transition from the lowest excited level of $^6P_{7/2}$ to the ground state of $^8S_{7/2}$. The $4f^65d^1$ state of Gd^{3+} is located in a relative high energy state, so the $4f^65d^1 \rightarrow 4f^7$ band transition emission is usually not observed. The luminescence of Gd^{3+} ion is mainly due to the $^6P \rightarrow {}^8S$ transition, which is located at 313 nm. A weaker electronic vibrational level can be observed at an energy lower than the emission energy of $^6P_{7/2} \rightarrow {}^8S_{7/2}$ transition, which is usually interpreted as a phonon subband [38].

(b) Eu^{3+} ($4f^6$): Usually the emission lines of Eu^{3+} are located at the red spectral region. Eu^{3+}-doped red phosphors have found important applications in lighting and display technologies. These emission lines of Eu^{3+} usually come from the transitions from the excited state $^5D_0$ level to the ground levels of ^7F_J (J = 0, 1, 2, 3, 4, 5, and 6) in the $4f^6$ electronic configuration. The $^5D_0 \rightarrow {}^7F_J$ emission of Eu^{3+} is very suitable for probing the site symmetry of the host. If the lattice sites occupied by Eu^{3+} ions have inversion symmetry, the electronic transitions of the $4f^n$ levels are parity forbidden. In this case very weak magnetic dipole transitions could occur. If Eu^{3+} ions are in the lattice position without inversion symmetry, the odd field term of the crystal field can mix with the opposite parity state into the $4f^n$ configuration level. At this point the electric dipole transition is no longer strictly forbidden, and a weak spectral line appears in the spectrum due to forced electric dipole transition. Therefore the $^5D_0 \rightarrow {}^7F_2$ transition is hypersensitive, while the $^5D_0 \rightarrow {}^7F_1$ transition is insensitive to the crystal field environment [39–41].

(c) Tb^{3+} ($4f^8$): Tb^{3+} ion with a $4f^8$ configuration has complicated energy levels and various possible transitions between 4f → 4f levels. The most important line emissions of Tb^{3+} originate from the $^5D_4/^5D_3 \rightarrow {}^7F_J$ ($J = 6, 5, 4, 3$, and 2) transitions, mainly located in the green spectral region. The line emissions of $^5D_3$ and higher energy levels are mainly located in the blue light region, but it is easily quenched due to cross-relaxation between two adjacent Tb^{3+} ion pairs at high Tb^{3+} doping concentration. There are two kinds of cross-relaxation energy transfer processes between Tb^{3+} ions: One is the well-known process by transitions $^5D_3 \rightarrow {}^5D_4$ and $^7F_6 \rightarrow {}^7F_0$, and the other is the process by transitions $^7D_J \rightarrow {}^5D_{1,\ 2,\ 3}$ and $^7F_6 \rightarrow {}^5D_4$. Both of these two processes are dependent on the Tb^{3+} concentration [42].

(d) Sm^{3+} ($4f^5$): Sm^{3+} ion has been widely used as an activator to produce reddish orange emission due to its efficient $^4G_{5/2} \rightarrow {}^6H_J$ ($J = 5/2, 7/2, 9/2$, and 11/2) transitions. Sm^{3+}-doped phosphors have the applications in lighting and display. The sharp characteristic excitation peaks of Sm^{3+} corresponding to the transitions from the ground state ($^6H_{5/2}$) to the different excited states of Sm^{3+} allow for efficiency excitation from UV to visible spectral region. Generally, Sm^{3+}-doped materials exhibit three major emission peaks in the orange-red region, corresponding to the $^4G_{5/2} \rightarrow {}^6H_{J/2}$ ($J = 5/2, 7/2$, and 9/2) transitions of Sm^{3+}, respectively. Among these emissions the transition of ($^4G_{5/2} \rightarrow {}^6H_{7/2}$) shows the strongest intensity. The strong red emission due to the transition of $^4G_{5/2} \rightarrow {}^6H_{7/2}$ ($\Delta J = \pm 1$) can be regarded as a partly magnetic dipole and partly electric dipole transition [43].

(e) Dy^{3+} ($4f^9$): Recently, trivalent dysprosium ion (Dy^{3+})–doped phosphors have been extensively studied due to the potential applications in white light LEDs, because of its intense blue (484 nm, $^4F_{9/2} \rightarrow {}^6H_{15/2}$) and yellow (575 nm, $^4F_{9/2} \rightarrow {}^6H_{13/2}$) emissions. The $^4F_{9/2} \rightarrow {}^6H_{13/2}$ transition is a forced electric dipole transition, which is allowed only in the case that the Dy^{3+} ions are located at lattice sites without inversion symmetry [40]. Thus the intensity ratio of the two dominant emission bands of Dy^{3+} arising from the transitions $^4F_{9/2} \rightarrow {}^6H_{15/2}$ (blue) and $^4F_{9/2} \rightarrow {}^6H_{13/2}$ (yellow) is host dependent. Therefore the emission color of Dy^{3+}-doped phosphors can be modified by changing the relative emission intensity. [44]

(f) Pr^{3+} ($4f^2$): The emission of Pr^{3+} is strongly dependent on the host matrix. The red emission is derived from the $^1D_2$ level of Pr^{3+}, the green emission originates from the $^3P_0$ level, the blue emission comes from the $^1S_0$ level, and the purple emission is derived from the 4f-5d state. For instance, Gd_2O_3/Pr^{3+} gives a green emission band due to the $^3P_0 \rightarrow {}^3H_4$ transition, while Pr^{3+}-doped $LiYF_4$ presents a red emission by the transitions of $^3P_0 \rightarrow {}^3H_6$ and $^3P_0 \rightarrow {}^3F_2$. The emission from the $^1D_2$ level is located in red and near-infrared region [45].

(g) Ce^{3+} (4f-5d): Ce^{3+} is the rare earth ion with the simplest energy level because it has only one 4f electron. The ground state $4f^1$ electronic configuration includes two energy levels of $^2F_{52}$ and $^2F_{7/2}$. Due to spin-orbit coupling, the energy difference between the doublets is 2000 cm^{-1}. The electronic configuration of the excited state is $5d^1$, which split by the crystal field into 2–5 components with a maximum splitting amplitude of approximately 15,000 cm^{-1}. For example, when Ce^{3+} is doped into the well-known yttrium aluminum garnet ($Y_3Al_5O_{12}$), it is capable of absorbing blue light and emitting yellow light by the d-f transition. Since the 5d orbital is strongly influenced by the crystal field and the polarization of the host lattice, the excitation and emission energy of the activator can be adjusted by changes in the chemical composition and structure of the host. The relationship between crystal structure and crystal-field splitting of Ce^{3+} has been summarized in several previous reports [46–48].

(h) The most well-known and widely used divalent lanthanide ion is Eu^{2+} ($4f^7$), which gives ultraviolet to red emission due to the 5d-4f transition. The reason for the relatively slow decay rate for the d-f transition can be understood by the spin selection rule [49].

(i) Divalent ytterbium (Yb^{2+}) ion also shows broadband emission due to the $4f^{13}5d$-$4f^{14}$ transition. From the spectral results obtained in different inorganic hosts, the difference in the 5d-4f emission energies of Eu^{2+} and Yb^{2+} is usually within 0.1 eV. However, thermal quenching is in general more prominent for Yb^{2+}, due to the closer proximity of the 5d excited state to the conduction band compared with the case of Eu^{2+}. In addition, the relatively long decay time of Yb^{2+} (1–10 ms) could lead to saturation under high excitation density. Therefore, in many compounds, the emission of Yb^{2+} is absent at room temperature or considerably red-shifted, due to interaction of the 5d excited state with the conduction band states and the formation of self-trapped excitons. The relatively strong thermal quenching for Yb^{2+}-doped phosphors makes them less suited for spectral converters used in LEDs [50].

1.4.3 Transition metal ions

Transition metal (TM) ions have their valence electrons orbiting in the incomplete d shell. Unlike *4f* electrons that are shielded by the outer d and s shell, *d* electrons in TM ions suffer large perturbation from the outer crystal field, compared with that of spin-orbit interaction. The crystal field of the coordination ligands dictates the emission profile of a TM compound as the ligands can remove the degeneracy of the orbitals and split them into different energy levels. There have been extensive studies on persistent phosphors doped with TM ions. In particular, for NIR persistent phosphorescence in the far-red to infrared regions, the afterglow properties of various TM ions—$3d^2$ (Cr^{4+}, Mn^{5+}, and Fe^{6+}), $3d^3$ (Cr^{3+} and Mn^{4+}), $3d^4$ (Mn^{3+}), and $3d^5$ (Mn^{2+} and Fe^{3+})—have been investigated in different host [51–64].

(a) Cr^{4+} and Mn^{5+}: Both Cr^{4+} and Mn^{5+} have the same outer layer electronic configuration, that is, $3d^2$. In this configuration the ground state is a spin triplet $^3A_2$ (e^2 electron configuration). Similar as the case of Cr^{3+} in weak and strong crystal fields (such as $KZnF_3$ and $A1_2O_3$), the energy levels of the Cr^{4+} in a weak crystal field has a spin triplet first excited state $^3T_2$ (t_2e electron configuration), which gives rise to broad tunable absorption and emission and allows for four-level laser action. For Mn^{5+} centers in a strong field, the first excited state is a spin singlet ^1E level (e^2 electron configuration) that gives sharp emission lines and allows for three-level laser operation. For instance, the weak bands observed in Cr/forsterite covering spectral region of 900–1100 nm can be assigned to spin-allowed parity-forbidden transitions to crystal field (CF) components (Cs local symmetry) of state $^3T_2$. In contrast the emission band in the visible region of 570–730 nm are assigned to spin- and parity-allowed transitions to CF components of state 3 T(t_2e). Other bands corresponding to spin- and parity-allowed transitions to state $^3T_1(e^2)$ could be expected at higher energy, but they are not observed, probably because they require a two-electron jump process. In this case it seems that for this configuration the selection rule is rather strict, a point that will be discussed again in the following section. On the other hand the low energy doublet line observed in $Ba_3(VO_4)_2/Mn^{5+}$, that is, 1180 nm (8467 cm^{-1}), is assigned to spin forbidden transitions to CF components (C_3, local symmetry) of state ~E. These bands corresponding to the transitions between the triplet states $^3T_2$ and ^3T (t_2e) are observed at around 900 and 700 nm, respectively [51–53].

(b) Cr^{3+} and Mn^{4+}: Cr^{3+} and Mn^{4+} ions present the same outer layer electronic configuration, that is, $3d^3$. They are the most popular TM ions used as the activators in persistent phosphors [54–57]. The spectral terms of d^3 in a strong crystal field split into $^4A_{2g}$, ^2E_g, $^2T_{1g}$, $^2T_{2g}$, etc. Energy levels and photoluminescence properties of Cr^{3+} and Mn^{4+} have been investigated in a great number of literatures. Whether the ^2E or $^4T_2$ energy level is the lowest excited states is determined by the strength of the crystal field. Commonly the emission of Cr^{3+} shows characteristic sharp R-line due to the ^2E-$^4A_2$ transition, such as the line-shaped emission in garnet crystals. In many hosts the transition of $^4T_2$-$^4A_2$ gives broadband emission. It is found that the crystal field splitting term, Dq, varies in different octahedral sites occupied by Cr^{3+} ions. Both the ^2E-$^4A_2$ (R-line) and $^4T_{2g} \rightarrow {}^4A_{2g}$ band are inhomogeneously broadened due to site-to-site disordering.

Luminescence bands originating from the ^2E_g-$^4A_{2g}$ transition of Mn^{4+} are located at 620–700 nm in most host materials. The spectrum has a structure consisting of several broad lines attributed to the d-d transitions and phonon lines. If Mn^{4+} occupies the octahedral site in a host, the $^4T_2$ level splits into two levels due to the low symmetric field. Generally the optical absorption bands originate from the Mn^{4+} intra-d shell transitions. The absorption band at ~480 nm is due to the transition of Mn^{4+} from the ground state $^4A_{2g}$ to the first excited state $^4T_{2g}$. The origin of the weak absorption band at ~400 nm is the spin-forbidden transition from the ground state $^4A_{2g}$ to the second excited quartet state $^2T_{2g}$; the strong absorption band at ~350 nm is attributed to the spin-allowed transition from $^4A_{2g}$ to $^4T_{1g}$.

(c) Mn^{2+} and Fe^{3+}: Mn^{2+} has an outer layer $3d^5$ electron configuration. The luminescence of Mn^{2+} and Fe^{3+} is attributed to the parity-forbidden d-d transitions from the lowest excited level $^4T_1$ (4G) to the ground state $^6A_1$ (6S). For Mn^{2+}, there are three emission peaks ascribed, respectively, to the transitions of $^6A_1(6S)$-$^4T_1(4G)$ (490 nm), $^6A_1(6S)$-$^4T_2(4G)$ (450 nm), and $^6A_1$ (6S)-$^4E(4G)/^4A(4G)$ (425 nm), which have been identified in different hosts. As an important activator in luminescent materials, Mn^{2+} ion has been extensively employed as the activator in a large number of important phosphors. The emission of Mn^{2+} usually shows a broad band located at spectra region from deep green to far red depending on the crystal field [58].

Fe^{3+} ion is isoelectronic with Mn^{2+}, with a d^5 electronic configuration. However, the optical spectra of Fe^{3+} differ greatly from that of Mn^{2+} due to the fact that weak crystal field bands of Fe^{3+} are superimposed on the tail of an intense charge transfer band, whereas for Mn^{2+} the charge transfer band is located in the UV region. The position of the charge transfer band is correlated with the coordination number of the Fe^{3+} ion; it shifts to higher energy for lower coordination number. In $LiGaTiO_4/Fe^{3+}$ the observed luminescence has a maximum intensity at 720 nm associated with $^4T(4G)$-$^6A_1$ transition of tetrahedrally coordinated Fe^{3+} ions. In accordance with the d-d transitions of Fe^{3+} ions in similar materials, these absorption bands can be tentatively assigned to the d-d transitions of tetrahedrally coordinated Fe^{3+} ions. The two stronger and sharper bands at 392 nm and 441 nm together with a shoulder peak at 465 nm are assigned to the crystal field–independent transitions of $^6A_1(6S) \rightarrow {}^4E(4D)$, $^6A_1(6S) \rightarrow {}^4T_2$ (4D), and $^6A_1(6S) \rightarrow {}^4A_1$, respectively. The two remnant broad bands at 520 nm and 630 nm are assigned to $^6A_1(6S) \rightarrow {}^4T_2(4G)$ and $^6A_1(6S) \rightarrow {}^4T_1$ (4G) transitions, and the broadening of the emission band can be explained by the crystal field effect [59, 60].

1.5 Codopants

It is often an effective method to optimize the persistent luminescent materials by codoping. The codoping ions, which can create more defects and serve as the electron or hole traps in the host, act as trapping centers. Choosing the appropriate codoping ions could notably increase the performance of persistent luminescence. The codoping of ions in a host can not only increase the number of trapping centers but also improve the capture efficiency [61]. For ions with a higher transition probability, electrons can be more easily pumped into the conduction band of the host via the energy levels of these ions, which result in the capture of more electrons by the traps.

For example, codopant ions, such as Dy^{3+}, Pr^{3+}, and Nd^{3+}, are considered to serve as charge carrier traps that can trap free electrons and holes. The trap depth can be adjusted by using different codopants, and it can be used to control the persistent intensity and lifetime. For instance, by codoping of trivalent lanthanide ions such as Dy^{3+}, Pr^{3+}, Ce^{3+}, and Nd^{3+}, which serve as electron traps in the diopside lattice, it is possible to improve the performance of persistent phosphors $CaMgSi_2O_6/Eu^{2+}$, Mn^{2+}, M^{3+}. Fig. 1.8 shows the proposed mechanism of persistent luminescence of $CaMgSi_2O_6/Mn^{2+}$, Ln^{3+}. Under UV excitation, Mn^{2+} ions, which occupy the Mg^{2+} sites, might capture holes from valence band to form Mn^{3+} ions; meanwhile the free electrons in conduction band are trapped by oxygen vacancies (V_O) and Ln^{3+} ions. After switching off the excitation, the trapped electrons are released and captured by Mn^{3+} ions, resulting in the emission by the $^4T_1 \rightarrow {}^6A_1$ transition. [62, 63]

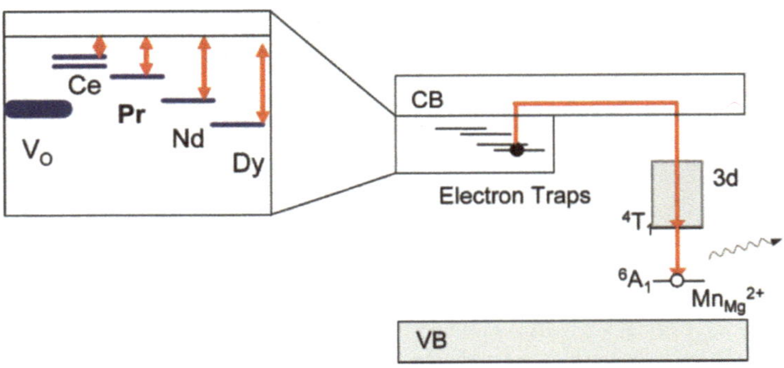

Fig. 1.8 Schematic energy level diagram of Mn^{2+} and Ln^{3+} in tremolite. The main hole traps are Mn^{2+} ions in the Mg^{2+} site, while electrons are trapped by oxygen vacancies (V_O) and Ln^{3+} ions. Thermal luminescence and persistent luminescence occur by thermally activated release of electrons captured by Mn^{3+} ions. The insert shows the relative positions of electron trap levels with respect to the conduction band edge.
Reproduced with permission from T. Maldiney, A. Bessiere, J. Seguin, E. Teston, S.K. Sharma, B. Viana, et al., The in vivo activation of persistent nanophosphors for optical imaging of vascularization, tumours and grafted cells, Nat. Mater. 13 (2014), 418–426, copyright 2014, Springer Nature Publication.

The depth of electron trap E, which is defined as the energy difference between the trapping level and the conduction band, can be roughly estimated by the following equation:

$$E \approx 0.002 T_M \tag{1.2}$$

where T_M is the peak temperature in the thermal luminescence spectra. The trap depths are calculated to be 0.45, 0.7, 0.92, and 0.96 eV for codoping ions of Ce^{3+}, Pr^{3+}, Nd^{3+}, and Dy^{3+}, respectively. It is worth noting that cooping of Pr^{3+} induces most suitable trap depth (0.7 eV), which produces the best persistent luminescence performance at room temperature [64].

1.6 Characterization of persistent luminescence

After excitation, all luminescent materials show afterglow with duration from nanosecond to over hours. Fig. 1.9 shows the afterglow intensity as a function of time after 1-min excitation with an unfiltered xenon arc lamp at 1000 lx. The decay curves follow straight lines in a log-log plot, implying that a power law can model them with a

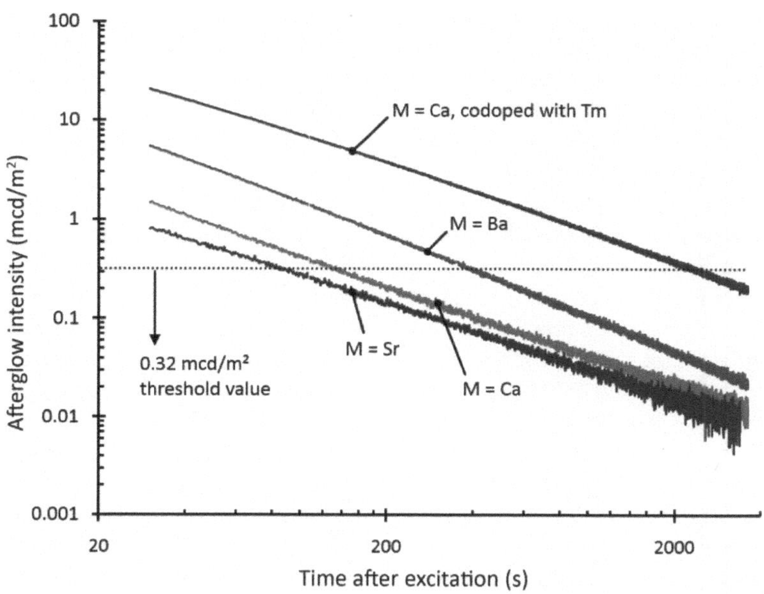

Fig. 1.9 Decay of the afterglow intensity of $MAl_2O_4:Eu^{2+}$ after 1-min excitation with a Xe arc lamp at 1000 lx.
Data from P.F. Smet, D.E.K. Van, D. Poelman, Persistent phosphors. In Handbook on the Physics and Chemistry of Rare Earths. Elsevier, 2015.

negative scaling exponent. In other words the decay is very fast initially but slows down over time. This is typical for most persistent luminescent materials [65–71].

Luminescent materials featuring afterglow curve as that shown in Fig. 1.9 are often referred as phosphorescent, but the definition of phosphorescence is rather ambiguous, since the term is also used for luminescence where a quasi-stable state is involved, causing an increased lifetime of the fluorescence decay. Nevertheless, the decay from such a quasi-stable state usually does not last for longer than a second. We are interested in materials where the afterglow is caused by the existence of suitable charge carrier traps in the crystal and remains visible for a reasonable duration of time. The borderline between visible and invisible is not sharply defined, and neither does there exist a consensus on the duration of observable afterglow. In this part, we will focus on materials that have an afterglow decay time longer than several minutes, where the decay time is defined as the time between the end of the excitation and the moment when the light intensity drops below 0.32 mcd/m^2, roughly 100 times the sensitivity of the dark-adapted human eye. This is a definition similar to the one used in the safety signage industry and by various researchers [65, 66].

The sensitivity of the human eye to light of certain intensity varies strongly over the wavelength range between 380 and 800 nm. Under daylight conditions the average normal sighted human eye is most sensitive at the wavelength of 555 nm, resulting in the fact that green light at this wavelength produces the impression of highest "brightness" when compared with light at other wavelengths. The spectral sensitivity function of the average human eye under daylight conditions (photopic vision) is defined by the CIE spectral luminous efficiency function $V(\lambda)$. Only in very rare cases, the spectral sensitivity of the human eye under dark-adapted conditions (scotopic vision), defined by the spectral luminous efficiency function $V'(\acute{e})$, becomes technically relevant. Conventionally, these sensitivity functions are normalized to a value of 1 with respect to their maximum [67].

As an example, the photopic sensitivity of the human eye to monochromatic light at 490 nm amounts to 20% of its sensitivity at 555 nm. Consequently, when a source of monochromatic light at 490 nm emits five times as much power (expressed in watts) as an otherwise identical source of monochromatic light at 555 nm, both sources produce the impression of same "brightness" to the human eye. A light source radiating 1 W of green light will appear much brighter than another source radiating the same amount of power of red light because the eye is more sensitive in the green region. The human eyes are sensible to photons at wavelengths roughly between 400 nm (violet) and 700 nm (red). Wavelengths shorter than 400 nm (ultraviolet, UV) or longer than 700 nm (infrared, IR) are not visible [68].

Human eyes behave differently in high and low brightness conditions: in daylight, for brightness levels above 3 cd/m^2, the vision is mainly done by the center of the retina, we can see colors, and the maximum sensitivity is at 555 nm (in the green region). This type of vision is called *photopic* vision. In low brightness conditions, for brightness levels below 30 μcd/m^2, the vision is mainly done by the peripheral region of the retina that is color-blind, while the center region is not sensitive enough to see any color. This type of vision is called *scotopic* vision. Maximum sensitivity is at 507 nm (in the blue-green region), and red light is almost invisible. The vision in-between photopic and scotopic vision is called *mesopic*. [68, 69]

The benefit of scotopic vision and its reduced sensitivity to red light is that you can use a red flashlight to illuminate an object (e.g., to read a map) without disturbing your night vision. It takes several minutes to get your vision used to the darkness; if you use a white flashlight instead, the eyes will switch back to photopic vision, and you'll have to wait a few minutes again. The center of the retina can directly pick up red light without affecting night vision. For this reason, astronomers often use red flashlights.

The standard eye response is the visible region given in Fig. 1.10 in both linear and logarithmic scale. This standard eye sensitivity is also called standard luminosity function $V(\lambda)$ and is used, for photopic vision, to define a conversion between the radiated energy (in Watts) and the luminous flux (in Lumen). The standard luminosity function $V'(\lambda)$ refers to scotopic vision, but it shouldn't be used to convert to and from photometric units [70].

For photopic vision, 1 W of radiant power at the wavelength of 555 nm is defined corresponding to a luminous flux of 683 lm. For scotopic vision the sensitivity of the eye is greater, and 1 W of radiant power at 507 nm corresponds to a luminous flux of 1700 lm. These two wavelengths correspond to the maximum of each curve. Of course, every eye is different; these data correspond to the standard eye, as defined by the CIE 1931 (photopic) and CIE 1951 (scotopic) standards [71].

1.7 Development history of persistent phosphors

The earliest record of persistent luminescence in ancient China can be date back to the 20th century (Fig. 1.11) [74].

In Europe, the well-documented history of persistent luminescence dates back to the beginning of the 17th century. In 1602 an Italian shoemaker, V. Casciarolo, observed strong luminescence from a mineral barite, $BaSO_4$, later to be known as the famous Bologna stone. It appears, however, that the material actually was not the barite mineral itself but rather the reduced product, barium sulfide, BaS. According to modern knowledge the reduction process of the sulfide cannot be held responsible for the luminescence since this luminescence occurred long after the process had been terminated. On the other hand, taking into account of the low purity of both the original barite and the charcoal used for the reduction, it is evident that many different (trace) impurities were present [72–77]. That is probably the reason for the varying color of the persistent emission reported, though mainly orange or red emission in the visible range prevails in the literature. Whatever the emission color, the cause for the persistent emission remained unknown then and, to certain extent, up to the 21st century. At that time, that is, during the decades following the discovery of the bright emission from the (reduced) Bologna stone, the phenomenon did not cease to arouse the interest of both scientists and laymen, and even several books were written on this miraculous phenomenon [72].

The matter clearly went beyond the knowledge of the scientists of the time, and, maybe for comprehensible reasons, even the famous interdisciplinary genius Galileo refused to get involved. Probably, Galileo had had enough trouble with the Catholic Church, and he preferred to leave the shine of the Bologna stone alone since it may have been of celestial origin. Since the 17th century, only scattered reports are

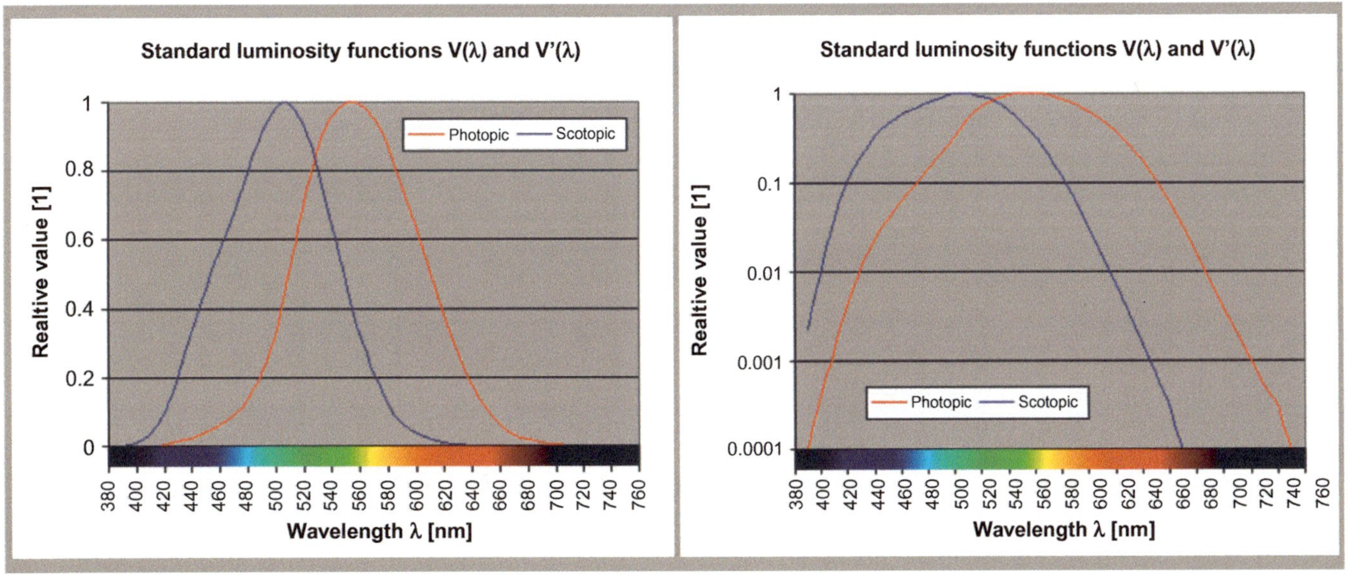

Fig. 1.10 Standard luminosity functions $V(\lambda)$ and $V'(\lambda)$.
Reproduced with permission from P.F. Smet, D.E.K. Van, D. Poelman, Persistent phosphors. In Handbook on the Physics and Chemistry of Rare Earths, Elsevier, 2015, Copyright 2015, Elsevier.

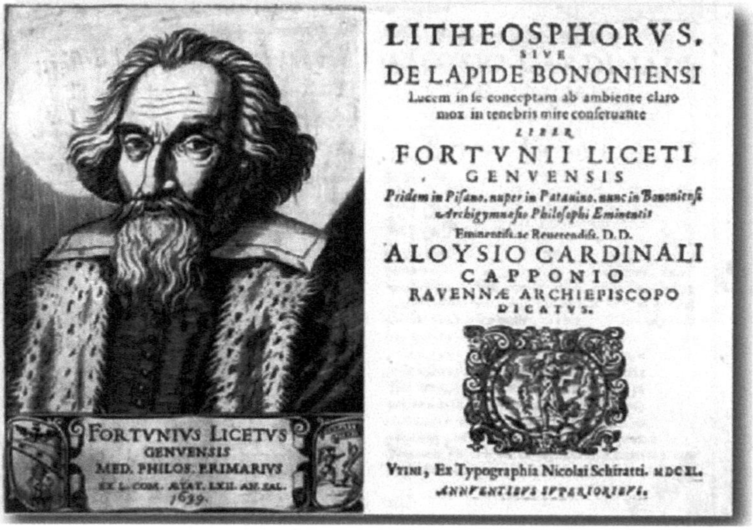

Fig. 1.11 The book Litheosphorus Sive de Lapide Bononiensi by Fortunius Licetus (Bologna, Italy, 1640) on the persistent luminescence of the Bologna stone.
Data from T. Aitasalo, J. Holsa, M. Kirm, T. Laamanen, M. Lastusaari, J. Niittykoski, J. Raud, R. Valtonen, Persistent luminescence and synchrotron radiation study of the $Ca_2MgSi_2O_7$: Eu^{2+}, R^{3+} materials, Radiat. Meas. 42 (2007) 644–647.

available on persistent luminescence, and the research receded considerably because no explanation was found for the persistent luminescence produced without any evident excitation source. From the previous text, it may well be concluded that persistent luminescence was one of the first genres of luminescence ever discovered and studied to some scientific extent.

Despite no explanation being found for the persistent luminescence, the applications were taken into use in the 20th century. Luminous paints were based on persistent luminescence from the different sulfides, such as ZnS doped with, for example, Cu. The emission was modified with the partial substitution of Zn with Cd. However, at its best, persistent luminescence from these materials was both weak and short, lasting for a few hours only. Some environmentally dubious "tricks," such as the doping of ZnS with radioactive elements (one of the few uses of the artificial radioactive lanthanide, promethium), were used to prolong the duration of persistence with external excitation. These tricks were only retarding the final rejection of the ZnS-based phosphors, also due to their pronounced instability against humidity [73, 74].

The leading role of persistent luminescence phosphorescence as the number one luminescence subject was finally lost with the advent of the introduction of the rare earth–based phosphors in the 1960s [75]. This state of affairs seemed to remain even while the other fields of luminescence flourished, for example, in the areas of lighting, cathode ray tubes, and scintillator material applications. The hectic research and development that followed the lanthanides now being commercially available in a pure form produced both new information about the different luminescence processes and ever better phosphor materials. None of this took place in the field of persistent phosphorescence. Because of the speeded-up introduction of many new phosphor materials, also some with serious problems with the resistance to the sometimes hostile operating conditions (e.g., mercury vapor and electron bombardment) appeared in the commercial market. Thus the negative side of persistent phosphorescence was introduced: the afterglow. Because of the formation of traps to store the excitation energy (e.g., UV radiation and electrons), the emission was retarded and in many cases significantly weakened. This was not always completely condemning the phosphor material since different ways to solve these problems were finally developed, for example, posttreatment in inert/reactive atmosphere or introduction of some additives. However, in general, the afterglow problem seemed to give persistent phosphorescence the final coup de grace, and practically nothing, neither in understanding the phenomenon nor in developing new materials, took place before the mid-1990s.

Then, out of the blue, the first generation of the modern persistent phosphors, the Eu^{2+}-doped and rare earth R^{3+}–codoped alkaline earth aluminates (MAl_2O_4: Eu^{2+}, R^{3+}; M: Ca and Sr) appeared in both the commercial market and in research laboratories as well. With this, the first new persistent phosphorescence era since 1995, there rapidly appeared not only new materials and mechanisms but also new methods of analysis and applications. Because of the increased research activity, there emerged also hundreds of papers, reports, and meeting communications [76, 77].

Unfortunately, since the phenomenon of persistent phosphorescence seemed to be rather complicated, the quality of the reports has not always been of the best caliber. As the new persistent phosphors have been developed, there are presently persistent phosphors for each of the main colors: blue, green, and red that fulfill—perhaps with the exception of red—the requirements not only for sufficiently strong and long persistent phosphorescence but also for the stability of a commercial phosphor. Fig. 1.12 shows the behavior of three different persistent phosphors in daylight, under UV radiation excitation and in dark after UV irradiation. Despite the high initial persistent phosphorescence intensity of the red-emitting material, Y_2O_2S: Eu^{3+}, Mg^{2+},Ti^{4+}, its long-term luminescence is much weaker than the duration of the blue ($Sr_2MgSi_2O_7$:Eu^{2+}, Dy^{3+}; 25+ hours) and green ($SrAl_2O_4$:Eu^{2+}, Dy^{3+}; 15+ hours) emitting counterparts [76–78].

The idea of mutating a well-established and commercially available phosphor for other luminescence applications such as Y_2O_2S: Eu^{3+}—though reinforced with codopants such as Mg^{2+} and Ti—does not always result in a superior persistent phosphorescent material, as was the case with $SrAl_2O_4$: Eu^{2+} [78]. The design of new efficient persistent phosphors requires more than this or a lot of luck. As a result of perhaps the latter, the most recent persistent phosphors, the Eu^{2+}-doped and rare earth R^{3+}–codoped alkaline earth disilicates ($M_2MgSi_2O_7$: Eu^{2+}, R^{3+}; M: Ca and Sr) were discovered to be much more efficient and stable than corresponding aluminates [78]. Their manufacture is also somewhat less complicated, as no flux material is needed. The host materials (both MAl_2O_4 and $M_2MgSi_2O_7$, but not Y_2O_2S) are rather cheap, though the high purity requirements increase the price tag. So far, there has not been any serious alternative for the Eu^{2+} dopant (the price of which is skyrocketing), mainly

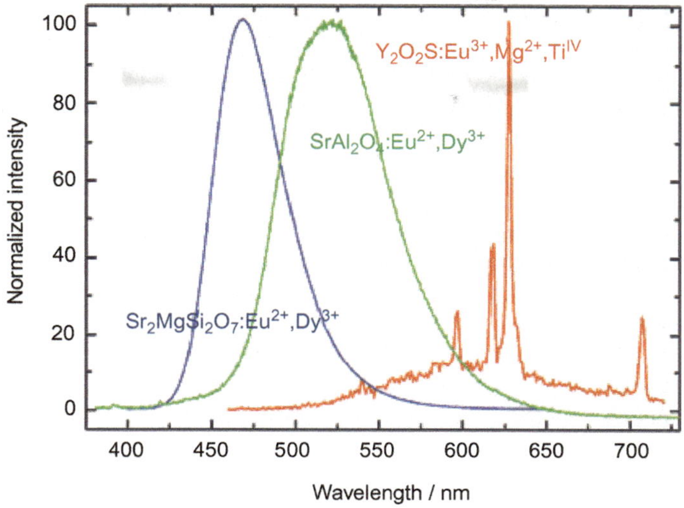

Fig. 1.12 The luminescence spectra of the blue emitting $Sr_2MgSi_2O_7$: Eu^{2+}, Dy^{3+}, green emitting $SrAl_2O_4$: Eu^{2+}, Dy^{3+}, and red emitting Y_2O_2S: Eu^{3+}, Mg^{2+},Ti^{4+} phosphors after UV excitation.
Reproduced with permission from D. Poelman, N. Avci and P. F. Smet, Measured luminance and visual appearance of multi-color persistent phosphors. Opt. Express 17 (2009) 358–364, Copyright 2009, The Optical Society.

because of the very favorable position of its ground electronic energy level vis-a-vis the host band structure. The low concentration of this element somewhat lowers the price, which is inherently high because europium is used in most luminescence applications, including security markers in Euro bank notes [78, 79].

The idea of making a white emitting persistent light source by combining the three individual blue, green, and red emitting phosphors, in a way similar to the tricolor fluorescent tubes, seems possible when the emission spectra of these phosphors are considered alone. However, the unbalanced duration of the three colors, especially due to the weak and short red persistent emission, will delay or even preclude this application.

1.8 Diverse applications of persistent phosphors

Persistent phosphors have now become an important class of multifunctional materials. [6] For a long time, they have been mainly used as luminous materials in the low-light display and lighting field. In recent years, persistent phosphors have been employed as a weak light source to drive photocatalysis in the dark environment. Furthermore, their high-temperature sensitivity and mechanoluminescence properties have been utilized for the measurement of temperature and pressure. In the biomedical field, persistent phosphors have been used as luminescent markers for tumor diagnosis and analysis and photodynamic therapy for cancer. For optoelectronics field, persistent phosphors have been used in optoelectronic nanoscale devices, fiber optic amplifiers, and photovoltaic materials. Persistent phosphors can also be used in agriculture as a light source to promote plant growth and to enhance plant photosynthesis in the dark. In addition to environmentally friendly and energy-saving products and industrial processes used for indication and lighting, with the development of some new persistent luminescent materials such as persistent luminescent glass, the main research direction of durable light-emitting materials might be shifted to high-tech fields, such as information storage, radiation detection, and imaging. [6, 18, 40] These will be discussed in the Chapter 7 in detail.

The key to the application of persistent phosphors is the close cooperation between materials research and development and subsequent processes, such as device fabrication. At present, biomedical application of persistent luminescent materials is an important direction due to their unique advantages. However, persistent phosphors in nanoscale sizes always show notable degradation of performance due to increase in the number of surface defects. In addition, surface modifications with organic moieties to stabilize the particles in aqueous solutions also create potential quenching channels that result in the reduction in afterglow time and intensity.

References

[1] S. Shionoya, W.M. Yen, H. Yamamoto, Phosphor Handbook, CRC Press, 2006.

[2] G. Blasse, B.C. Grabmaier, Luminescent Materials, Springer, Berlin, Heidelberg, 1994.

[3] R.S. Becker, R.S. Becker, Theory and Interpretation of Fluorescence and Phosphorescence, Wiley Interscience, New York, 1969.

[4] Z. Pan, Y.Y. Lu, F. Liu, Sunlight-activated long-persistent luminescence in the near-infrared from Cr^{3+}-doped zinc gallogermanates, Nat. Mater. 11 (1) (2012) 58.

[5] R. Kabe, C. Adachi, Organic long persistent luminescence, Nature 550 (7676) (2017) 384.

[6] Y. Li, M. Gecevicius, J.R. Qiu, Long persistent phosphors-from fundamentals to applications, Chem. Soc. Rev. 45 (8) (2016) 2090–2136.

[7] Y. Li, Y.Y. Li, R.C. Chen, K. Sharafudeen, S.F. Zhou, M. Gecevicius, H.H. Wang, G.P. Dong, Y.L. Wu, Tailoring of the trap distribution and crystal field in Cr^{3+}-doped non-gallate phosphors with near-infrared long-persistence phosphorescence, NPG Asia Mater. 7 (5) (2015) e180.

[8] Y. Li, M. Gecevicius, J.R. Qiu, Long persistent phosphors-from fundamentals to applications, Chem. Soc. Rev. 45 (8) (2016) 2090–2136.

[9] A.H. Wako, B.F. Dejene, H.C. Swart, Roles of doping ions in afterglow properties of blue $CaAl_2O_4$: Eu^{2+}, Nd^{3+} phosphors, Physica B 439 (2014) 153–159.

[10] H. Yamamoto, T. Matsuzawa, Mechanism of long phosphorescence of $SrAl_2O_4$: Eu^{2+}, Dy^{3+} and $CaAl_2O_4$: Eu^{2+}, Nd^{3+}, J. Lumin. 72 (1997) 287–289.

[11] Y. Wang, Z. Wang, Characterization of Y_2O_2S: Eu^{3+}, Mg^{2+}, Ti^{4+} long-lasting phosphor synthesized by flux method, J. Rare Earths 24 (1) (2006) 25–28.

[12] A. Abdukayum, J.T. Chen, Q. Zhao, X.P. Yan, Functional near infrared-emitting Cr^{3+}/Pr^{3+} co-doped zinc gallogermanate persistent luminescent nanoparticles with superlong afterglow for in vivo targeted bioimaging, J. Am. Chem. Soc. 135 (38) (2013) 14125–14133.

[13] K.W. Böer, Handbook of the Physics of Thin-Film Solar Cells, Springer Science & Business, 2014.
[14] W. Lu, Y. Fu, Optical properties of fluorescent colloidal quantum dots, in: Spectroscopy of Semiconductors, Springer, Chem, 2018.
[15] T.M. Tritt (Ed.), Thermal Conductivity: Theory, Properties, and Applications, Springer Science & Business Media, 2005.
[16] J. Hinks, C. Geoff, The Technology of Building Defects, Routledge, 2002.
[17] M.D. McCluskey, E.H. Eugene, Dopants and Defects in Semiconductors, CRC Press, 2018.
[18] R.J.D. Tilley, Principles and Applications of Chemical Defects, Routledge, 2019.
[19] J. Pelleg, Defects in Materials. Diffusion in Ceramics, Springer, Cham, 2016.
[20] E. Finley, M.T. Aria, B. Jakoah, Intrinsic defects drive persistent luminescence in monoclinic $SrAl_2O_4$: Eu^{2+}, J. Phys. Chem. C 122 (2018) 16309–16314.
[21] D.E.K. Van, P.F. Smet, D. Poelman, Persistent luminescence in Eu^{2+}-doped compounds: a review, Materials 3 (4) (2010) 2536–2566.
[22] K. Van den Eeckhout, A.J. Bos, D. Poelman, P.F. Smet, Revealing trap depth distributions in persistent phosphors, Phys. Rev. B 87 (4) (2013) 045126.
[23] O. Vazquez-Robaina, A.F. Cabrera, M. Meyer, R.M. Romano, A.F. Cruz, C.E.R. Torres, Room-temperature ferromagnetism induced by high-pressure hydrogenation of ZnO, J. Phys. Chem. C 123 (2019) 19851–19861.
[24] A. Oguri, A.C. Hewson, Higher-order Fermi-liquid corrections for an Anderson impurity away from half filling: equilibrium properties, Phys. Rev. B 97 (2018) 17.
[25] G.A. Craig, M. Mark, 3d single-ion magnets, Chem. Soc. Rev. 44 (2015) 2135–2147.
[26] L. Ungur, Introduction to the electronic structure, luminescence, and magnetism of lanthanides, in: Lanthanide-Based Multifunctional Materials, Elsevier, 2018, pp. 1–58.
[27] L.E. Orgel, Spectra of transition-metal complexes, J. Chem. Phys. 23 (1955) 1004–1014.
[28] H. Ge, G. Lei, W. Sheng, Z. Shuang, An oxadiazole-functionalized ligand and its yellow-emitting re(I) complex for organoelectronic application, Opt. Mater. 34 (2012) 1303–1309.
[29] D. Xu, W. Chen, M. Zeng, H. Xue, Y. Chen, X. Sang, Y. Xiao, T. Zhang, R.U. Raymond, K. Xiao, L. Fu, Crystal-field tuning of photoluminescence in two-dimensional materials with embedded lanthanide ions, Angew. Chem. Int. Ed. 57 (2018) 755–759.
[30] K.V. Lamonova, E.S. Zhitlukhina, R.Y. Babkin, S.M. Orel, S.G. Ovchinnikov, Y.G. Pashkevich, Intermediate-spin state of a 3d ion in the octahedral environment and generalization of the Tanabe-Sugano diagrams, J. Phys. Chem. A 115 (2011) 13596–13604.
[31] P. Dorenbos, Mechanism of persistent luminescence in Eu^{2+} and Dy^{3+} codoped aluminate and silicate compounds, J. Electrochem. Soc. 152 (2005) H107–H110.
[32] K. Ogasawara, F. Alluqmani, H. Nagoshi, Multiplet energy level diagrams for Cr^{3+} and Mn^{4+} in oxides with Oh site symmetry based on first-principles calculations, ECS J. Solid State Sci. Technol. 5 (1) (2016) R3191–R3196.
[33] M.J. Weber, T.E. Varitimos, B.H. Matsinger, Optical intensities of rare-earth ions in yttrium orthoaluminate, Phys. Rev. B 8 (1973) 47.
[34] M. Vollmer, K.P. Möllmann, Infrared Thermal Imaging: Fundamentals, Research and Applications, John Wiley & Sons, 2017.
[35] M. Pawlak, T. Szidarovszky, G.J. Halasz, A. Vibok, Robust field-dressed spectra of diatomics in an optical lattice, Phys. Chem. Chem. Phys. 22 (2020) 3715–3723.
[36] J.A. Jimenez, Silicon-induced UV transparency in phosphate glasses and its application to the enhancement of the UV type B emission of Gd^{3+}, ACS Appl. Mater. Interfaces 9 (2017) 15599–15604.

[37] C.M. Dodson, R. Zia, Magnetic dipole and electric quadrupole transitions in the trivalent lanthanide series: calculated emission rates and oscillator strengths, Phys. Rev. B 86 (2012) 125102.

[38] J. Liao, H. Huang, H. You, X. Qiu, Y. Li, B. Qiu, H.R. Wen, Photoluminescence properties of $NaGd(MoO_4)_2:Eu^{3+}$ nanophosphors prepared by sol-gel method, Mater. Res. Bull. 45 (2010) 1145–1149.

[39] T.O. Sales, R.J. Amjad, C. Jacinto, M.R. Dousti, Concentration dependent luminescence and cross-relaxation energy transfers in Tb^{3+} doped fluoroborate glasses, J. Lumin. 205 (2019) 282–286.

[40] A. Hooda, S.P. Khatkar, A. Khatkar, R.K. Malik, M. Kumar, S. Devi, V.B. Taxak, Reddish-orange light emission via combustion synthesized $Ba_3Y_4O_9:Sm^{3+}$ nanocrystalline phosphor upon near ultraviolet excitation, J. Lumin. 217 (2019) 116806.

[41] J.L. Zhang, Q.F. Guo, L.B. Liao, Y.J. Wang, M.Y. He, H. Ye, L.F. Mei, H.K. Liu, T.S. Zhou, B. Ma, Structure and luminescence properties of $La_6Ba_4(SiO_4)(6)F\text{-}2:Dy^{3+}$ phosphor with apatite structure, RSC Adv. 8 (2018) 38883–38890.

[42] S. Kaewjaeng, C. Jumpee, S. Kothan, J. Kaewkhao, N. Srisittipokakun, H.J. Kim, White emission from NaO-BaO-Bi_2O_3-SiO_2 glass system doped with Dy^{3+}, Mater. Today: Proc. 17 (2019) 1774–1779.

[43] S.J. Park, H.W. Jang, J.Y. Park, J.W. Chung, H.K. Yang, B.K. Moon, Gd_2O_3: Pr^{3+} nanospheres as bi-functional contrast agents for optical and magnetic resonance imaging properties, Ceram. Int. 45 (2019) 5958–5964.

[44] R.Y. Yunusov, O.V. Solovyev, Nonadiabatic effects in the $4f^1$-$5d^1$ absorption spectrum of a $LiYF_4$: Ce^{3+} crystal, Opt. Spectrosc. 116 (2014) 764–768.

[45] T. Wang, Z. Xia, Q. Xiang, S. Qin, Q. Liu, Relationship of 5d-level energies of Ce^{3+} with the structure and composition of nitride hosts, J. Lumin. 166 (2015) 106–110.

[46] S. Khiari, M. Velázquez, R. Moncorgé, J.L. Doualan, P. Camy, A. Ferrier, M. Diaf, Red-luminescence analysis of Pr^{3+} doped fluoride crystals, J. Alloys Compd. 451 (1–2) (2008) 128–131.

[47] D.E.K. Van, P.F. Smet, D. Poelman, Persistent luminescence in Eu^{2+}-doped compounds: a review, Materials 3 (4) (2010) 2536–2566.

[48] S. Wang, Z. Song, Y. Kong, Z. Xia, Q. Liu, Crystal field splitting of 4f-5d-levels of Ce^{3+} and Eu^{2+} in nitride compounds, J. Lumin. 194 (2018) 461–466.

[49] S. Wang, Z. Song, Y. Kong, Z. Xia, Q. Liu, 5d-level centroid shift and coordination number of Ce^{3+} in nitride compounds, J. Lumin. 200 (2018) 35–42.

[50] H. Terraschke, C. Wickleder, UV, blue, green, yellow, red, and small: newest developments on Eu^{2+}-doped nanophosphors, Chem. Rev. 115 (20) (2015) 11352–11378.

[51] W. Zhang, R. Hua, T. Liu, J. Zhao, L. Na, B. Chen, Preparation, photoluminescent properties and luminescent dynamics of $BaAlF_5$: Eu^{2+} nanophosphors, Mater. Res. Bull. 60 (2014) 247–251.

[52] J.M.P.J. Verstegen, J.L. Sommerdijk, Line emission of $SrBe_2Si_2O_7$: Eu^{2+} and $BaBe_2Si_2O_7$: Eu^{2+}, J. Lumin. 9 (1974) 297–301.

[53] P. Dorenbos, Anomalous luminescence of Eu^{2+} and Yb^{2+} in inorganic compounds, J Phys-Condens. Mater. 15 (2003) 2645.

[54] H. Watanabe, N. Kijima, Crystal structure and luminescence properties of $Sr_xCa_{1\text{-}x}AlSiN_3$: Eu^{2+} mixed nitride phosphors, J. Alloys Compd. 475 (2009) 434–439.

[55] C. Chudoba, J.G. Fujimoto, E.P. Ippen, H.A. Haus, U. Morgner, F. Kärtner, T. Tschudi, All-solid-state Cr: forsterite laser generating 14-fs pulses at 1.3 μm, Opt. Lett. 26 (2001) 292–294.

[56] L.D. Merkle, A. Pinto, H.R. Verdún, B. McIntosh, Laser action from Mn^{5+} in $Ba_3(VO_4)_2$, Appl. Phys. Lett. 61 (1992) 2386–2388.
[57] M.G. Brik, A.M. Srivastava, Critical review-a review of the electronic structure and optical properties of ions with d^3 electron configuration (V^{2+}, Cr^{3+}, Mn^{4+}, Fe^{5+}) and main related misconceptions, ECS J. Solid State Sci. 7 (1) (2018) R3079–R3085.
[58] Y. Katayama, Red to near-infrared persistent luminescence in transition metal ion activated phosphors, J. Ceram. Soc. Jpn. 125 (11) (2017) 793–798.
[59] J.M. Liu, Y.Y. Liu, D.D. Zhang, G.Z. Fang, S. Wang, Synthesis of $GdAlO_3$: Mn^{4+}, Ge^{4+}@au core-shell nanoprobes with plasmon-enhanced near-infrared persistent luminescence for in vivo trimodality bioimaging, ACS Appl. Mater. Interfaces 8 (44) (2016) 29939–29949.
[60] Q. Ye, Y. Wang, H. Guo, X. Zhou, P. Feng, S. Ding, Design a novel red to near-infrared persistent phosphors of $CaMgGe_2O_6$: Mn^{2+}, Sm^{3+} based on vacuum referred binding energy diagram, Dalton Trans. 48 (2019) 11052–11062.
[61] R. Kang, X. Dou, H. Lian, Y. Li, $SrAl_{12}O_{19}$: Fe^{3+}@3-aminopropyl triethoxysilane: ambient aqueous stable near-infrared persistent luminescent nanocomposites, J. Am. Ceram. Soc. 103 (2020) 258–265.
[62] L.P. Sosman, T. Abritta, M.R. Amaral, N. Cella, H. Vargas, Optical properties of $LiGaTiO_4$: Fe^{3+}, Solid State Commun. 105 (2) (1998) 135–138.
[63] T. Maldiney, A. Bessiere, J. Seguin, E. Teston, S.K. Sharma, B. Viana, A.J.J. Bos, P. Dorenbos, M. Bessodes, D. Gourier, D. Scherman, C. Richard, The in vivo activation of persistent nanophosphors for optical imaging of vascularization, tumours and grafted cells, Nat. Mater. 13 (2014) 418–426.
[64] X. Li, P. Li, Z. Wang, S. Liu, Q. Bao, X. Meng, Z. Yang, Color-tunable luminescence properties of Bi^{3+} in $Ca_5(BO_3)_3F$ via changing site occupation and energy transfer, Chem. Mater. 29 (20) (2017) 8792–8803.
[65] L. Li, J. Cao, B. Viana, S. Xu, M. Peng, Site occupancy preference and antithermal quenching of the Bi^{2+} deep red emission in β-$Ca_2P_2O_7$: Bi^{2+}, Inorg. Chem. 56 (11) (2017) 6499–6506.
[66] S. Asano, N. Yamashita, Y. Nakao, Luminescence of the Pb^{2+}-ion dimer center in CaS and CaSe phosphors, Phys. Status Solidi B 89 (2) (1978) 663–673.
[67] H.H. Chen, W.T. Zhu, Z.B. Zhang, W.X. Cai, X.F. Zhou, Er and Mg co-doped TiO_2 nanorod arrays and improvement of photovoltaic property in perovskite solar cell, J. Alloys Compd. 771 (2019) 649–657.
[68] P.F. Smet, D.E.K. Van, D. Poelman, Persistent phosphors, in: Handbook on the Physics and Chemistry of Rare Earths, Elsevier, 2015.
[69] S. Grangeon, A. Vinsot, C. Tournassat, C. Lerouge, E. Giffaut, S. Heck, N. Groschopf, M. A. Denecke, S. Wechner, T. Schafer, The influence of natural trace element distribution on the mobility of radionuclides. The example of nickel in a clay-rock, Appl. Geochem. 52 (2015) 155–173.
[70] Y. Jin, X.Y. Long, Y.N. Zhu, M.Q. Ge, Optical performance study of $Sr_2ZnSi_2O_7$: Eu^{2+}, Dy^{3+}, $SrAl_2O_4$:Eu^{2+}, Dy^{3+} and Y_2O_2S: Eu^{3+}, Mg^{2+}, Ti^{4+} ternary luminous fiber, J. Rare Earths 34 (2016) 1206–1212.
[71] S.Y. Liu, J.S. Zhang, J.S. Sun, B.J. Chen, X.P. Li, S. Xu, L.H. Cheng, Synthesis and long afterglow characteristics of $Sr_2MgSi_2O_7$:Eu^{2+}, Dy^{3+} by experimental optimization design, Acta Phys. Sin. 68 (2019) 7.
[72] L. Wang, Z.J. Shang, M.M. Shi, P.Y. Cao, B.B. Yang, J. Zou, Preparing and testing the reliability of long-afterglow $SrAl_2O_4$:Eu^{2+}, Dy^{3+} phosphor flexible films for temperature sensing, RSC Adv. 10 (2020) 11418–11425.

[73] B.Y. Qu, J.T. Wang, K. Liu, R.L. Zhou, L. Wang, A comprehensive study of the red persistent luminescence mechanism of Y_2O_2S: Eu, Ti, Mg, Phys. Chem. Chem. Phys. 21 (2019) 25118–25125.

[74] T. Aitasalo, J. Holsa, M. Kirm, T. Laamanen, M. Lastusaari, J. Niittykoski, J. Raud, R. Valtonen, Persistent luminescence and synchrotron radiation study of the $Ca_2MgSi_2O_7$: Eu^{2+}, R^{3+} materials, Radiat. Meas. 42 (2007) 644–647.

[75] R.M. Calderón-Olvera, E.A. Albanés-Ojeda, M. García-Hipólito, J.M. Hernández-Alcántara, M.A. Álvarez-Perez, C. Falcony, O. Álvarez-Fregoso, Characterization of luminescent $SrAl_2O_4$ films doped with terbium and europium ions deposited by ultrasonic spray pyrolysis technique, Ceram. Int. 44 (2018) 7917–7925.

[76] Z. Chen, Y.X. Pan, L.Q. Xi, R. Pang, S.M. Huang, G.K. Liu, Tunable yellow-red photoluminescence and persistent afterglow in phosphors $Ca_4LaO(BO_3)_3$:Eu^{3+} and $Ca_4EuO(BO_3)_3$, Inorg. Chem. 55 (2016) 11249–11257.

[77] Y. Katayama, J. Ueda, S. Tanabe, Effect of Bi_2O_3 doping on persistent luminescence of $MgGeO_3$:Mn^{2+} phosphor, Opt. Mater. Express 4 (2014) 613–623.

[78] Z.L. Wang, J.C. Zhang, G.S. Zheng, X.J. Peng, H.X. Dai, Violet-blue afterglow luminescence properties of non-doped $SrZrO_3$ material, J. Lumin. 144 (2013) 30–33.

[79] D. Poelman, N. Avci, P.F. Smet, Measured luminance and visual appearance of multicolor persistent phosphors, Opt. Express 17 (2009) 358–364.

Synthesis methods

2

Synthetic method of persistent phosphors plays an important role for the determination of the features of microstructure, afterglow performance, fluorescent quantum efficiency, and defect distribution. It is also closely related to applications of persistent phosphors, for example, bioimaging and photocatalysis. Conventional method to synthesize persistent phosphors mainly relies on solid-state reaction, during which a powdery oxide or carbonate raw materials are used as starting material usually. Since these powder particles are comparatively rough, a high sintering temperature is required to obtain persistent phosphors with designed compositions and structures. However, the high sintering temperature leads to a heterogeneous distribution of particle size and the formation of large crystallites that are considered unfavorable for applications such as biolabeling. In recent years, persistent phosphors have attracted extensive attention that was stimulated by using persistent phosphors as biolabels and photocatalysts. These applications require high homogeneity, monodispersity, and small particle size of persistent phosphors to improve the permeability, fluidity, and biocompatibility of persistent phosphors in organisms. Conventional synthesis methodology seemingly poses a conflict between the optimized morphology and afterglow time, that is, a high synthesis temperature is advantageous to improve the persistent duration, but results in the formation of agglomerated products with irregular morphologies [1, 2]. Therefore new synthesis approaches for preparing persistent phosphors, including sol-gel process, hydrothermal method, and template method, have been proposed to control their morphology in nanoscale and manage luminescent features.

The investigations on the synthesis of high-quality persistent phosphors with controlled crystalline phases, shapes, and sizes as well as long persistent duration could be divided into two main directions:

(1) synthesis of the nanoparticles and then activation of the persistent phosphors;
(2) synthesis of the persistent phosphors and then reduce the size by a top-down approach.

For the first direction, ordinary routes are based on wet-chemical methods, including sol-gel, combustion synthesis, hydro(solvo)thermal route, coprecipitation, microwave synthesis, and template methods. By comparison, some researchers also propose another distinctive way of preparing nanoparticles via laser ablation method. By choosing the appropriate parameters for laser ablation (wavelength, repetition rate, peak power, intensity, irradiation time, etc.), it is possible to obtain the desired nanoparticles with different morphology and size. The as-synthesized nanoparticles are then activated by, for example, a thermal treatment process to induce long persistent phosphorescence.

Table 2.1 gives a comparison of synthesis methods in terms of particle sizes, required temperatures, homogeneity, morphology, and afterglow times. High-temperature annealing generally produces persistent phosphors with large sizes and

Persistent Phosphors. https://doi.org/10.1016/B978-0-12-818637-4.00002-1

Table 2.1 Comparison of synthesis methods of persistent phosphors in terms of the particle sizes, homogeneity, morphology control, synthesis temperature, and afterglow time.

Synthesis method	Particle size	Homogeneity	Morphology	Synthesis temperatures (°C)	Afterglow time	Reference
Solid-solid reaction	Micrometer scale	Bad	Bad	>1000	Long	[3, 4]
Sol-gel method	Nanometer to micrometer scale	Medium	Medium	<1000	Short	[5–7]
Combustion method	Nanometer to micrometer scale	Medium	Medium	<1000	Short	[8–11]
Hydrothermal/ solvothermal method	Nanometer to micrometer scale	Good	Medium	<1000	Short	[12–17]
Coprecipitation	Nanometer to micrometer scale	Medium	Medium	<1000	Short	[18, 19]
Template method	Nanometer to micrometer scale	Good	Medium	<1000	Short	[20–27]
Electrospinning	Micrometer scale	Good	Medium	>1000	Short	[28]
Reverse micelle method	Nanometer to micrometer scale	Medium	Medium	>1000	Short	[29]
Microwave synthesis	Nanometer to micrometer scale	Medium	Medium	>1000	Short	[30]
Laser technique	Nanometer to micrometer scale	Medium	Bad	None	Medium	[31, 32]
Electron beam irradiation	Nanometer to micrometer scale	Medium	Bad	None	Medium	[33]

poor dispersibility and cannot meet the requirement for the biomedical applications, which need the persistent phosphorescent nanoparticles with small sizes, good monodispersity, and easy modification. Although significant progress has been achieved, it is still a challenge to improve the monodispersity and size distribution for particles synthesized at a high temperature.

2.1 Solid-state reaction

Solid-state reaction, which occurs between powders in the solid state, has been the most widely used approach for the preparation of persistent phosphors including oxides, (oxy)sulfide, and (oxy)nitrides since it is comparatively simple, very suitable for mass production, and easy to obtain a long persistent luminescent duration. Since high sintering temperatures can generally increase the number density of intrinsic thermal defects, high-temperature sintering is an effective way for activating the afterglow. These lattice defects and distortions in the form of cation vacancy, anion vacancy, interstitial, etc. (i.e., Schottky imperfections, anti-Schottky imperfections, and Frenkel imperfections of the point defects) are believed to be beneficial to improve the afterglow properties. There are many factors influencing the afterglow properties of persistent phosphors prepared by solid-state reaction. The main factors affecting quality of persistent phosphors include (1) preparation temperature, (2) preparation duration, (3) ambient atmosphere, and (4) ambient pressure. Generally the preparation temperature of the persistent phosphors is controlled at 1000–1500°C. A temperature too low might suppresses the formation of defects, while an excessively high temperature is also undesirable for the preparation of persistent phosphors. Though solid-state method has many advantages in synthesizing the persistent phosphors with remarkable afterglow property, it still has the unavoidable disadvantages. On the one hand the grain size of phosphor powders prepared by this method is several tens of micrometers. On the other hand, high calcination temperature usually leads to poor homogeneity and morphology control. These shortcomings have greatly limited the development and application of the synthesized persistent phosphors in the areas of biomedicine and bioscience [34–40].

2.2 Wet-chemical approach

2.2.1 Sol-gel method

Sol-gel method is a well-known method for preparing inorganic nanoparticles. The sol-gel process can be carried out in different ways, including sol-gel route (based on the hydrolysis and condensation of molecular precursors), gelation route (based on concentration of aqueous solutions containing metal chelates), and polymerizable complex route. Generally, this process uses metal salts or alkoxide as precursors, citric acid as a chelating ligand, and alcohol as a cross-linking agent to form a polymeric resin on the molecular level. The further required heat treatment is beneficial for the formation of defects, and small nanocrystallites are packed into the discrete space offered by complexing agent, thereby limiting the growth of nanocrystallites under such thermal treatment. Moreover, compared with the solid-state reaction, the persistent phosphors prepared by the sol-gel method have a smaller size, better dispersibility, and improved morphology uniformity. Much attention has been directed to the inorganic salt-based sol-gel approach in the synthesis of strontium aluminate persistent phosphors [41, 42]. A representative synthesis process is illustrated in Fig. 2.1.

Chermont et al. first proposed a bioimaging method by using near-infrared persistent phosphorescence nanoparticles, $Ca_{0.2}Zn_{0.9}Mg_{0.9}Si_2O_6$: Eu^{2+}, Dy^{3+}, Mn^{2+} as the biolabel, which was prepared by the sol-gel method. This work opened a completely new era for applications of persistent phosphors [5]. Briefly, metal salt precursors and tetraethoxysilane (TEOS) were reacted at 70°C to induce the sol-to-gel transition, and the dried gel was further calcinated at 1050°C (for 10 h in a weakly reductive atmosphere). The resulting persistent phosphors were ground with a mortar, and the smallest particles were separated by selective sedimentation. To be used for subsequent bioconjugation, the obtained persistent phosphorescence nanoparticles were further treated with NaOH solution to introduce hydroxyl groups onto their surface. Electron microscopy analysis showed that the sizes of the persistent phosphorescence nanoparticles were ranging from 50 to 200 nm. The persistent luminescence was in the near-infrared region, and the maximum luminescence intensity was centering at around 690 nm. The persistent phosphorescence in the persistent phosphorescence nanoparticles was recorded with a photon-counting system based on a cooled GaAs intensified charge-coupled device camera. The persistent phosphorescence was detectable for over 24 h in the dark. In the following years, other persistent phosphorescence nanoparticles based on different hosts and activation centers prepared by sol-gel technique were reported, such as $SrAl_2O_4$: Eu, Dy; $Sr_2MgSi_2O_7$: Eu, Dy; ZrO: Ti; $CaTiO_3$: Pr [6]; $ZnSiO_4$: Mn [43]; $CaTiO_3$: Eu; Ga_2O_3: Cr; ZrO_2: Yb, Er; $Sr_4Al_{14}O_{25}$: Ce; $BaCa_2Al_8O_{15}$: Eu; Ca_2SnO_4: Sm; $CaMgSi_2O_6$: Mn [44]; $CdSiO_3$: Mn, Eu; Y_2O_2S: Dy, Mg, Ti; $CaAl_2O_4$: Eu, Nd; $BaZrO_3$: Mg; $ZnGa_2O_4$:Cr [45]; and $Zn_3Ga_2GeO_8$:Yb, Er, Cr [46]. Many of these persistent phosphorescence nanoparticles were prepared and widely employed in biomedicine application (Fig. 2.2) [5, 45, 46].

Xue et al. reported the preparation of $ZnGa_2O_4$:Cr^{3+} persistent phosphorescence nanoparticles by the sol-gel method followed by calcination at 1000°C for 3 h. The calcined phosphors were further grinded and then dispersed into NaOH solution for

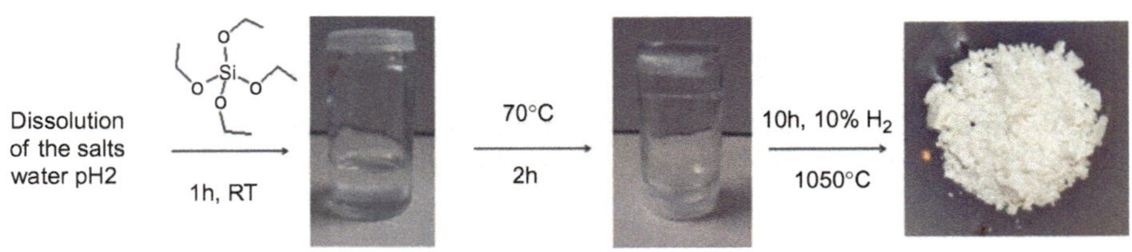

Fig. 2.1 Sol-gel synthesis of persistent phosphors.
From T. Maldiney, et al., In vivo imaging with persistent luminescence silicate-based nanoparticles, Opt. Mater. 35 (10) (2013) 1852–1858.

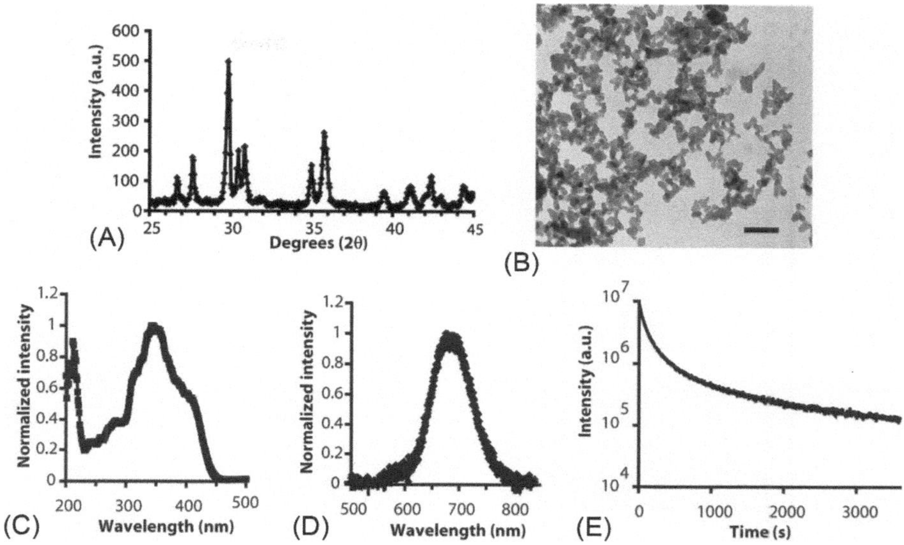

Fig. 2.2 Physical characteristics of persistent phosphorescence nanoparticles. (A) Persistent phosphorescence nanoparticles showed a clinoenstatite-like structure. (B) Transmission electronic microscopy images of the synthesized persistent phosphorescence nanoparticles. (Scale bar: 200 nm.) (C) Excitation spectrum. (D) Long afterglow emission spectrum. (E) Time dependence of the luminescence intensity of the persistent phosphorescence nanoparticles. Persistent phosphorescence nanoparticles (10 mg) was put in 96-well plates under direct exposure to a 6-W UV lamp for 5 min. The luminous decay data were fit by a power law function for time > 100 s.

From Q. le Masne de Chermont, C. Chanéac, J. Seguin, F. Pellé, S. Maîtrejean, J.P. Jolivet, D. Gourier, M. Bessodes, D. Scherman, Nanoprobes with near-infrared persistent luminescence for *in vivo* imaging, Proc. Natl. Acad. Sci. U S A 104 (2007), 9266–9271, Copyright 2007, National Academy of Sciences, USA.

sonication. The persistent phosphorescence nanoparticles with the desired size were collected by selective sedimentation and functionalized with PEG for bioimaging applications. The $ZnGa_2O_4:Cr^{3+}$ persistent phosphorescence nanoparticles show a size of less than 200 nm, and the hydrodynamic diameter of the persistent phosphorescence nanoparticles before/after PEG modification is about 141/190 nm (Fig. 2.3) [45].

In addition to the size distribution, sol-gel method also enables the control of the morphology (nanosphere [47], nanowire [48], etc.) of as-prepared materials by varying the synthesis condition, including precursors, solvents, ligands, pH value, and treatment temperature. Zhang et al. synthesized $LiGa_5O_8$: Cr^{3+} persistent phosphorescence nanoparticles by a sol-gel method. They investigated the influence of synthesis temperature (700–800–900–1100°C) on the morphology and size distribution of $LiGa_5O_8$: Cr^{3+} nanoparticles [49]. To inhibit the growth of nanocrystallites, some

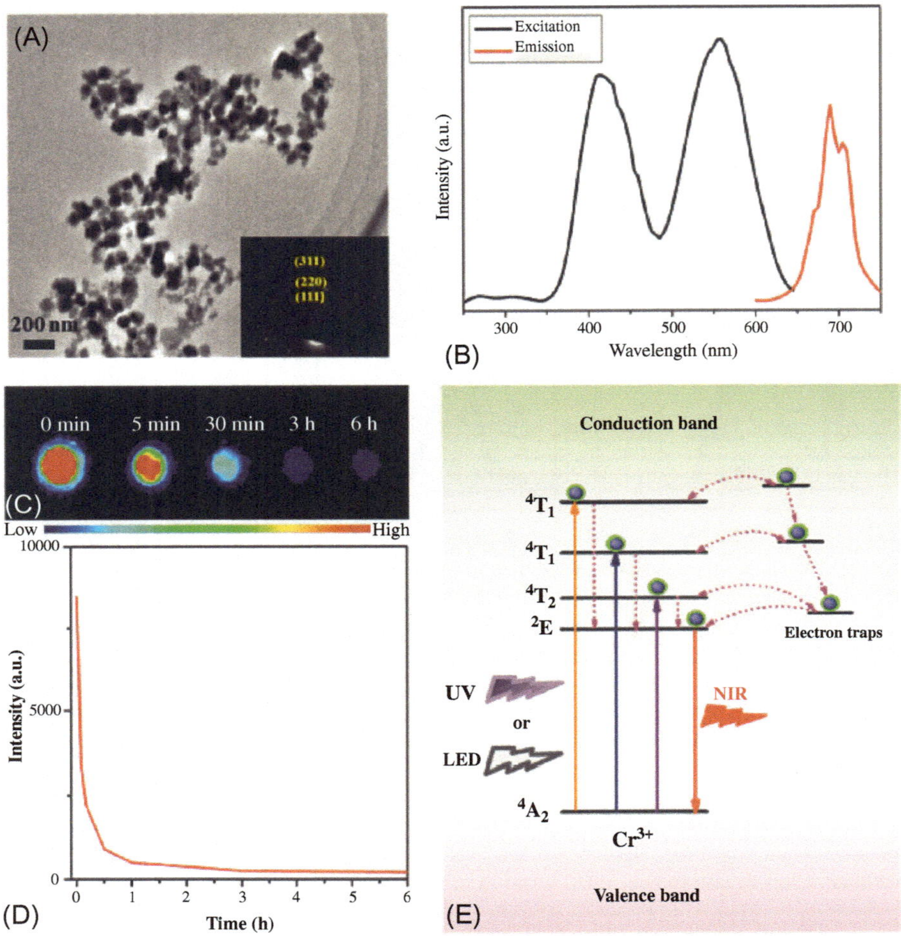

Fig. 2.3 (A) TEM image of $ZnGa_2O_4:Cr^{3+}$ persistent phosphorescence nanoparticles, inset: SAED. (B) Excitation (*black curve*) and emission (*red curve*) spectra of $ZnGa_2O_4:Cr^{3+}$ persistent phosphorescence nanoparticles at room temperature. (C) In vitro decay near-infrared persistent phosphorescence imaging of P $ZnGa_2O_4:Cr^{3+}$ persistent phosphorescence nanoparticles after ceasing x-ray (5 min, 45 kV) irradiation. (D) The corresponding decay curve of persistent luminescence intensity of (C). (E) Energy schematic diagram of the persistent luminescence mechanism under UV/LED excitation; the *dotted line arrows* and *solid line arrows* represent electron transfer processes and optical transitions, respectively. Reproduced with permission from Z. Xue, X. Li, Y. Li, M. Jiang, H. Liu, S. Zeng, et al., X-ray-activated near-infrared persistent luminescent probe for deep-tissue and renewable *in vivo* bioimaging. ACS Appl. Mater. Interfaces 9 (2017), 22132–22142, Copyright 2017, American Chemical Society.

compositional approaches have been proposed to synthesis persistent phosphorescence nanoparticles, including microwave-assisted sol-gel route [30], sol-gel combustion processing [50], and sol-gel template technology (which will be discussed in Section 2.5). Red-emitting $Sr_3Al_2O_6$: Eu^{2+}, Dy^{3+} nanorods with a width of about 80 nm were synthesized by using the microwave-assisted sol-gel method [48]. The advantage of microwave irradiation is that it can effectively shorten the reaction time, accordingly decreasing the particle size. In the sol-gel combustion synthesis of $SrAl_2O_4$: Eu^{2+}, Dy^{3+}, raw materials were dissolved in concentrated nitric acid. After the mixed solution was heated up to 90°C, a small amount of ethanol was added dropwise into the solution until no rufous gases (NO_2) evaporated out. Finally, spontaneous ignition occurred, which is accompanied by combustion with enormous swelling [49]. Electron microscopy analysis showed that the nanoparticles exhibit a size distribution with the diameter ranging from 90 to 250 nm.

The sol-gel method has the following advantages: (1) The obtained nanoparticles have good uniformity and high purity, (2) it is energy saving due to the low annealing temperature, and (3) it can be used to prepare coatings, fibers, and composite materials. But there are still some problems: first, some raw materials are organic and toxic; second the entire sol-gel process usually takes a long time, from several days to weeks.

2.2.2 *Hydrothermal method*

Hydrothermal synthesis refers to material synthesis by chemical reaction performed within a sealed vessel filled with aqueous of reagents under temperatures above 100°C and autogenously high pressure. Different from the sol-gel and the coprecipitation methods (which will be described in S2.4), hydrothermal synthesis is usually carried out within specialized reaction vessels known as autoclaves, which provide a high-pressure and closed environment to facilitate solvation of the reagents and reactions between liquid precursors. Compared with other synthesis methods, this method can synthesize high-crystalline nanomaterials under relatively mild conditions.

Hydrothermal method is a promising method that allows facile control of reaction parameters such as reaction time and temperature to obtain highly homogeneous products. In a typical hydrothermal synthesis procedure for synthesizing persistent phosphorescence nanoparticles, organic additives and/or surfactants with specific functional groups like oleic acid (OA), polyethylenimine (PEI), EDTA, and cetyltrimethylammonium bromide (CTAB) are generally added along with the reaction precursors to achieve simultaneous control over the crystalline phases, sizes, and morphologies. A high reaction temperature is usually considered indispensable to form the target persistent phosphorescence nanoparticles [51, 52].

Persistent phosphorescence nanoparticles with different morphologies can be synthesized by the hydrothermal process, such as $SrAl_2O_4$: Eu^{2+}, Dy^{3+} nanosheets [53]; β-Ga_2O_3:Cr^{3+} nanowire; Y_2O_2S: Eu^{3+} M^{2+} (M = Mg, Ca, Sr, and Ba), Ti^{4+} nanotubes [54]; $SrAl_2O_4$: Eu^{2+}, Dy^{3+} nanoparticles; needlelike $SrAl_2O_4$: Eu^{2+} Dy^{3+} nanophosphors; and Zn_2GeO_4:Mn^{2+} nanorods [12]. Maldiney and coworkers demonstrated a successful example for hydrothermal synthesis of $ZnGa_{1.995}O_4$: 0.5%Cr persistent phosphorescence nanoparticles with sizes between 50 and 100 nm [13].

The experimental results showed that the critical factors for synthesizing pure products are pH value, the initial composition, and the reaction temperature (Fig. 2.4A–C). Li et al. reported the hydrothermal synthesis of $ZnGa_2O_4$: 0.4%Cr^{3+} persistent

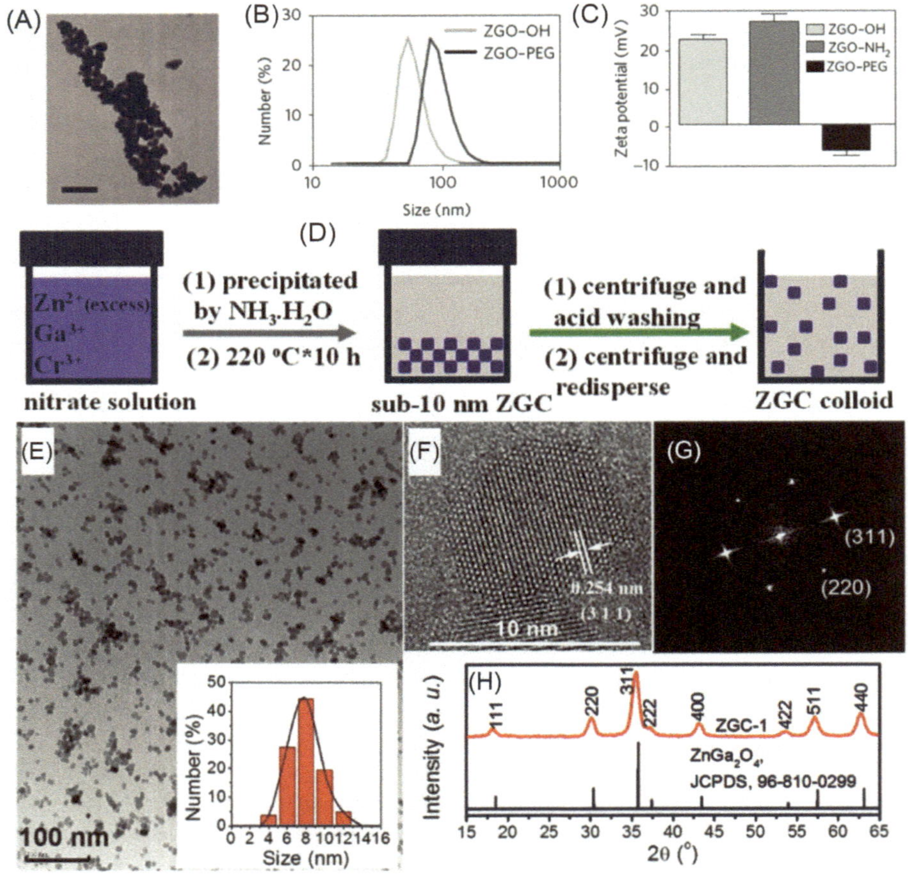

Fig. 2.4 (A) Transmission electron micrograph of $ZnGa_{1.995}O_4$: 0.5%Cr nanoparticles. Scale bar, 100 nm. (B) The hydrodynamic diameter measured by dynamic light scattering before and after PEG coverage. (C) The evolution of zeta potential as a function of surface coverage. (D) Schematic illustration of the synthesis of $ZnGa_2O_4$: 0.4%Cr^{3+} PPNPs. (E) TEM image and size distribution of $ZnGa_2O_4$: 0.4%Cr^{3+} PPNPs. (F) HRTEM image of $ZnGa_2O_4$: 0.4%Cr^{3+} PPNPs. (G) SAED. (H) x-Ray diffraction pattern of $ZnGa_2O_4$: 0.4%Cr^{3+} NPPs.
A–C: Reproduced with permission from T. Maldiney, A. Bessiere, J. Seguin, E. Teston, S.K. Sharma, B. Viana, et al., The *in vivo* activation of persistent nanophosphors for optical imaging of vascularization, tumours and grafted cells, Nat. Mater. 13 (2014), 418–426, Copyright 2014, Nature Publishing Group; D–G: Reproduced with permission from Z. Li, Y. Zhang, X. Wu, L. Huang, D. Li, W. Fan, G. Han, Direct aqueous-phase synthesis of sub-10 nm "luminous pearls" with enhanced *in vivo* renewable near-infrared persistent luminescence, J. Am. Chem. Soc. 137 (2015), 5304–5307, Copyright 2015, American Chemical Society.

phosphorescence nanoparticles. The specific solution-phase reaction procedure was outlined in Fig. 2.4D, and this method produced monodisperse persistent phosphorescence nanoparticles as small as circa 8 nm. They demonstrated that such sub-10-nm nanocrystals could be readily dispersed and functionalized and formed stable colloidal solutions in aqueous solution (Fig. 2.4E–H) [14].

The morphology of the phosphors is controlled by adjusting the reaction conditions, including reaction temperature, heating rate, holding time, modifier, and pH. Wang et al. reported that the size of the $Zn_{1+x}Ga_{2-2x}Ge_xO_4$: Cr, $0 \leq x \leq 0.5$ nanoparticles gradually increases with the increase of x in the formula. In their synthesis, metal nitrates were dissolved as raw materials in deionized water, and pH was adjusted to 8.5 with ammonium hydroxide. Then the mixture was transferred to a Teflon-lined autoclave and reacted at 220°C for 6 h. The as-prepared persistent phosphorescence nanoparticles were collected by centrifugation and washed with deionized water. As shown in Fig. 2.5, all of the persistent phosphorescence nanoparticles are well dispersed with a uniform shape and size. As the value of x in the chemical formula increases, the size of the nanoparticles increases from about 7 nm to about 80 nm [15]. In addition, Wang et al. also reported that the length and persistent phosphorescence of Zn_2GeO_4: Mn^{2+} nanorods can be finely tuned by simply changing the pH (6.0–9.5) of the hydrothermal reaction system (Fig. 2.6) [12].

Although the hydrothermal method has the advantages of high yield, easy size control, uniform morphology, and excellent dispersion of product, it also has its shortcomings as follows: (1) Even the method is relatively simple, it needs a longer reaction duration. (2) It is not possible to directly observe the reaction process carried out within a closed system. (3) Current research on hydrothermal synthesis is generally limited to the preparation of oxide powders. (4) The hydrothermal synthesis has to be carried out within pressurized vessels that may raise safety concerns and hinders its large-scale applications.

2.2.3 Template method

Template method is another well-developed synthesis method to fabricate the persistent phosphorescence nanoparticles [55]. A typical process is to deposit the persistent phosphors' layer on the surface of the template, that is, mesoporous silica nanospheres (MSNs), which results in the formation of the MSNs@persistent phosphors composite material (Fig. 2.7A) [20].

This material combines the mesoporous structure of MSNs and the phosphorescence of the persistent phosphorescence nanoparticles to allow for novel dual-functional biolabels and drug delivery. Up to now, MSNs have been used as templates for the synthesis of many materials, such as noble metals, metal oxides, carbon materials, and phosphors. [20] The templating process is usually combined with a high-temperature treatment, and it is regarded as the most convenient method for synthesizing high-quality persistent phosphorescence nanoparticles. Zhang and Han et al. recently reported the synthesis of persistent phosphorescent nanospheres (SiO_2/$SrMgSi_2O_6$: Eu, Dy, SiO_2/$ZnGa_2O_4$: Cr, and SiO_2/$CaMgSi_2O_6$: Eu, Pr, Mn), SiO_2/$Gd_3Ga_5O_{12}$:Cr, Nd with a narrow size distribution (Fig. 2.7B and C) [20–24]. In

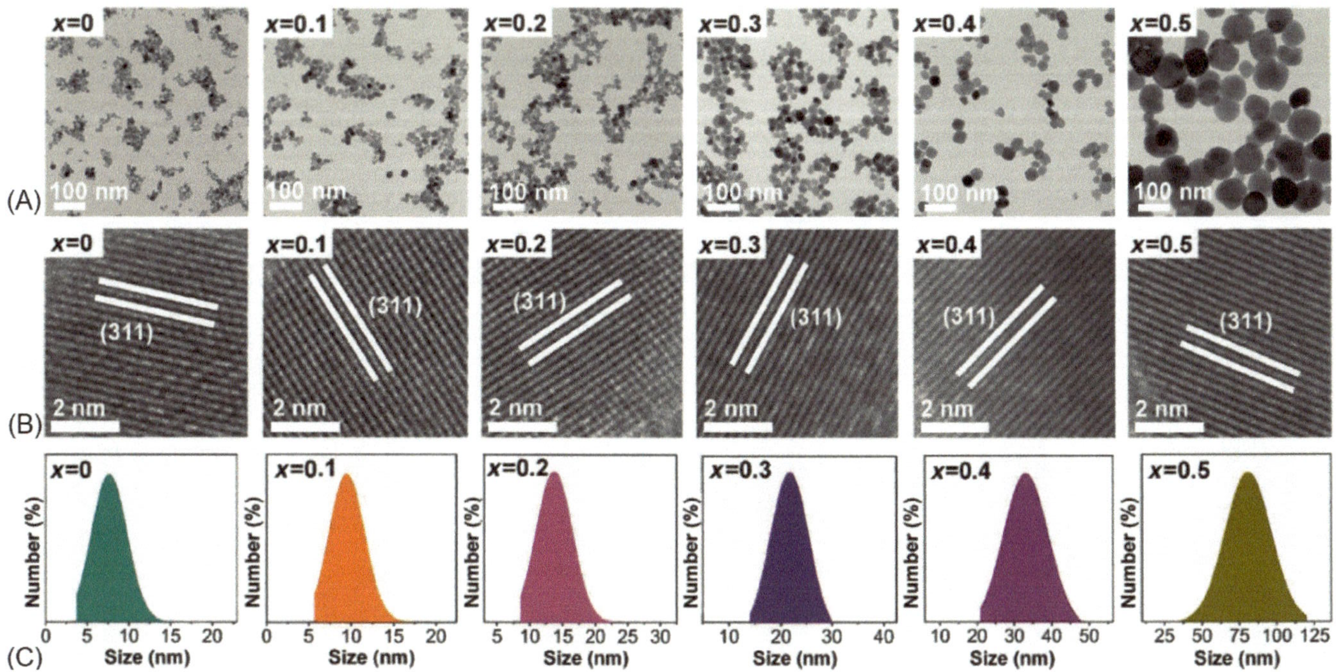

Fig. 2.5 (A) TEM images, (B) HRTEM images, and (C) corresponding size distributions of the $Zn_{1+x}Ga_{2-2x}Ge_xO_4$: Cr nanoparticles with x from 0 to 0.5.

Reproduced with permission from J. Wang, Q. Ma, X.X. Hu, H. Liu, W. Zheng, X. Chen, Q. Yuan, W. Tan, Autofluorescence-free targeted tumor imaging based on luminous nanoparticles with composition-dependent size and persistent luminescence, ACS Nano 11 (2017), 8010–8017, Copyright 2017, American Chemical Society.

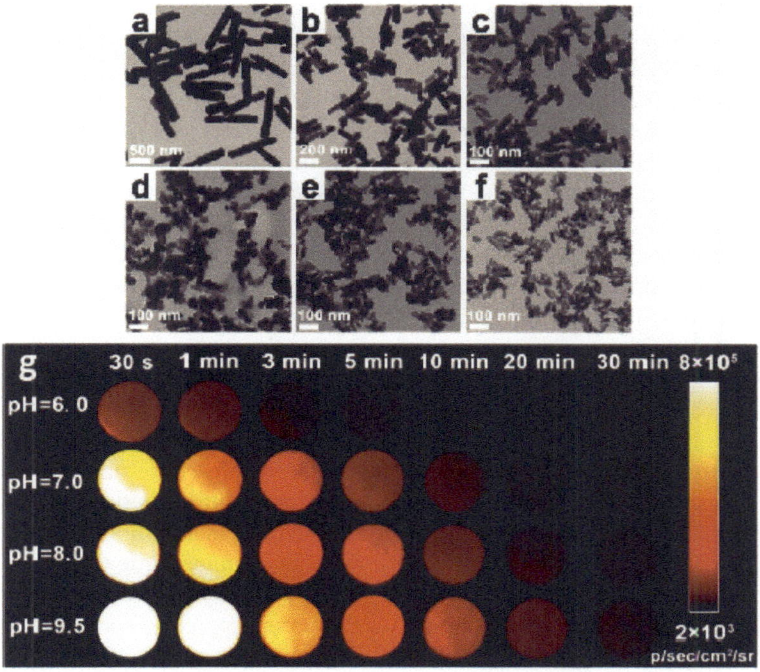

Fig. 2.6 TEM images of the Zn_2GeO_4: Mn^{2+} nanorods prepared at different pH values (A, pH = 6.0; B, pH = 6.5, C, pH = 7.0; D, pH = 7.5; E, pH = 8.5; and F, pH = 9.5). (G) Persistent phosphorescence decay images of the Zn_2GeO_4: Mn^{2+} nanorods.
Reproduced with permission from J. Wang, Q. Ma, W. Zheng, H. Liu, C. Yin, F. Wang, X. Chen, Q. Yuan, W. Tan, One-dimensional luminous nanorods featuring tunable persistent luminescence for autofluorescence-free niosensing, ACS Nano 11 (2017), 8185–8191, Copyright 2017, American Chemical Society.

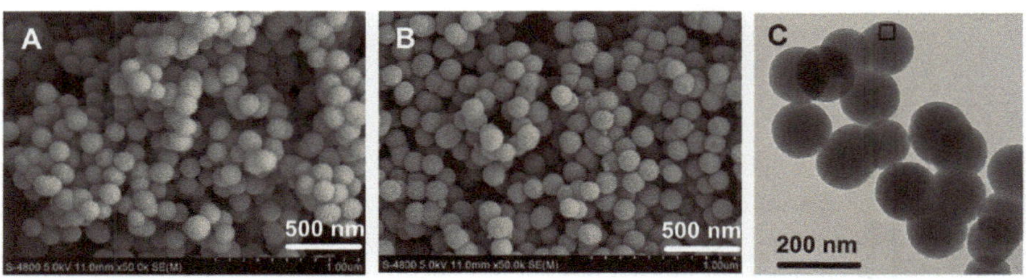

Fig. 2.7 FESEM images of (A) SiO_2 template and (B) $SiO_2/ZnGa_2O_4$: Cr^{3+} prepared by template method. (C) TEM of $SiO_2/ZnGa_2O_4$: Cr^{3+}.
Reproduced with permission from Z. Li, Y. Zhang, X. Wu, X. Wu, R. Maudgal, H. Zhang, G. Han, *In vivo* repeatedly charging near-infrared-emitting mesoporous $SiO_2/ZnGa_2O_4$: Cr^{3+} persistent luminescence nanocomposites, Adv. Sci. 2 (2015), 1500001, Copyright 2015, Wiley.

the synthesis process, metal ions were first impregnated into the mesopores of MSN templates by immersing MSNs into an aqueous solution of metal ions. Afterward, ion-impregnated MSNs were annealed at high temperature to form phosphorescent nanomaterials with a narrow size distribution. The key feature of this method was that mesoporous silica nanoparticles act not only as the silicon source but also as hard template to control the morphology and the size of nanoparticles (50–100 nm) [20].

Lin et al. synthesized $SiO_2@ZnGa_2O_4:Cr^{3+}@SiO_2$ nanoparticles with kiwifruit-like structure that can tolerate high-temperature calcination. The $SiO_2@ZnGa_2O_4$: Cr^{3+} nanoparticles with a core-shell structure were prepared by the template method in hydrothermal process. The obtained nanoparticles were then coated with silica and used for in vivo bioimaging (Fig. 2.8) [25]. The agglomeration of particles in solid-state reaction is circumvented by regulating the reaction on the surfaces and the pores of MSNs. Importantly the structure of silica spheres is inherited by resultant persistent phosphorescence nanoparticles. In addition to silica sphere, the use of a purgeable carbon spheres as the templates has also been reported. Wang et al. reported the synthesis of hollow structured $ZnGa_2O_4$: Cr^{3+} persistent phosphorescence nanoparticles with tunable sizes by using purgeable carbon spheres as template. This strategy yields persistent phosphorescence nanoparticles with large and size tunable hollow cavities after calcination. Due to the hollow cavity of NPPs, high chemical drug (DOX) or photosensitizer (Si-Pc) loading can be achieved. The DOX/Si-Pc-loaded hollow persistent phosphorescence nanoparticles exhibit efficient tumor suppression benefited from the large cavities, and this process can be monitored optically by the afterglow of persistent phosphorescence nanoparticles (Fig. 2.9) [26].

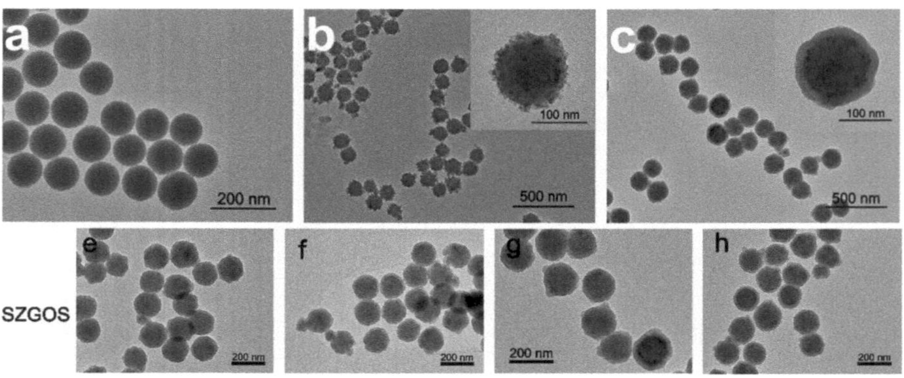

Fig. 2.8 TEM images of (A) SiO_2 nanospheres, (B) $SiO_2@ZnGa_2O_4$: Cr^{3+} nanoparticles, and (C) $SiO_2@ZnGa_2O_4$: $Cr^{3+}@SiO_2$ nanoparticles. (E) $SiO_2@ZnGa_2O_4:Cr^{3+}@SiO_2$ nanoparticles without calcination and (F–H) $SiO_2@ZnGa_2O_4$: $Cr^{3+}@SiO_2$ nanoparticles after calcination for 4 h in air at 600, 800, and 950°C, respectively.

Reproduced with permission from X.H. Lin, L. Song, S. Chen, X.F. Chen, J.J. Wei, J. Li, et al., Kiwifruit-like persistent luminescent nanoparticles with high-performance and in situ activatable near-infrared persistent luminescence for long-term *in vivo* bioimaging, ACS Appl. Mater. Interfaces 9 (2017), 41181–41187, Copyright 2017, American Chemical Society.

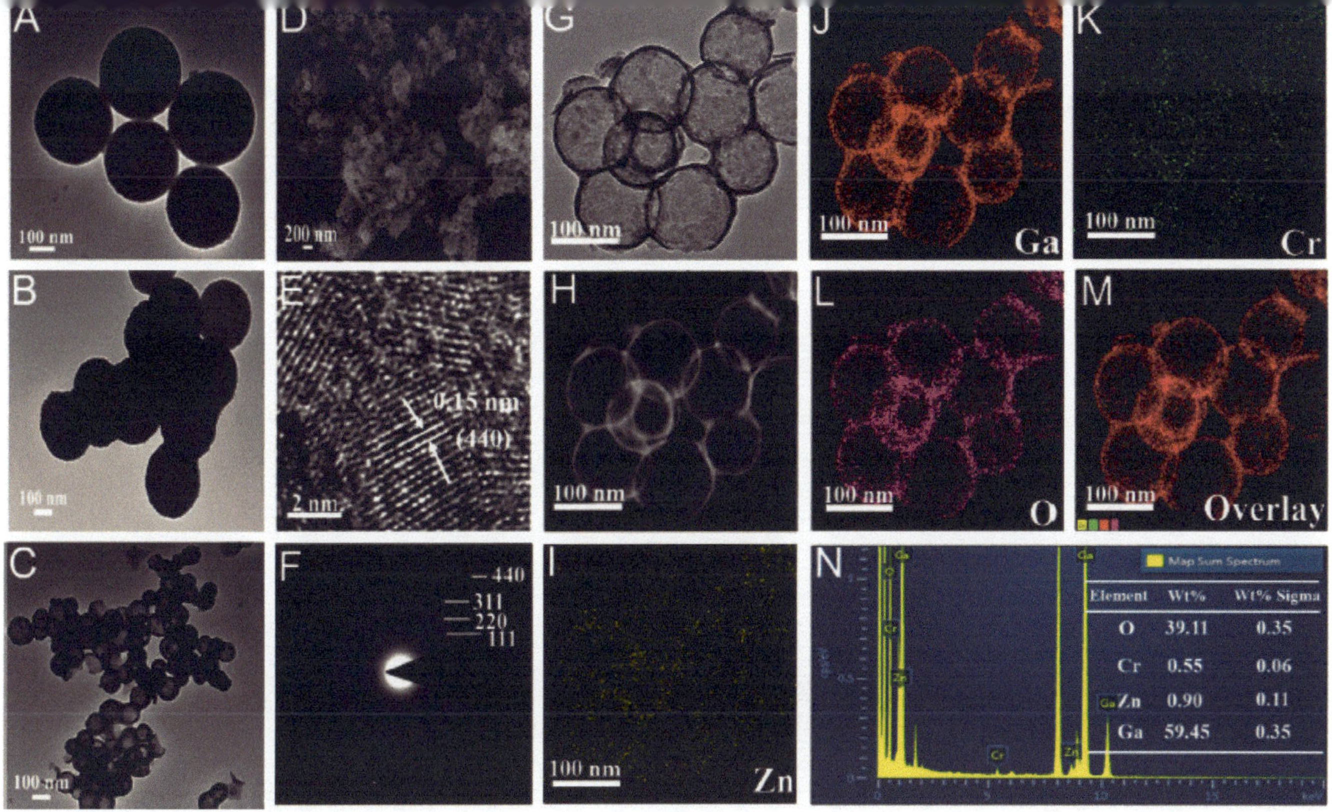

Fig. 2.9 Synthesis of hollow persistent phosphorescence nanoparticles by using 300-nm-sized carbon spheres as templates. TEM images of the carbon spheres with a size of 300 nm (A). Precursor ions labeled carbon spheres (B). The hollow persistent phosphorescence nanoparticles (C). SEM image of the hollow persistent phosphorescence nanoparticles (D). High-resolution TEM image of the hollow persistent phosphorescence nanoparticles (E). SAED pattern of the hollow persistent phosphorescence nanoparticles (F). Element mapping (G–M). EDS of the hollow persistent phosphorescence nanoparticles (N). Reproduced with permission from J. Wang, J. Li, J. Yu, H. Zhang, B. Zhang, Large hollow cavity luminous nanoparticles with near-infrared persistent luminescence and tunable sizes for tumor afterglow imaging and chemo-/photodynamic therapies, ACS Nano 12 (2018), 4246–4258, Copyright 2018, American Chemical Society.

2.2.4 Coprecipitation

In the coprecipitation process, all the raw materials are dissolved in solutions, and the products are precipitated through adding of certain ions or changing of, for example, pH and temperature. The precipitates are then separated, heat-treated to decompose the precursor, and finally annealed to form crystallized powder. Coprecipitation methods can generally be divided into direct precipitation methods and homogeneous precipitation methods. In the direct precipitation method, a reactant directly contacts a precipitating agent, resulting in precipitation of the product. The disadvantage of this method is that the reaction is too rapid, which tends to cause agglomeration of the product particles. The homogeneous precipitation method usually uses urea as a precipitating agent, and by hydrolysis of urea, anions such as OH^- and CO_3^{2-} are slowly generated in the solution, which then combine with metal cations to form precipitated product. In other words the precipitant does not directly react with the cation. It is well known that urea hydrolysis is a slow process occurring in aqueous solution. Compared with the direct precipitation method, the slow decomposition of urea allows the homogeneous mixing of anions and cations in the molecular level, resulting in the formation of product with facile size control. This approach is an efficient technique for the preparation of the inorganic nanomaterials as a homogeneous reaction environment favors the formation of product particles with narrow size distribution. Now this technology has also been used to fabricate fluoride-based and oxide-based luminescent nanoparticles. After a higher-temperature treatment, this modified coprecipitation method certainly can be employed to prepare different NPPs, such as Mn-doped calcium phosphates; $Ca_5(PO_4)_3(OH)$ hydroxyapatite; β-$Ca_3(PO_4)_2$ and $Ca_{9.5}Mn(PO_4)_7$ whitlockite; $(Y_{1-x}Gd_x)_2O_3$: Eu^{3+}, Sm^{3+}, Si^{4+}, Mg^{2+} [56]; and $Sr_2MgSi_2O_7$: Eu^{2+}, Dy^{3+}.

Cheng et al. synthesized the hexagonal $SrAl_2O_4$: Eu^{2+}, Dy^{3+} and $BaAl_2O_4$: Eu^{2+}, Dy^{3+} nanotubes with persistent phosphorescence by a coprecipitation method followed by heat treatment in a reducing atmosphere. TEM results revealed the presence of large amount of nanotubes, with about 30–60 nm wall thickness and about 80–200 nm diameter, in the sample. But the tube-form nanomaterials are not suitable as the biolabels for optical imaging [18, 57]. Recently, Li et al. reported a two-step synthesis route to prepare morphology and afterglow controllable persistent luminescent oxycarbonates ($La_2O_2CO_3$: Eu^{3+}, Ho^{3+}). In the first step, Li et al. employed a precipitation pretreatment to ensure uniform precursor nanorods. Then a morphology evolution from bamboo-like nanorods to nanoparticles occurred, and the control of afterglow time can be realized by simply changing posttreatment conditions, such as reaction duration and reaction temperature. In particular the nanoparticles can be easily produced via cutting the bamboo-like nanorod section by section by mechanical grinding (Fig. 2.10) [58].

2.2.5 Combustion method

Another popular synthesis technique for the preparation of persistent phosphorescence nanoparticles is combustion method. This technique is based on exothermic redox reactions that undergo self-sustaining combustion. For instance, mixtures of metal

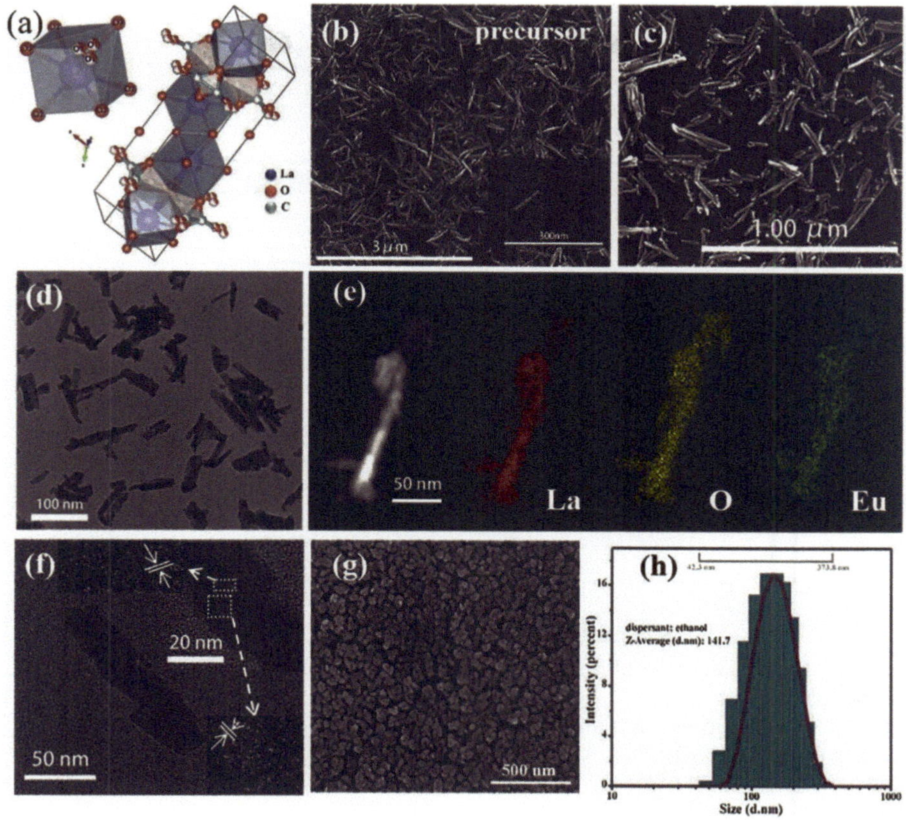

Fig. 2.10 (A) Crystal structure of $La_2O_2CO_3$. (B) SEM image of precursor nanorods. (C) SEM image of $La_2O_2CO_3$: 1%Eu^{3+} nanorods after thermal treatment at 600°C for 0.5 h. (D) TEM image of $La_2O_2CO_3$: 1%Eu^{3+} bamboo-like nanorods. (E) EDS mapping of $La_2O_2CO_3$: 1%Eu^{3+} bamboo-like nanorods. (F) HRTEM image of $La_2O_2CO_3$: 1%Eu^{3+} bamboo-like nanorods. (G) SEM image of $La_2O_2CO_3$: 1%Eu^{3+} nanoparticles after mechanical grinding. (H) DLS of $La_2O_2CO_3$: 1%Eu^{3+} nanoparticles.
Reproduced with permission from X. Doua, Y. Li, T. Vaneckova, R. Kang, Y. Hu, H. Wen, et al., Versatile persistent luminescent oxycarbonates: morphology evolution from nanorods through bamboo-like nanorods to nanoparticles, J. Lumin. 215 (2019), 116635, Copyright 2019, Elsevier.

nitrates (oxidizers) and a fuel (reducer; e.g., urea, citric acid, or glycine) undergo spontaneous combustion under heating, and the chemical energy from the exothermic reaction heats the precursor mixture to high temperatures. The detailed synthesis process can be found in Fig. 2.11 [8]. The combustion method can be further divided into a low-temperature combustion synthesis and gas combustion synthesis.

Low-temperature combustion synthesis, also known as sol-gel combustion synthesis, is a strong redox reaction between organic salt gel or organic salt and metal nitrate gel during heating, which generates a large amount of gas, and it is usually used for

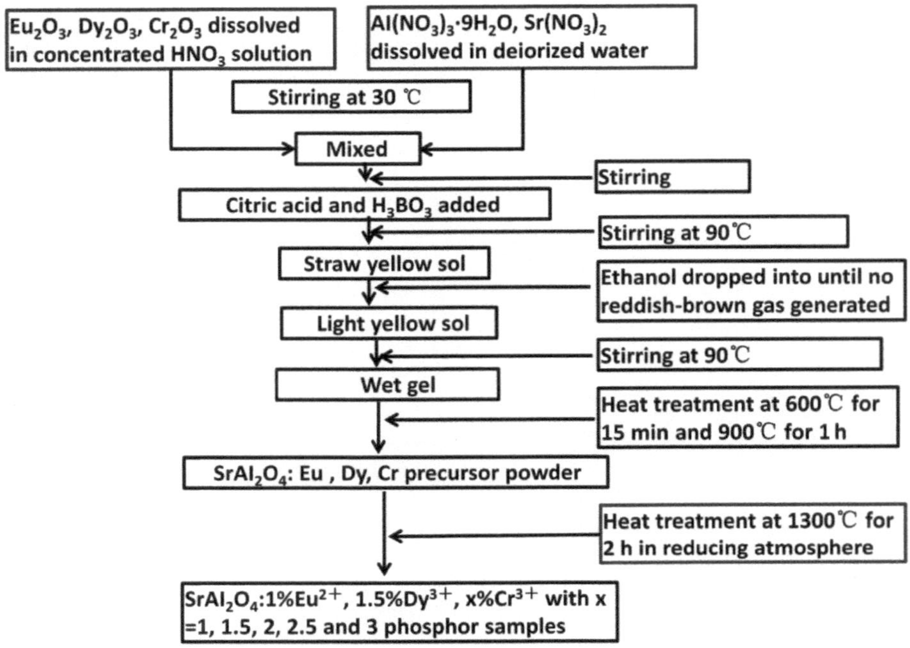

Fig. 2.11 Synthesis process of $SrAl_2O_4$: Eu^{2+}, Dy^{3+}, Cr^{3+} persistent phosphors via a combustion method.
Reproduced with permission from Y. Teng, J. Zhou, S.N. Khisro, S. Zhou, J. Qiu, Persistent luminescence of $SrAl_2O_4$: Eu^{2+}, Dy^{3+}, Cr^{3+} phosphors in the tissue transparency window, Mater. Chem. Phys. 147 (2014), 772–776, Copyright 2014, Elsevier.

synthesizing metal oxide powders. This combustion reaction is characterized by a low ignition temperature (150–200°C). The principle of the gas combustion synthesis is to use a gas (CO or CH_4 or H_2) and a raw material gas ($SiCl_4$ or $TiCl_4$) to be burned under the protection of an inert gas into a high-temperature oxygen-rich environment. Finally, persistent phosphorescence nanoparticles are obtained after cooling the product.

After obtaining the precursors, a high-temperature treatment must be carried out. Such a high temperature leads to the formation and crystallization of persistent phosphorescence nanoparticles. By carefully adjusting the experimental variables (such as nature of solvents, concentration of precursors, reaction temperature, and time), it is possible to obtain persistent phosphorescence nanoparticles with narrow size distribution, good crystallinity, and afterglow properties. Cheng et al. investigated the influence of organic fuel and postannealing temperature on the variation of morphology and crystallographic structure. They demonstrated that triethanolamine (TEA) can not only be used as an organic fuel but also could control the morphology of nanostructured combustion products. With an increase of postannealing temperature, their morphology transformed from nanosheets into semicurled nanobelts due to the crystal growth, whose transition exhibited the following rule: amorphous/hexagonal β-$SrAl_2O_4$ (1000°C)/monoclinic $SrAl_4O_7$ (1100°C)/orthorhombic $Sr_4Al_{14}O_{25}$ (1300°C). In fact, this method has been widely accepted nowadays for the synthesis of main aluminate-based persistent phosphorescence nanoparticles. There are some reports on the synthesis of other persistent phosphorescence nanoparticles, such as $SrAl_2O_4$: Eu, Dy; $CaAl_2O_4$: Eu, Dy; $BaAl_2O_4$: Eu, Dy; $CaAl_4O_7$: Eu, Dy; $CaAl_2O_4$: Eu; $Sr_4Al_{14}O_{25}$: Ce; Y_2O_2S: Sm; $BaCa_2Al_8O_{15}$: Eu, Dy; ZrO_2: Yb, Er; $SrAl_2O_4$: Eu, Dy, Cr; and $SrAl_2O_4$: Eu, Dy, Er, using a high-temperature combustion process. Although the combustion method has become a common method for preparing nanomaterials at present, the obtained products usually show large particle size, uneven morphology, and poor dispersion in solvents [9, 10, 59].

2.2.6 Other methods

While high-temperature coprecipitation, hydrothermal synthesis, combustion synthesis, template method, and sol-gel are currently well established for the synthesis of monodisperse persistent phosphorescence nanoparticles, a few groups have also attempted to develop other synthetic strategies like solvothermal [60–62], electrospinning [28, 55], hydrothermal ion exchange reaction [63], microwave synthesis [29, 64], and reverse micelle method [29].

In recent years, Qiu et al. reported the synthesis of a hybrid nanocluster composed of upconversion nanoparticles (UCPL-NC), β-$NaYbF_4$: Tm@$NaYF_4$, and persistent nanoparticles, $Zn_{1.1}Ga_{1.8}Ge_{0.1}O_4$: 0.5%Cr, which can be activated by a 980-nm laser. This hybrid nanoparticle exhibits an afterglow at 700 nm that can be used for near-infrared (NIR) to NIR UCPL bioimaging. In their process, core-shell β-$NaYbF_4$:0.5%Tm@$NaYF_4$ nanoparticles (Tm-UCNPs) with an average diameter of 21 nm were first synthesized using a well-established solvothermal method [62]. $Zn_{1.1}Ga_{1.8}Ge_{0.1}O_4$:0.5%Cr nanoparticles (ZGGO persistent phosphorescence nanoparticles) were synthesized via a modified hydrothermal method. The synthesis of hybrid nanoparticle was carried out with an evaporation-induced self-assembly method, in which spherical oleic acid (OA)-coated ZGGO persistent phosphorescence nanoparticles and Tm-UCNPs were simultaneously encapsulated within micelles by dodecyltrimethylammonium bromide (DTAB) (Fig. 2.12) [61].

Qiu et al. successfully synthesized the red $CaTiO_3$: Pr^{3+}; green $SrAl_2O_4$: Eu^{2+}, Dy^{3+}; and blue $CaAl_2Si_2O_8$: Eu^{2+}, Dy^{3+} nanofibers by electrospinning (Fig. 2.13D–F) [28]. The schematic diagram of electrospinning equipment was illustrated in Fig. 2.13A, and the average diameter was estimated to be 800 nm (Fig. 2.13B and C). After the as-prepared nanofibers were calcined at 1100°C, the size of nanofibers decreased owing to the degradation and combustion of PVP polymer and organic solvents. They further pointed out that by controlling the electrospinning parameters (the concentration of emission activator, the feeding rate of the solution, the dielectric constant of the precursor solution, etc.), the emission intensity of $CaTiO_3$: Pr^{3+}; $SrAl_2O_4$: Eu^{2+}, Dy^{3+}; and $CaAl_2Si_2O_8$: Eu^{2+}, Dy^{3+} nanofibers can be easily controlled. Consequently the emitting color of microstructural fibers can be widely tuned from blue, green, and red light to white. Moreover, regular arrays can be built by using the ternary microstructural fibers as the basic

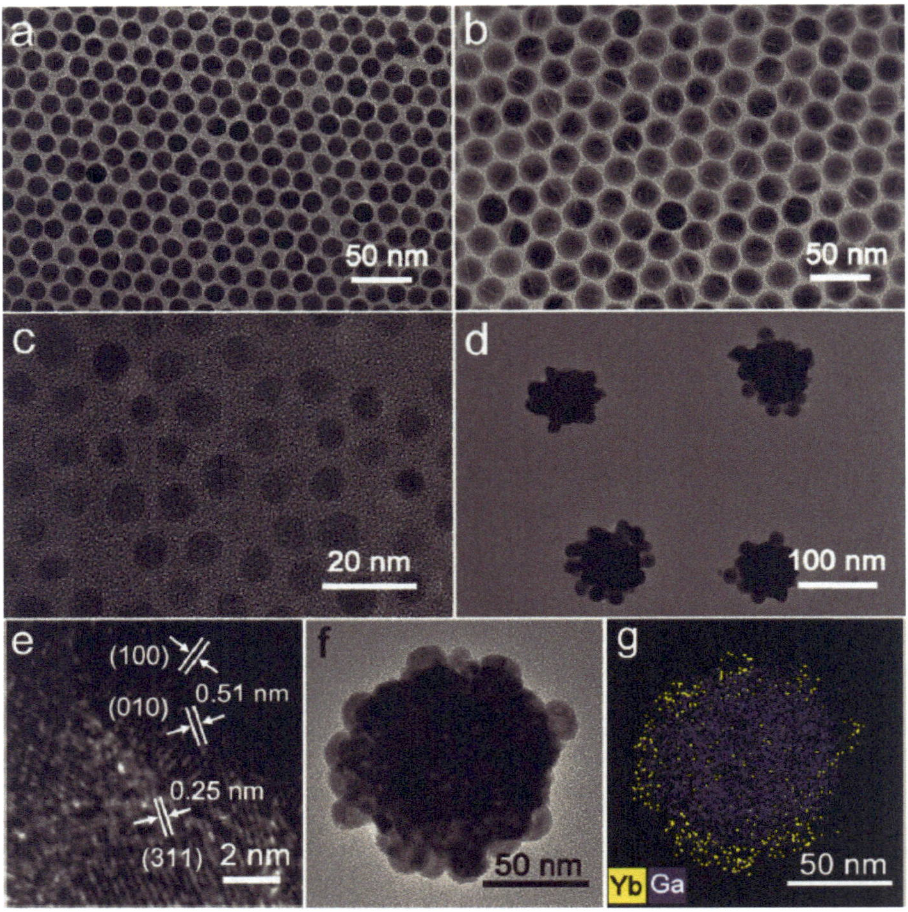

Fig. 2.12 TEM image of (A) as-prepared β-$NaYbF_4$:0.5%Tm nanoparticles, (B) β-$NaYbF_4$:0.5%Tm@$NaYF_4$ nanoparticles, (C) $Zn_{1.1}Ga_{1.8}Ge_{0.1}O_4$ nanoparticles, and (D) UCPL-NCs. (E) High-resolution TEM image of UCPL-NCs. (F and G) TEM image and the corresponding EDS elemental mapping of UCPL-NCs.
Reproduced with permission from X. Qiu, X. Zhu, M. Xu, W. Yuan, W. Feng, F. Li, Hybrid nanoclusters for near-infrared to near-infrared upconverted persistent luminescence bioimaging, ACS Appl. Mater. Interfaces 9 (2017), 32583–32590, Copyright 2017, American Chemical Society.

building blocks. By controlling the fraction of red-, green-, and blue-emitting nanofibers, the emitting colors of electrospun array mats can also be conveniently tuned (Fig. 2.13G).

Yan et al. synthesized zinc gallogermanate persistent phosphor by hydrothermal ion exchange reaction [63]. The precursor, $K_{0.9}Ga_{0.9}Ge_{0.1}O_2$, was first prepared by solid-state reaction and dispersed in water to form a colloidal suspension. After

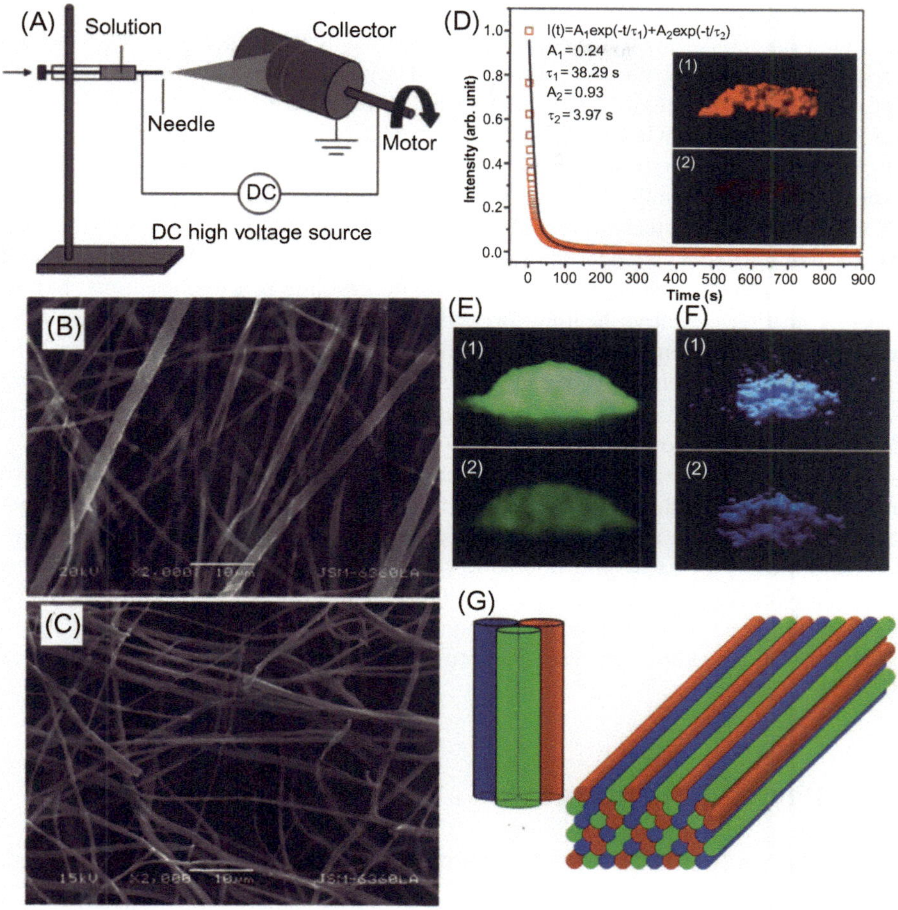

Fig. 2.13 (A) Schematic diagram of the electrospinning equipment. (B) SEM images of $CaTiO_3$: Pr^{3+} electrospun nanofibers calcined at 1100°C in air. (C) SEM images of $SrAl_2O_4$: Eu^{2+}, Dy^{3+} electrospun nanofibers calcined at 1100°C in air. (D) Afterglow decay curve for the 615 nm emission of $CaTiO_3$: Pr^{3+} electrospun nanofibers. Inset shows the digital photographs of electrospun nanofibers (after grinding) taken (1) upon irradiation with a 365 nm LED, (2) 1 min after the removal of irradiating light. (E) The digital photographs of $SrAl_2O_4$: Eu^{2+}, Dy^{3+} electrospun nanofibers (after grinding) taken (1) upon irradiation with a 365 nm LED, (2) 1 min after the removal of irradiating light. (F) The digital photographs of $CaAl_2Si_2O_8$: Eu^{2+}, Dy^{3+} electrospun nanofibers (after grinding) taken (1) upon irradiation with a 365 nm LED, (2) 1 min after the removal of irradiating light. (G) Schematic diagrams of microstructural fibers and array.
Reproduced with permission from G.P. Dong, X.D. Xiao, L.L. Zhang, Z.J. Ma, X. Bao, M.Y. Peng, et al., Preparation and optical properties of red, green and blue afterglow electrospun nanofibers, J. Mater. Chem. 21 (2011), 2194–2203, Copyright 2011, The Royal Society of Chemistry.

Zn^{2+} ions (from $Zn[CH_3COO]_2$) aqueous solution was introduced into the $K_{0.9}Ga_{0.9}Ge_{0.1}O_2$ colloidal suspension, hydrothermal treatment was applied to the dispersion to synthesis the zinc gallogermanate. They also suggested that the cubic zinc gallogermanate formed via the ion exchange reaction of $10K_{0.9}Ga_{0.9}Ge_{0.1}O_2 + 6.5Zn(CH_3COO)_2 \rightarrow 4.5(ZnGa_2O_4){:}(Zn_2GeO_4) + 9CH_3COOK + 4CH_3COOH$. The high-resolution TEM image showed that the microstructures can be related with the aggregation of the nanocrystals with a particle size of about 10 nm.

Microwave heating technique is also an effective method to synthesize a wide variety of homogeneous nanoparticles. It has several advantages over conventional heating techniques, such as high heating rate, no grinding process, less time, homogeneous heating, selective heating, no contact between the heating source and the reactants and/or solvents, and improved reproducibility. Microwave chemistry definitely satisfies the promise of being a fast synthesis technique. Microwave radiation has been applied to the synthesis of nanomaterial thin films and porous ceramics. This method was applied to the synthesis of $SrAl_2O_4$: Eu: Dy, $CaTiO_3$: Pr, etc. [64]

Reverse micelle or microemulsion process constrains the reaction in the micron- or nanosized droplet, such that the size of the product particles always shows narrow distribution and agglomeration can be easily prevented. This method can be used to synthesize a variety of oxide and nonoxide particles from micron to nanoscale sizes. Many simple oxide-based phosphors can be prepared by this method, while a low reaction temperature usually leads to poor crystallinity of the products, such that a postsynthesis annealing process is often applied to improve the luminescence properties [29].

2.3 Physical-based routes

2.3.1 Laser technique (ablation/deposition)

Laser ablation/deposition is an efficient technique for synthesis of various functional materials. The representative equipment for laser ablation in liquid is shown in Fig. 2.14C. The target is first placed in a cell filled with water or ethyl alcohol and then irradiated with a focused laser beam. By controlling parameters of the laser source (wavelength, repetition rate, peak power, focusing condition, and irradiation duration), it is easy to obtain the desired particles with different morphology and size. This method has already been reported for the synthesis of noble metal and semiconductor nanoparticles. In this route the morphology and dispersity of prepared persistent phosphorescence nanoparticles can be improved by plasma etching process followed by laser ablation/deposition, where phosphors are formed or "broken off" from a larger structure (Fig. 2.14A) [65]. Compared with other synthesis techniques, laser ablation has many advantages in terms of the size distribution of nanoparticles. The most important feature is that the size of particles is reduced while maintaining excellent crystallinity and afterglow properties.

Yoshimura et al. prepared the $Sr_2MgSi_2O_7$: Eu^{2+}, Dy^{3+} persistent phosphorescence nanoparticles by using this technique [31]. They demonstrated that the size of

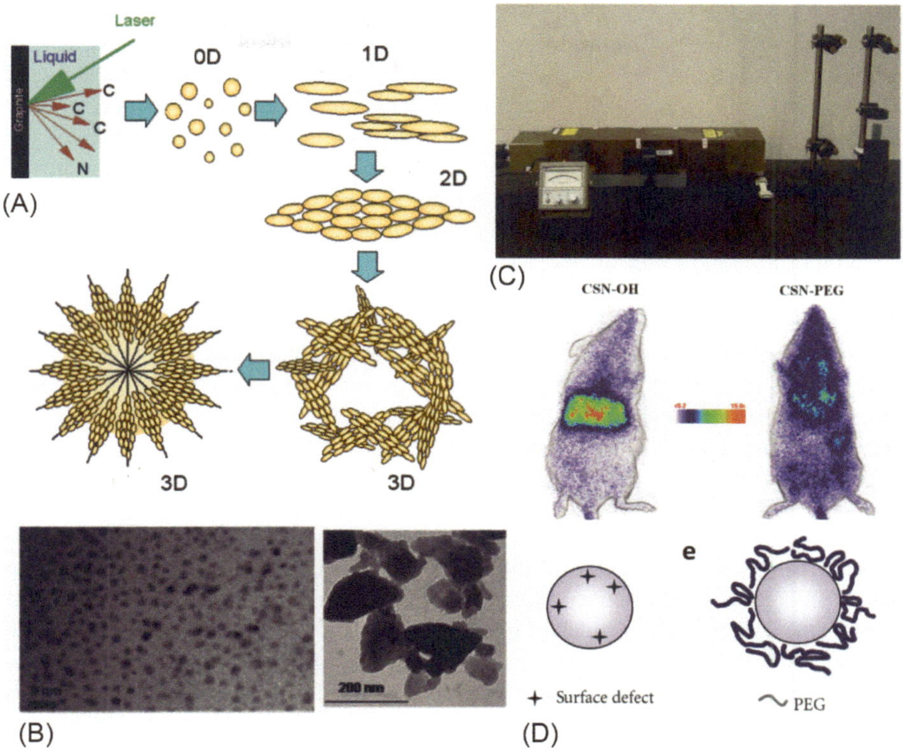

Fig. 2.14 (A) Proposed sequential growth pathway model involving structural and morphology modifications by using laser ablation. (B) Transmission electron micrograph (TEM) of the obtained nanoparticles from the laser ablation technique. (C) Equipment for laser ablation in liquid. (D) Distribution of $Ca_2Si_5N_8$: Eu^{2+}, Tm^{3+}-OH and $Ca_2Si_5N_8$: Eu^{2+}, Tm^{3+}-PEG, 15 min after tail vain injection. (E) Schematic of an afterglow nanoparticle and of PEG molecules. Left panel shows an afterglow nanoparticle without the addition of PEG, and right panel shows an afterglow nanoparticle capped with PEG molecules at a high PEG concentration.
A: Reproduced with permission from L. Yang, P.W. May, L. Yin, R. Brown, T.B. Scott, Direct growth of highly organized crystalline carbon nitride from liquid-phase pulsed laser ablation, Chem. Mater. 18 (2006), 5058–5064, Copyright 2006, American Chemical Society; B: Reproduced with permission from T. Maldiney, G. Sraiki, B. Viana, D. Gourier, C. Richard, D. Scherman, et al., *In vivo* optical imaging with rare earth doped $Ca_2Si_5N_8$ persistent luminescence nanoparticles. Opt. Mater. Express 2 (2012), 261–268, Copyright 2012, The Optical Society; C: From F. Yoshimura, M. Ishizaki, F. Wakai, M. Hara, O. Odawara, H. Wada, Optical properties of afterglow nanoparticles $Sr_2MgSi_2O_7$: Eu^{2+}, Dy^{3+} capped with polyethylene glycol, Adv. Opt. Technol. 2012 (6) (2012), https://doi.org/10.1155/2012/814745; D: Reproduced with permission from L. Yang, P.W. May, L. Yin, R. Brown, T.B. Scott, Direct growth of highly organized crystalline carbon nitride from liquid-phase pulsed laser ablation, Chem. Mater. 18 (2006) 5058–5064; E: From F. Yoshimura, M. Ishizaki, F. Wakai, M. Hara, O. Odawara, H. Wada, Optical properties of afterglow nanoparticles $Sr_2MgSi_2O_7$: Eu^{2+}, Dy^{3+} capped with polyethylene glycol, Adv. Opt. Technol. 2012 (6) (2012), https://doi.org/10.1155/2012/814745.

nanoparticles could be controlled by the energy density of laser. During this process, some byproducts were also produced, depending on the liquid used. The amount of byproduct was reduced by a decrease in the energy density, which also reduced the particle size of the nanoparticles. To improve the quantum yield and biocompatibility needed for bioimaging, they added PEG as a surface capping agent. An increase in the amount of PEG increased the quantum yield of the nanoparticles and improved the afterglow property at the same time (Fig. 2.14E). This technique is a versatile technique for synthesizing high-quality persistent phosphorescence nanoparticles, regardless of the high cost of the laser systems.

Maldiney et al. demonstrated the synthesis of $Ca_2Si_5N_8$: Eu^{2+}, Tm^{3+} biolabel with persistent phosphorescence: first, preparing the bulk-form persistent phosphors; second, synthesizing nanoparticles with the pulsed laser ablation method in liquid; and finally, the surface modification. Meanwhile, they revealed the influence of surface functionalization on the biodistribution of biomaker after systematic injection and discussed the potential application in in vivo imaging of the functionalized nanoparticles (Fig. 2.14D) [66].

Popov et al. reported the size-tunable synthesis of TiN nanoparticles by femtosecond (fs) laser ablation in liquids. A Yb: KGW pulsed laser (ns pulse) and an fs laser were used in their experiment for ablation (Fig. 2.15A and B). They found that the nanoparticles produced in different liquids by laser ablation show different morphologies. Most nanoparticles produced in water were amorphous and nonspherical with hollow cavities (Fig. 2.15G). In comparison, when acetone was used as the solvent, the produced particles were mostly spherical, and polycrystalline with very few particles contain cavities (Fig. 2.15C–F), and the mean size of nanoparticles depended on laser pulse energy (Fig. 2.15H) [32]. Such technique should be applicable to preparation of nitride with excellent persistent phosphorescence properties.

2.3.2 Swift heavy ion irradiation and electron beam bombardment

A synthetic method that allows synchronously activation of afterglow and size control in synthesis of persistent phosphorescence nanoparticles is highly desirable, while it is admittedly difficult. Swift heavy ion (SHI) irradiation and electron beam bombardment have found wide applications in many fields such as biology, medicine, physics, and materials science [67, 68]. They are always used to evaporate materials that are hard to evaporate by thermal evaporation due to high melting points of these materials. Compared with the conventional ionizing radiation processes, the intense energy deposition in the target material is several orders of magnitude higher, resulting in loss of atoms and formation of vacancies during the process. After SHI or *E*-beam irradiation, formation of defects has been observed in various materials, including semiconductors, insulators, and various metals. Therefore SHI irradiation or electron beam bombardment can be used for the synthesis of persistent phosphorescence nanoparticles by simultaneous inducing suitable lattice defects. Xu et al. reported an approach for improving the afterglow of long-lasting phosphors with SHI

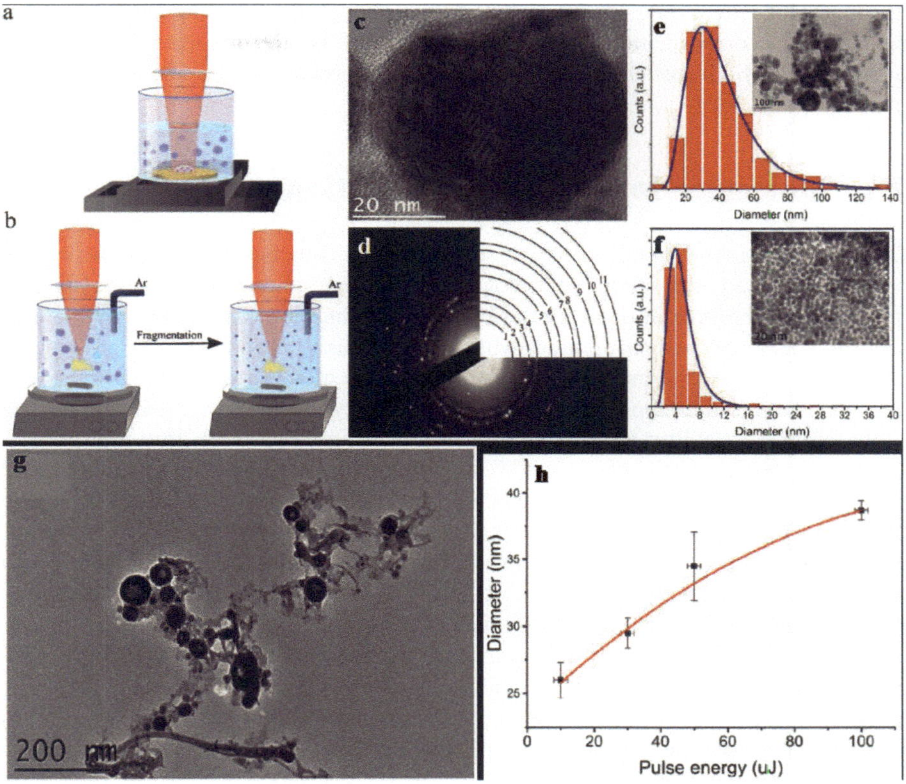

Fig. 2.15 (A) Schematics of laser ablation setup. A laser beam is focused onto the surface of the TiN target, which is placed in a vessel filled with a liquid. The vessel is mounted on a moving translation stage to avoid ablation from the same area of the target. (B) Schematic of laser ablation setup. Ar bubbling was applied before ablation to remove dissolved oxygen. (C) HRTEM image of TiN nanoparticle synthesized by pulsed laser ablation in acetone. (D) SAED pattern of TiN nanoparticles with ring sequence typical for FCC lattice. (E) Size distribution of TiN nanoparticles synthesized by laser ablation in acetone at 100 μJ pulse energy and (F) after laser ablation in acetone. Insets show typical TEM images of corresponding TiN nanoparticles. (G) TEM image of TiN nanoparticles synthesized by laser ablation in water having multiple interior cavities and linked by amorphous nonspherical flakes. (H) Mean diameter of TiN nanoparticles, obtained by laser ablation in acetone at different pulse energies. Reproduced with permission from A.A. Popov, G. Tselikov, N. Dumas, C. Berard, K. Metwally, N. Jones, et al., Laser-synthesized TiN nanoparticles as promising plasmonic alternative for biomedical applications, Sci. Rep. UK 9 (2019), 1194, Copyright 2019, Nature Publishing Group.

irradiation. The results showed that the afterglow time of $SrAl_2O_4$: Eu^{2+} persistent phosphorescence nanoparticles notably increased after SHI treatment [69]. Smet et al. found that electron beam annealing was a viable option to reduce the europium ions in $CaAl_2O_4$: Eu, Nd persistent phosphors and it led to well-crystallized nanoparticles with a long and bright afterglow. Though this method was highly efficient, free of contamination and energy saving, and uses or produces no hazardous byproduct, the high cost of the equipment strongly limits the wide application of these methods [33].

2.4 Molecules for surface modification

Due to their biocompatibility and biodegradability, biodegradable polymers have been widely used in the fields of biological science and biomedicine. Biodegradable polymers can be broken down in the body, producing nontoxic natural byproducts such as water and carbon dioxide, which can be easily eliminated. At present the use of degradable polymers to modify the surface of inorganic nanoparticles has become a hot research topic. The aforementioned polymer-inorganic nanocomposites show improved stability in aqueous solutions and can be used for drug transport.

Great progress has been made in the preparation of polymer-inorganic nanocomposites. Compared with naked inorganic compounds, nanocomposites have the following advantages: (1) Since inorganic compounds are mainly composed of unhydrolyzed structures, which are usually insoluble in water, the combination with water-soluble polymers can significantly improve their hydrophilicity; (2) compound with polymer can improve the metabolism of inorganic compounds and the elimination of organisms; and (3) polymers provide additional functional groups for the combination of targeted and therapeutic moieties. With the rapid development of medicine and materials science, various kinds of polymer materials have been used. To be suitable for stabilizing the persistent phosphorescence nanoparticles, the selected polymers should meet the following requirements: (1) biocompatibility, (2) water solubility, (3) biodegradable, and (4) functional groups capable of organic reactions and binding with other molecules (such as drugs and targeted molecules).

2.4.1 (3-Aminopropyl)triethoxysilane (APTES)

APTES has a high reactivity, which can fix effective groups on the surface of inorganic materials, change their surface structure and properties, and effectively improve the interface compatibility between inorganic powders and organic phases. Moreover, after covering APTES onto the surface of inorganic materials, water or alcohol products are generated in the reaction process, which has the advantages of no cytotoxicity and no physiological inflammation and can be eliminated in vitro with metabolism.

Li et al. have designed and synthesized a new kind of near-infrared persistent luminescent nanocomposite with aqueous solution stability. In the first step, persistent phosphorescence nanoparticles were hydroxylated by grinding and stirring with NaOH solution. NaOH solution was selected for grinding and stirring as the hydroxylation step of persistent phosphorescence nanoparticles by taking full advantages of its easy operation and simple experimental setup. In the second step, DMF (the reader might be NOT clear to the meaning of DMF) and APTES were selected and used as the solvent and amination modifier, respectively. Subsequently, $SrAl_{12}O_{19}$: Fe^{3+}

@APTES nanocomposites were obtained by stirring at 80°C for 24 h (Fig. 2.16A). After encapsulating the persistent phosphorescence nanoparticles with APTES, $SrAl_{12}O_{19}$: Fe^{3+} @APTES nanocomposites stayed stable for over 20 days in aqueous solution (Fig. 2.16B). This result demonstrated that the nanocomposites are promising in studying the biological functions of biomolecules and monitoring cellular networks in their native contexts [70].

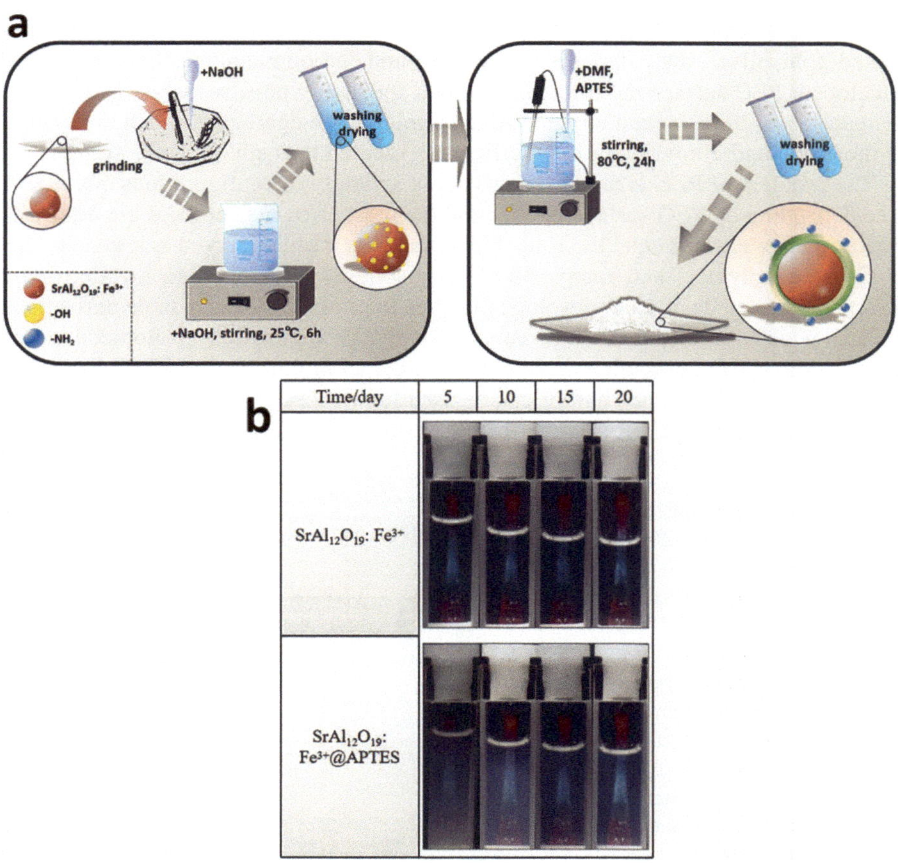

Fig. 2.16 (A) Demonstration of ambient aqueous stability. After the surface functionalization, $SrAl_{12}O_{19}$: Fe^{3+} @APTES nanocomposites were dispersed in ethanol (0.3 mg/mL) for storage period of 20 days. (B) Schematic illustration for the surface functionalization of nanocomposites.

Reproduced with permission from R. Kang, X. Dou, H. Lian, Y. Li, $SrAl_{12}O_{19}$: Fe^{3+} @3-aminopropyl triethoxysilane: ambient aqueous stable near-infrared persistent luminescent nanocomposites, J. Am. Ceram. Soc. 103 (2020), 258–265, Copyright 2020, Wiley.

2.4.2 Poly(ethylene glycol) (PEG)

Supramolecular amphiphilic copolymers consisting of amphiphilic blocks can form micelles with core/shell-like nanostructures in aqueous solution, where the hydrophobic blocks aggregate to create an inner core surrounded by an outer corona made of hydrophilic blocks. By varying the length and number of blocks, the copolymers can form diverse micellar structures. The most popular hydrophilic block used in constructing polymer micelles is poly(ethylene glycol) (PEG), which confers the micelles with good biocompatibility and reduces toxicity for cells. PEG coating is widely used in surface modification for creating renal clearable nanoparticles with a neutral surface charge. Since the pioneer work of renal clearable QDs in 2007, other inorganic nanoparticles such as ultrasmall fluorescent silica nanoparticles; C-dots; glutathione-coated gold, silver, and copper nanoparticles; and carbon nanotubes have also been developed. PEG surface modification changes the highly negatively charged surface to neutral, which prevents inorganic from forming large aggregates when circulating in the blood and allows them to be efficiently cleared through the renal system.

The problem of PEG is that its chains adopt a variety of configurations on the surface depending on PEG surface density and the most effective densities are often difficult to achieve. Although PEG is not biodegradable, it can be served as a standard for PEG-based polymers and many other water-soluble polymers. Meanwhile, many efforts have been devoted to develop strategies to produce biodegradable derivatives of PEG, which are believed to be promising for diverse applications in biomedicine. In the design of PEG-inorganic nanocomposites, the amino group of APTES-inorganic nanocomposites mentioned earlier can be used for synthesis. Moreover the PEG surface groups can be further used for folic acid packaging (Fig. 2.17) [71].

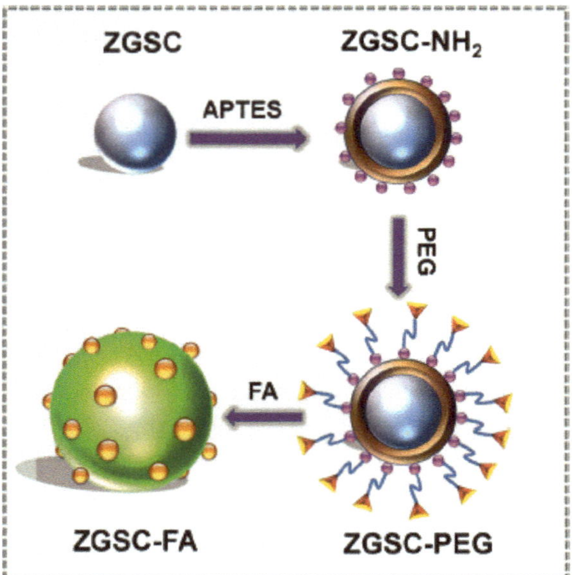

Fig. 2.17 Schematic illustration for the synthesis of FA-ZGSC nanoparticles.
Reproduced with permission from Y. Li, R. Chen, Y. Li, K. Sharafudeen, S. Liu, D. Wu, Y. Wu, X. Qin and J. Qiu, Folic acid-conjugated chromium(III) doped nanoparticles consisting of mixed oxides of zinc, gallium and tin, and possessing near-infrared and long persistent phosphorescence for targeted imaging of cancer cells. Microchimica Acta, 2015, 182, 1827-1834, Copyright 2015, Springer.

2.4.3 Other molecules

Polyesters, such as polylactic acid (PLA), polyglycolic acid (PGA), and polycaprolactone (PCL), also have been used for the preparation of biocompatible hydrogels. They are attractive materials for the design of drug delivery systems, tissue-engineering scaffolds, implants, and surgical materials. Polyester-based polymer coatings offer inherent biodegradability due to ester hydrolysis under physiological conditions. Thus, using combinations of polyesters with other synthetic or natural inorganics, the rate of hydrogel degradation can be tuned as per application requirements.

Chitosan, the deacetylated derivative of chitin, is a linear polycationic polysaccharide composed of randomly distributed b-(1–4)-linked D-glucosamine and *N*-acetyl-D-glucosamine. It degrades by various mechanisms, including surface erosion, enzymatic degradation through chitosanase and lysozyme, and dissolution. By using appropriate cross-linking chemistries and densities, the degradation kinetics can be tuned. Chitosan has excellent cell compatibility, biodegradability, minimal foreign body reactivity, and antibacterial properties, which makes chitosan hydrogel promising for applications, including wound healing, bioactive molecule delivery, and soft tissue engineering.

Liposome is a spherical self-closed structure made up of one or several lipid bilayer (s), which is similar to membrane of cell that contains an aqueous core. The inorganic structure can be loaded within the lipid bilayer or in the aqueous core or at the interface between them. Since the lipid is an essential biomolecule for most living tissues and has an amphiphilic nature, that is, ability to spontaneously self-assemble into a variety of microstructures, liposome is widely used as a temperature or pH-sensitive drug delivery vehicle particularly for cytotoxic anticancer drugs.

2.5 Synthesis of glassy, glass-ceramic, and composite persistent phosphors

The usually used PPs involve persistent phosphorescent powders, nanoparticles, glasses, glass ceramics, and composites. Persistent luminescent powders are usually prepared by solid-state reaction method, while PPNPs are synthesized via the aforementioned chemical route and laser technique. Following, we describe the synthesis routes of glassy, glass-ceramic, and composite PPs.

2.5.1 Persistent luminescent glasses (PPGs) and glass ceramics (PPGCs)

PPGs and PPGCs are known to show many advantages as the medium for the optical applications such as lighting, display, and decoration. Here, glasses refer to the activator-doped glasses, not the composites including persistent luminescent powders and glasses. The latter will be discussed in the next section.

Qiu et al. first reported the persistent phosphorescence in Eu^{2+}-doped CaO-Al_2O_3-B_2O_3 glasses [72]. After irradiation by a white color fluorescent lamp with a power density of 2×10^4 Ix for 30 min, the glass samples emit a visible light peaking at 510 nm that can be seen with the naked eye in the dark even 8 h after the removal of the activating light. Based on the x-ray absorption spectra, the long-lasting phosphorescence in the Eu^{2+}-doped glasses is considered to result from the recombination of electrons and holes at shallow traps in the glass matrix that can be thermally released at room temperature. Qiu et al. also observed persistent phosphorescence in Eu^{2+}-doped SrO-Al_2O_3-SiO_2 glasses [73]. A visible light (peaking at 510 nm) can be clearly seen with the naked eye in the dark even 24 h after removal of the activating light. They measured x-ray absorption near edge structure (XANES) and electron spin resonance (ESR) spectra and suggested that the long-lasting phosphorescence in the Eu^{2+}-doped glasses results from the recombination of electrons and holes at shallow traps that are not Eu^{2+} and Dy^{3+} ions in the glass matrix. These glasses may be promising materials for energy-saving display and useful for the improvement of the conversion efficiency of solar cells. Hosono et al. discussed the PP in Tb^{3+}-doped glasses [74]. A green PP was found in Tb^{3+}-doped calcium aluminate glasses prepared under a strongly reducing atmosphere. An electron-trapped center giving an ESR signal at $g = 1.999$, an analogue of the F^+ center in CaO, was induced after irradiation with UV light and decayed after the illumination was stopped. They observed UV light–induced oxidation of Tb^{3+} ions to Tb^{4+} ions. A mechanism of the persistent phosphorescence was suggested to consist of recombination of an electron released thermally from the F^+-like center with the photooxidized Tb^{3+} (i.e., Tb^{4+}). Peng et al. observed red to near-infrared persistent luminescence from Mn^{2+}-doped sodium gallium aluminum germanate glasses, covering the spectral region of 600–800 nm. The glass of $8Na_2O$-$12Ga_2O_3$-$80GeO_2$ was thoroughly mixed and filled into alumina crucibles for high-temperature calcination, and $MnCO_3$ was added for Mn^{2+} doping in nominal quantities. They show that the afterglow of Mn^{2+}-doped sodium gallium germanate glass can persist longer than 48 h at a dopant content of 0.1% Mn^{2+}. Surprisingly the nanocrystalline Na(*Ga,Al*)Ge_3O_8 phase can be precipitated from the supercooled melt, and it is composed of the sodium octahedral sites and the aluminum/germanium/gallium sites. Mn^{2+}-doped glass ceramics can preserve the red emission of Mn^{2+} ions that substitute for sodium sites in Na(*Ga,Al*)Ge_3O_8, and in the meanwhile the emission intensity can be enhanced, while the persistence of luminescence can be improved longer than 100 h [75, 76].

Tanabe et al. reported a Mn^{2+}-activated glass and transparent glass ceramics containing spinel-type $Zn_{1+x}Ga_{2-2x}GexO_4$ $(0 \leq x \leq 1)$ nanocrystals, which show multicolor persistent luminescence for more than 10 min. The color of persistent phosphorescence showed a remarkable change from red to orange, greenish, and green with respect to the increase of annealing temperature [77].

Therefore PPGs can be considered as the precursor of PPGCs. Crystallization of glass, a unique technique for precipitating crystal, allows designing the morphology of precipitated crystals by controlling crystal nucleation and growth rate in glasses. By using this technique, various optically transparent glass ceramics containing nanocrystals have been developed so far. For example, from these advantages, fabrication of glass ceramics with $SrAl_2O_4$ has been attempted in SrO-Al_2O_3-B_2O_3 systems, where the addition of B_2O_3 is necessary for glass formation. Crystallization of glass with related (stoichiometric) composition of precipitated crystal has been received much attention as a new technique to synthesize ceramics because of its formability and morphology controllability of precipitated crystal.

Furthermore, it should be pointed out that glasses have a potential for precipitating metastable crystal phase because of its nonequivalent transition. A levitation (containerless) melting technique has also been carried out to develop metastable crystal phase and new glasses without any traditional glass forming components such as SiO_2 and B_2O_3. In this technique, melts are taken into deep supercooled state and solidified in noncontact process with inhibiting surface nucleation, where whether to form glass or crystals depends on the composition and cooling rate.

Therefore these high-temperature phases of nanocrystals were synthesized via two paths: (1) heat treatment on glass (a conventional glass crystallization technique) and (2) solidification of supercooled melts in melt-cooling process with noncontact process, where a levitation melting technique was used to obtain precursor glass and ceramic beads.

Note that both process of crystallization of glass and solidification of supercooled melt have a large nonequivalence transition, that is, precipitated crystal structure and formation of defects, which are the origin of energy trap center in afterglow crystals, could be different from other synthesis roots based on equivalent transition. For instance, formation of metastable phase by these techniques has been reported so far such as β'-$Gd_2(MoO_4)_3$, $BiBO_3$ (phase II), $SrAl_2O_4$, $LiGa_5O_8$, and $BaTi_2O_5$.

The afterglow in PPGs is known to occur when the charge carriers, namely, electrons and holes, created inter glass by the excitation process are trapped in and then recombined radiatively around the trapping centers. For the silicate glasses, it has been reported that the presence of the cerium may significantly reduce the afterglow. To understand the mechanism of the afterglow, Blasse et al. considered that the doped cerium is present mainly in tetravalent state in the oxidized sample and suggested that the presence of Ce^{4+} in $Gd_3Ga_5O_{12}$: Cr is the main reason for suppression of the afterglow [78].

Recent researches have demonstrated their developments in PPGs and PPGCs. Allix et al. adopt the method of controlling the crystallization to synthesize the PLGCs [79]. In their experiment, SiO_2-ZnO-Ga_2O_3-Na_2O glasses doped with 0.1% Cr^{3+} were prepared by conventional melt-quenching technique. The mixture was subsequently melted at 1600°C for 30 min in an electrical muffle furnace in air. The glass samples were then annealed at 550°C for 1 h to release thermal residual stress and finally cooled down to room temperature. To obtain the final glass-ceramic materials, the glass was introduced at room temperature in an electrical furnace, heated under air at 10°C min^{-1} up to a temperature between 900°C and 1200°C, and maintained for 10 min before being cooled down to room temperature at 10°C min^{-1}. Highly transparent $ZnGa_2O_4$ glass-ceramic materials are elaborated via a simple heat treatment of a $55SiO_2$-$5Na_2O$-$17ZnO$-$23Ga_2O_3$ parent glass composition, which presents nanoscale spinodal phase separation. This optimized glass ceramic exhibits 50 wt% of $ZnGa_2O_4$ nanocrystals showing a homogeneous and tunable size. Remarkably, red persistent phosphorescence arising from the entire sample volume is observed in the Cr^{3+}-doped transparent glass ceramics, opening the route to a wider range of performing applications for this famous zinc gallate persistent phosphor.

Cr^{3+}-doped transparent glass ceramics of SiO_2-Ga_2O_3-Li_2O also were fabricated via the similar crystallization route. The glass precursors were heated to and maintained for 2 h at different temperature at 700, 800, and 900°C to obtain transparent glass ceramics through crystallization. Cr^{3+}(0.5 mol%):$LiGa_5O_8$ (LGO) polycrystals were prepared by a high-temperature solid-state reaction method. These Cr^{3+}-doped transparent glass ceramics embedding $LiGa_5O_8$ nanocrystals show the near-infrared persistent phosphorescence lasting for more than 2 h [80].

The similar route also can be used to synthesize other PPGCs. Wang et al. report a silicate transparent oxyfluoride glass and glass ceramic with multicolor persistent luminescence lasting for ~1 h visible by the naked eye. By adjusting the crystallization duration, controlled growth of 5–20 nm α-Zn_2SiO_4 nanocrystals in a glass matrix is achieved. The coordination environment of Mn^{2+} evolves from octahedra in the glass host to tetrahedra in the α-Zn_2SiO_4 nanophase, resulting in the PP color changing from red to yellow, and then to green, as the crystallization of Mn^{2+}:α-Zn_2SiO_4 proceeds [81].

Besides to the heat treatment on glass (a conventional glass crystallization technique), solidification of supercooled melts in melt-cooling process with noncontact process is the other path to synthesize these high-temperature phase of nanocrystals. For example, hexagonal $SrAl_2O_4$ (β'-phase) were formed in both samples, not stable phase of monoclinic phase (α-phase), which is well known as one of the best afterglow materials. Both samples with hexagonal phase showed brighter broad green emission, higher intensity of PLE peak in visible region ($\lambda = 400$–500 nm), and longer duration of afterglow compared with monoclinic $SrAl_2O_4$: Eu^{2+}, Dy^{3+} sintered by solid-state reaction. Nakanishi et al. suggested so-called Frozen sorbet (FS) method to fabricate glass ceramics consisting of Eu^{2+}, Dy^{3+}-doped $SrAl_2O_4$ in the SrO-Al_2O_3-B_2O_3 system. It is found that Eu^{2+}, Dy^{3+}-doped $SrAl_2O_4$ in the glass ceramics show excellent long persistent phosphorescent properties. Via the crystallization the Al^{3+} sites in $SrAl_2O_4$ crystals are partially substituted by B^{3+} ions. This study proposes that the Frozen sorbet method can be a useful technique for the design of new $SrAl_2O_4$ PPGCs [82].

2.5.2 Persistent phosphorescent composites (PPCPs)

Persistent phosphorescent composites (PPCPs) usually are composed of persistent luminescent powders and the medium used to achieve the encapsulation of persistent phosphor powders. The medium includes glass, polymer, and even liquid slurry.

The conventional synthesis route includes the respective preparation of the persistent phosphorescent powders and the encapsulation medium. Of course, that is the feasible solution to directly use commercial products as raw materials. The synthesis technique depends on the encapsulation medium. For the glasses, all the raw materials are then melted at a secondary high temperature; for the polymers a conventional ultraviolet light polymerization is applicative.

Hupa et al. propose a technique to prepare glass composites with persistent phosphorescence properties using the melting process that has the advantage of being able to produce glass composites with microparticles, the composition of which does not depend on the glass matrix composition while the Frozen sorbet method can only be used for the processing of $SrAl_2O_4:Eu^{2+}$, Dy^{3+} glass composites within the SrO-Al_2O-B_2O_3 glass system. The glasses with the composition 50P_2O_5-10Na_2O-40SrO were elaborated using a standard melting method. Glasses were prepared with 2 ($x = 2\%$) and 3 ($x = 3\%$) wt% of commercial $Sr_4Al_{14}O_{25}:Eu^{2+}$, Dy^{3+} microparticles. The $Sr_4Al_{14}O_{25}:Eu^{2+}$, Dy^{3+} microparticles were added in the glass batch that was then melted at 1000°C for 10 min in a quartz crucible in air. The melt was poured on a brass mold, and the resulting glass was annealed at 40°C below the glass transition temperature [83].

Obviously, this method just simply mixes the powder with the glass substrate and then adopts the secondary melting technique for glass. Therefore these materials should be named persistent phosphorescent glass composites. These PiGs are another technique to prepare the persistent phosphorescent glass composites.

Xie et al. developed a series of persistent phosphorescent materials $Y_3Al_{5-x}Ga_xO_{12}$: Ce^{3+}, V^{3+} with a garnet structure and developed novel phosphor-in-glass (PiG) films containing these phosphors. The PiG films in their experiment were fabricated by using a blade-coating method as reported in the literature with a minor modification. Briefly the synthesized phosphors were crushed into fine powers. The phosphors were blend with glass powders and organic vehicles to form viscous phosphor slurries. The phosphor slurries were carefully flowed on a thin layer of single-crystal sapphire substrate. The upper surfaces of the phosphor slurries were slicked smooth using a sheet of glass. They heated the phosphor slurries first on a hot stage at 120°C for 10 min for air bubble removing and preliminary shaping. The films were sintered in a muffle furnace at 850°C for 30 min to eliminate the organics. After cooling to RT naturally, PiG films were finally obtained. The synthesized PiG films exhibited sufficiently deep traps, narrow trap depth distributions, high trap density, high quantum efficiency, and excellent chemical stability, which solved the problem of chemical stability at high temperatures in the reported phosphor-in-silicone films. The emission color was simultaneously changed from green to yellow due to the variation of crystal field strength [84].

These PiGs can also be fabricated via a simpler low-temperature cosintering technique. Chen et al. prepared transparent bulk Ce/Cr:$Y_3Al_2Ga_3O_{12}$ phosphor in glass by a facile low-temperature cosintering route at 850°C for 30 min in an ambient atmosphere. Afterward the prepared glass was milled to powder and mixed with 5–20 wt% Ce/Cr:$Y_3Al_2Ga_3O_{12}$ phosphors thoroughly and sintered at 570°C for 20 min to form Ce/Cr:YAGG PiGs [85].

2.6 Summary

In summary, with the continuous development and improvement of nanomaterial synthesis technology, PPs with different composition, sizes, and structure have been prepared by various physical and chemical methods, and their optical properties have been examined by various characterization methods. Unfortunately, up to now, there is still no perfect synthesis technology to prepare the nanoparticles with long afterglow time, small particle size, high homogeneity, and highly controllable morphology. Solid-state reaction, with its highest synthetic temperature, is certainly beneficial for improving the persistent duration, but the products are often in the micron size. Liquid-phase reaction is more suitable for synthesizing nanomaterials, but small particle size, high homogeneity, and highly controllable morphology can be only obtained under moderate synthesis or sintering temperatures. Meanwhile, these features inevitably link to a shorter persistent time. In comparison, by SHI or *E*-beam irradiation of the liquid phase–derived PPs, long persistent phosphorescence can be induced or enhanced without influencing the morphology or size distribution of the particles. The only limitation is the high cost of the equipment, which limits its wide application. Furthermore, to broaden the application of inorganic nanoparticles in the field of biology, surface modification of inorganic nanoparticles is a necessary and important step. Therefore all these factors have to be considered in the development of techniques for synthesizing nanoparticles with longer persistent duration, smaller particle size, higher homogeneity, and highly controllable morphology. Glassy, glass-ceramic, and composite persistent phosphors are easily prepared; wide applications are expected for these materials.

References

[1] J. Xu, S. Tanabe, A.D. Sontakke, J. Ueda, Near-infrared multi-wavelengths long persistent luminescence of Nd^{3+} ion through persistent energy transfer in Ce^{3+}, Cr^{3+} co-doped $Y_3Al_2Ga_3O_{12}$ for the first and second bio-imaging windows, Appl. Phys. Lett. 107 (2015), 081903.

[2] H. Kanno, K. Noda, K. Matsui, New long-afterglow phosphors: Yb^{2+}-doped strontium aluminates, Chem. Phys. Lett. 580 (2013) 103–107.

[3] H. Luo, A.J.J. Bos, A. Dobrowolska, P. Dorenbos, Low-temperature VUV photoluminescence and thermoluminescence of UV excited afterglow phosphor $Sr_3Al_xSi_{1-x}O_5$: Ce^{3+},Ln^{3+} (Ln=Er, Nd, Sm, Dy and tm), Phys. Chem. Chem. Phys. 17 (2015) 15419–15427.

[4] Z. Xia, Q. Li, G. Li, M. Xiong, L. Liao, Crystal growth of $Ca_3SiO_4Br_2$: new photoluminescence bromosilicate host, J. Cryst. Growth 318 (2011) 958–961.

[5] Q. le Masne de Chermont, C. Chanéac, J. Seguin, F. Pellé, S. Maîtrejean, J.P. Jolivet, D. Gourier, M. Bessodes, D. Scherman, Nanoprobes with near-infrared persistent luminescence for *in vivo* imaging, Proc. Natl. Acad. Sci. U. S. A. 104 (2007) 9266–9271.

[6] A. Abdukayum, J.T. Chen, Q. Zhao, X.P. Yan, Functional near infrared-emitting Cr^{3+}/Pr^{3+} co-doped zinc gallogermanate persistent luminescent nanoparticles with superlong afterglow for *in vivo* targeted bioimaging, J. Am. Chem. Soc. 135 (2013) 14125–14133.

[7] D. Meroni, L. Porati, F. Demartin, D. Poelman, Sol-gel synthesis of $CaTiO_3$:Pr^{3+} red phosphors: tailoring the synthetic parameters for luminescent and afterglow applications, ACS Omega 2 (2017) 4972–4981.

[8] Y. Teng, J. Zhou, S.N. Khisro, S. Zhou, J. Qiu, Persistent luminescence of $SrAl_2O_4$: Eu^{2+}, Dy^{3+}, Cr^{3+} phosphors in the tissue transparency window, Mater. Chem. Phys. 147 (2014) 772–776.

[9] W. Xie, C. Zou, M. Xie, Effect of H_3BO_3 on phases, micromorphology and persistent luminescence properties of $SrAl_2O_4$: Eu^{2+}, Dy^{3+} phosphors, J. Mater. Sci. Mater. Electron. 28 (2017) 6328–6334.

[10] D. Verma, R.P. Patel, M. Verma, Optical properties of Sr_2SiO_4: Eu^{2+}, Dy^{3+} phosphors prepared by combustion method, Mater. Sci. Pol. 36 (2018) 387–396.

[11] B. Cheng, Z. Zhang, Z. Han, Y. Xiao, S. Lei, $SrAl_xO_y$: Eu^{2+}, Dy^{3+} ($x = 4$) nanostructures: structure and morphology transformations and long-lasting phosphorescence properties, CrystEngComm 13 (2011) 3545–3550.

[12] J. Wang, Q. Ma, W. Zheng, H. Liu, C. Yin, F. Wang, X. Chen, Q. Yuan, W. Tan, One-dimensional luminous nanorods featuring tunable persistent luminescence for autofluorescence-free biosensing, ACS Nano 11 (2017) 8185–8191.

[13] T. Maldiney, A. Bessiere, J. Seguin, E. Teston, S.K. Sharma, B. Viana, A.J.J. Bos, P. Dorenbos, M. Bessodes, D. Gourier, D. Scherman, C. Richard, The *in vivo* activation of persistent nanophosphors for optical imaging of vascularization, tumours and grafted cells, Nat. Mater. 13 (2014) 418–426.

[14] Z. Li, Y. Zhang, X. Wu, L. Huang, D. Li, W. Fan, G. Han, Direct aqueous-phase synthesis of sub-10 nm "luminous pearls" with enhanced *in vivo* renewable near-infrared persistent luminescence, J. Am. Chem. Soc. 137 (2015) 5304–5307.

[15] J. Wang, Q. Ma, X.X. Hu, H. Liu, W. Zheng, X. Chen, Q. Yuan, W. Tan, Autofluorescence-free targeted tumor imaging based on luminous nanoparticles with composition-dependent size and persistent luminescence, ACS Nano 11 (2017) 8010–8017.

[16] Y. Lv, D. Ding, Y. Zhuang, Y. Feng, J. Shi, H. Zhang, T.L. Zhou, H. Chen, R.J. Xie, Chromium-doped zinc gallogermanate@zeolitic imidazolate framework-8: a multifunctional nanoplatform for rechargeable *in vivo* persistent luminescence imaging and pH-responsive drug release, ACS Appl. Mater. Interfaces 11 (2019) 1907–1916.

[17] Y. Wang, C.X. Yang, X.P. Yan, Hydrothermal and biomineralization synthesis of a dual-modal nanoprobe for targeted near-infrared persistent luminescence and magnetic resonance imaging, Nanoscale 9 (2017) 9049–9055.

[18] B.C. Cheng, L.T. Fang, Z.D. Zhang, Y.H. Xiao, S.J. Lei, $BaAl_2O_4$: Eu^{2+}, Dy^{3+} nanotube synthesis by heating conversion of homogeneous coprecipitates and afterglow characteristics, J. Phys. Chem. C 115 (2011) 1708–1713.

[19] M. Pellerin, E. Glais, T. Lecuyer, J. Xu, J. Seguin, S. Tanabe, C. Chanéac, B. Viana, C. Richard, $LaAlO_3$: Cr^{3+}, Sm^{3+}: Nano-perovskite with persistent luminescence for *in vivo* optical imaging, J. Lumin. 202 (2018) 83–88.

[20] Z. Li, Y. Zhang, X. Wu, X. Wu, R. Maudgal, H. Zhang, G. Han, *In vivo* repeatedly charging near-infrared-emitting mesoporous $SiO_2/ZnGa_2O_4$: Cr^{3+} persistent luminescence nanocomposites, Adv. Sci. 2 (2015) 1500001.

[21] J. Shi, X. Sun, S. Zheng, J. Li, X. Fu, H. Zhang, A new near-infrared persistent luminescence nanoparticle as a multifunctional nanoplatform for multimodal imaging and cancer therapy, Biomaterials 152 (2018) 15–23.

[22] Z. Li, J. Shi, H. Zhang, M. Sun, Highly controllable synthesis of near-infrared persistent luminescence $SiO_2/CaMgSi_2O_6$ composite nanospheres for imaging *in vivo*, Opt. Express 22 (2014) 10509–10518.

[23] Z.K. Xu, G.T. Duan, H.W. Zhang, Y.Y. Wang, L. Xu, W.P. Cai, In situ synthesis of porous array films on a filament induced micro-gap electrode pair and their use as resistance-type gas sensors with enhanced performances, Nanoscale 7 (2015) 14264–14271.

[24] Z.J. Li, Y.J. Zhang, H.W. Zhang, H.X. Fu, Long-lasting phosphorescence functionalization of mesoporous silica nanospheres by $CaTiO_3$: Pr^{3+} for drug delivery, Microporous Mesoporous Mater. 176 (2013) 48–54.

[25] X.H. Lin, L. Song, S. Chen, X.F. Chen, J.J. Wei, J. Li, G. Huang, H.H. Yang, Kiwifruit-like persistent luminescent nanoparticles with high-performance and in situ activatable near-infrared persistent luminescence for long-term *in vivo* bioimaging, ACS Appl. Mater. Interfaces 9 (2017) 41181–41187.

[26] J. Wang, J. Li, J. Yu, H. Zhang, B. Zhang, Large hollow cavity luminous nanoparticles with near-infrared persistent luminescence and tunable sizes for tumor afterglow imaging and chemo –/photodynamic therapies, ACS Nano 12 (2018) 4246–4258.

[27] C. Hou, Y. Wang, H. Zhu, L.C. Zhou, Formulation of robust organic-inorganic hybrid magnetic microcapsules through hard-template mediated method for efficient enzyme immobilization, J. Mater. Chem. B 3 (2015) 2883–2891.

[28] G.P. Dong, X.D. Xiao, L.L. Zhang, Z.J. Ma, X. Bao, M.Y. Peng, Q.Y. Zhang, J.R. Qiu, Preparation and optical properties of red, green and blue afterglow electrospun nanofibers, J. Mater. Chem. 21 (2011) 2194–2203.

[29] E. Finley, A.S. Paterson, A. Cobb, R.C. Willson, J. Brgoch, Reducing particle size of persistent luminescent $SrAl_2O_4$: Eu^{2+},Dy^{3+} via microwaveassisted, reverse micelle synthesis, Opt. Mater. Express 7 (2017), 2598.

[30] Y. Tian, P. Zhang, Z. Zheng, Y. Chai, A novel approach for preparation of $Sr_3Al_2O_6$: Eu^{2+}, Dy^{3+} nanoparticles by sol-gel-microwave processing, Mater. Lett. 73 (2012) 157–160.

[31] T. Maldiney, G. Sraiki, B. Viana, D. Gourier, C. Richard, D. Scherman, M. Bessodes, K. Van den Eeckhout, D. Poelman, P.F. Smet, *In vivo* optical imaging with rare earth doped $Ca_2Si_5N_8$ persistent luminescence nanoparticles, Opt. Mater. Express 2 (2012) 261–268.

[32] A.A. Popov, G. Tselikov, N. Dumas, C. Berard, K. Metwally, N. Jones, A. Al-Kattan, B. Larrat, D. Braguer, S. Mensah, A.D. Silva, M.A. Estève, A.V. Kabashin, Laser-synthesized TiN nanoparticles as promising plasmonic alternative for biomedical applications, Sci. Rep. UK 9 (2019) 1194.

[33] P.F. Smet, N. Avci, K. Van den Eeckhout, D. Poelman, Extending the afterglow in $CaAl_2O_4$: Eu, Nd persistent phosphors by electron beam annealing, Opt. Mater. Express 2 (2012) 1306–1313.

[34] J.H. Lim, B.N. Kim, Y. Kim, S. Kang, R.J. Xie, I.S. Chong, K. Morita, H. Yoshida, K. Hiraga, Non-rare earth white emission phosphor: Ti-doped $MgAl_2O_4$, Appl. Phys. Lett. 102 (2013), 031104.

[35] L. Jin, H.Y. Zhang, R.Q. Pan, P. Xu, J.C. Han, X.H. Zhang, Q. Yuan, Z.H. Zhang, X.J. Wang, Y. Wang, B. Song, Observation of the long afterglow in AlN helices, Nano Lett. 15 (2015) 6575–6581.

[36] W. Sun, R. Pang, H. Li, D. Li, L. Jiang, S. Zhang, J. Fu, C. Li, Investigation of a novel color tunable long afterglow phosphor $KGaGeO_4$: Bi^{3+}: luminescence properties and mechanism, J. Mater. Chem. C 5 (2017) 1346.

[37] X. Lin, R. Zhang, X. Tian, Y. Li, B. Du, J. Nie, Z. Li, L. Chen, J. Ren, J. Qiu, Y. Hu, Coordination geometry-dependent multi-band emission and atypically deep-trap-dominated NIR persistent luminescence from chromium-doped aluminates, Adv. Optical Mater. (2018), 1701161.

[38] Z. Zou, C. Cao, T. Zhang, L. Feng, J. Zhang, Z. Ci, Z. Zhang, Y. Wang, Structure, long persistent luminescent properties and mechanism of a novel efficient red emitting $Ca_2Ga_2GeO_7$: Pr^{3+} phosphor, J. Alloys Compd. 680 (2016) 397–405.

[39] X. Fu, S. Zheng, J. Shi, Y. Li, H. Zhang, Long persistent luminescence property of anovel green emitting $SrLaGaO_4$: Tb^{3+} phosphor, J. Lumin. 184 (2017) 199–204.

[40] X. Zhang, J. Nie, S. Liu, Y. Li, J. Qiu, Deep-red photoluminescence and long persistent luminescence in double perovstkite-type La_2MgGeO_6: Mn^{4+}, J. Am. Ceram. Soc. 101 (2018) 1576–1584.

[41] P. Gluchowski, W. Strek, M. Lastusaari, J. Holsa, Optically stimulated persistent luminescence of europium-doped $LaAlO_3$ nanocrystals, Phys. Chem. Chem. Phys. 17 (2015) 17246–17252.

[42] N. Basavaraju, K.R. Priolkar, D. Gourier, A. Bessiere, B. Viana, Order and disorder around Cr^{3+} in chromium doped persistent luminescent AB_2O_4 spinels, Phys. Chem. Chem. Phys. 17 (2015) 10993–10999.

[43] M. Milde, S. Dembski, A. Osvet, M. Batentschuk, A. Winnacker, G. Sextl, Polymer-assisted sol-gel process for the preparation of photostimulable core/shell structured SiO_2/Zn_2SiO_4: Mn^{2+} particles, Mater. Chem. Phys. 148 (2014) 1055–1063.

[44] A. Bessie're, A. Lecointre, K.R. Priolkar, D. Gourier, Role of crystal defects in red long-lasting phosphorescence of $CaMgSi_2O_6$: Mn diopsides, J. Mater. Chem. 22 (2012) 19039–19046.

[45] Z. Xue, X. Li, Y. Li, M. Jiang, H. Liu, S. Zeng, J. Hao, X-ray-activated near-infrared persistent luminescent probe for deep-tissue and renewable *in vivo* bioimaging, ACS Appl. Mater. Interfaces 9 (2017) 22132–22142.

[46] Z. Xue, X. Li, Y. Li, M. Jiang, G. Ren, H. Liu, S. Zeng, J. Hao, A 980 nm laser-activated upconverted persistent probe for NIR-to-NIR rechargeable *in vivo* bioimaging, Nanoscale 9 (2017) 7276.

[47] D. Wei, L. Qin, Y. Huang, H.J. Seo, Photoluminescence properties of Eu^{2+}-activated $BaCa_2Al_8O_{15}$ nanophosphors, Ceram. Int. 39 (2013) 2383–2387.

[48] Y.Y. Lu, F. Liu, Z. Gu, Z. Pan, Long-lasting near-infrared persistent luminescence from β-Ga_2O_3: Cr^{3+} nanowire assemblies, J. Lumin. 131 (2011) 2784–2787.

[49] X. Fu, C. Liu, J. Shi, H. Man, J. Xu, H. Zhang, Long persistent near infrared luminescence nanoprobes $LiGa_5O_8$: Cr^{3+}-PEG-OCH_3 for *in vivo* imaging, Opt. Mater. 36 (2014) 1792–1797.

[50] P. Huang, Q. Zhang, C. Cui, J. Li, Influence of excitation wavelengths on luminescent properties of $Sr_3Al_2O_6$: Eu^{2+}, Dy^{3+} phosphors prepared by sol-gel-combustion processing, Opt. Mater. 33 (2011) 1252–1257.

[51] D.C. Rodriguez Burbano, S.K. Sharma, P. Dorenbos, B. Viana, J.A. Capobianco, Persistent and photostimulated red emission in CaS: Eu^{2+}, Dy^{3+} Nanophosphors, Adv. Opt. Mater. 3 (2015) 551–557.

[52] J.T. Kong, W. Zheng, Y.S. Liu, R.F. Li, E. Ma, H.M. Zhu, X.Y. Chen, Persistent luminescence from Eu^{3+} in SnO_2 nanoparticles, Nanoscale 7 (2015) 11048–11054.

[53] Y.F. Xu, D.K. Ma, M.L. Guan, X.A. Chen, Q.Q. Pan, S.M. Huang, Controlled synthesis of single-crystal $SrAl_2O_4$: Eu^{2+}, Dy^{3+} nanosheets with long-lasting phosphorescence, J. Alloys Compd. 502 (2010) 38–42.

[54] D. Liu, C.e. Cui, P. Huang, L. Wang, G. Jiang, Luminescent properties of red long-lasting phosphor Y_2O_2S: Eu^{3+}, M^{2+} (M= Mg, ca, Sr, Ba), Ti^{4+} nanotubes via hydrothermal method, J. Alloys Compd. 583 (2014) 530–534.

[55] S.Y. Xin, Y.H. Wang, P.Y. Dong, W. Zeng, J. Zhang, Preparation, characterization, and luminescent properties of $CaAl_2O_4$: Eu^{2+}, Nd^{3+} nanofibers using core-sheath $CaAl_2O_4$: Eu^{2+}, Nd^{3+}/carbon nanofibers as templates, J. Mater. Chem. C 1 (2013) 8156–8160.

[56] Y. Liu, S. Liu, M. Wang, W. Li, T. Zhang, X. Zhang, Synthesis and luminescence properties of Eu^{3+}, Sm^{3+} doped $(Y_xGd_{1-x})_2O_3$: Si^{4+}, Mg^{2+} long-lasting phosphor, Int. J. Miner. Metall. Mater. 17 (2010) 347–352.

[57] B. Cheng, H. Liu, M. Fang, Y. Xiao, S. Lei, L. Zhang, Long-persistent phosphorescent $SrAl_2O_4$: Eu^{2+}, Dy^{3+} nanotubes, Chem. Commun. (2009) 944–946.

[58] X. Doua, Y. Li, T. Vaneckova, R. Kang, Y. Hu, H. Wen, X. Gao, S. Zhang, M. Vaculovicova, G. Han, Versatile persistent luminescent oxycarbonates: morphology

evolution from nanorods through bamboo-like nanorods to nanoparticles, J. Lumin. 215 (2019), 116635.
[59] N.J. Zúniga-Rivera, P. Salas-Castillo, V. Chernov, L.A. Díaz-Torres, R. Melendrez, R. García-Gutierrez, R.C. Carrillo-Torres, M. Barboza-Flores, Thermally and optically stimulated luminescence in long persistent orthorhombic strontium aluminates doped with Eu, Dy and Eu, Nd, Opt. Mater. 67 (2017) 91–97.
[60] H.H. Li, S. Yin, T. Sato, Persistent photocatalysis of continuously flowing gas-phase nitric oxide using $CaAl_2O_4$: (Eu,Nd)/rutile phase $TiO_{2-x}Ny$ composites, Res. Chem. Intermed. 39 (2013) 1501–1507.
[61] X. Qiu, X. Zhu, M. Xu, W. Yuan, W. Feng, F. Li, Hybrid Nanoclusters for near-infrared to near-infrared upconverted persistent luminescence bioimaging, ACS Appl. Mater. Interfaces 9 (2017) 32583–32590.
[62] Z. Li, Y. Zhang, An efficient and user-friendly method for the synthesis of hexagonal-phase $NaYF_4$:Yb, Er/Tm nanocrystals with controllable shape and upconversion fluorescence, Nanotechnology 19 (2008), 345606.
[63] S. Yan, J. Wang, H. Gao, N. Wang, H. Yu, Z. Li, Y. Zhou, Z. Zou, Zinc gallogermanate solid solution: a novel photocatalyst for efficiently converting CO_2 into solar fuels, Adv. Funct. Mater. 23 (2013) 1839–1845.
[64] M. Elsagh, M. Rajabi, E. Amini, Characterization of $SrAl_2O_4$: Eu^{2+}, Dy^{3+} phosphor nanopowders produced by microwave synthesis route, J. Mater. Sci. Mater. Electron. 25 (2014) 1612–1619.
[65] L. Yang, P.W. May, L. Yin, R. Brown, T.B. Scott, Direct growth of highly organized crystalline carbon nitride from liquid-phase pulsed laser ablation, Chem. Mater. 18 (2006) 5058–5064.
[66] F. Yoshimura, K. Nakamura, F. Wakai, M. Hara, M. Yoshimoto, O. Odawara, H. Wada, Preparation of long-afterglow colloidal solution of $Sr_2MgSi_2O_7$: Eu^{2+}, Dy^{3+} by laser ablation in liquid, Appl. Surf. Sci. 257 (2011) 2170–2175.
[67] T. Zhan, C. Xu, H. Yamada, Y. Terasawa, L. Zhang, H. Iwase, M. Kawai, Enhancement of afterglow in $SrAl_2O_4$: Eu^{2+} long-lasting phosphor with swift heavy ion irradiation, RSC Adv. 2 (2012) 328–332.
[68] L.L. Noto, O.M. Ntwaeaborwa, J.J. Terblans, H.C. Swart, Dependence of luminescence properties of $CaTiO_3$: Pr^{3+} on different TiO_2 polymorphs, Powder Technol. 256 (2014) 477–481.
[69] T.Z. Zhan, C.N. Xu, H. Yamada, Y. Terasawa, L. Zhang, H. Iwase, M. Kawai, Beam profile indicator for swift heavy ions using phosphor afterglow, AIP Adv. 2 (2012), 032116.
[70] R. Kang, X. Dou, H. Lian, Y. Li, $SrAl_{12}O_{19}$: Fe^{3+} @3-aminopropyl triethoxysilane: ambient aqueous stable near-infrared persistent luminescent nanocomposites, J. Am. Ceram. Soc. 103 (2020) 258–265.
[71] Y. Li, R. Chen, Y. Li, K. Sharafudeen, S. Liu, D. Wu, Y. Wu, X. Qin, J. Qiu, Folic acid-conjugated chromium(III) doped nanoparticles consisting of mixed oxides of zinc, gallium and tin, and possessing near-infrared and long persistent phosphorescence for targeted imaging of cancer cells, Microchim. Acta 182 (2015) 1827–1834.
[72] J. Qiu, K. Hirao, Long lasting phosphorescence in Eu^{2+}-doped calcium aluminoborate glasses, Solid State Commun. 106 (1998) 795–798.
[73] J. Qiu, M. Kawasaki, K. Tanaka, Y. Shimizugawa, K. Hira, Phenomenon and mechanism of long-lasting phosphorescence in Eu^{2+}-doped aluminosilicate glasses, J. Phys. Chem. Solids 59 (1998) 1521–1525.

[74] H. Hosono, T. Kinoshita, H. Kawazoe, M. Yamazaki, Y. Yamamoto, N. Sawanobori, Long lasting phosphorescence properties of Tb^{3+}-activated reduced calcium aluminate glasses, J. Phys. Condens. Matter 10 (1998) 9541–9547.

[75] S. Ye, F. Xiao, Y.Y. Pan, Y.Y. Ma, Q.Y. Zhang, Phosphors in phosphor-converted white light-emitting diodes: recent advances in materials, techniques and properties, Mater. Sci. Eng. R 71 (2010) 1–34.

[76] P.F. Li, M.Y. Peng, M. Lothar, Y.Q. Zhao, V. Bruno, Red to near infrared ultralong lasting luminescence from Mn^{2+}-doped sodium gallium aluminum germanate glasses and (Al, Ga)-albite glass-ceramics, J. Mater. Chem. C 3 (2015) 3406.

[77] Y.X. Zhuang, J. Ueda, S. Tanabe, Multi-color persistent luminescence in transparent glass ceramics containing spinel nano-crystals with Mn^{2+} ions, Appl. Phys. Lett. 105 (2014), 191904.

[78] Blasses, Study of the vibronic transitions of Gd^{3+} and Eu^{3+} in crystalline materials and glasses of the same composition, Phys. Rev. B 39 (1989) 3408.

[79] A. Mathieu, A. Salaheddine, F. Franck, S. Matthew, P. Florence, C. Thierry, M. Guy, Highly transparent $BaAl_4O_7$ polycrystalline ceramic obtained by full crystallization from glass, Phys. Rev. B 24 (2012) 5570.

[80] T. Suzuki, G.S. Murugan, Y. Ohishi, Optical properties of transparent Li_2O-Ga_2O_3-SiO_2 glass-ceramics embedding Ni-doped nanocrystals, Appl. Phys. Lett. 86 (2005), 131903.

[81] T. Hu, H. Lin, J. Xu, B. Wang, J.M. Wang, Y.S. Wang, Color-tunable persistent luminescence in oxyfluoride glass and glass ceramic containing Mn^{2+}:alpha-Zn_2SiO_4 nanocrystals, J. Mater. Chem. C 5 (2017) 479.

[82] K. Shinozaki, T. Honma, M. Affatigato, T. Komatsu, Long afterglow in hexagonal $SrAl_2O_4$:Eu^{2+}, Dy^{3+} synthesized by crystallization of glass and solidification of supercooled melts, J. Lumin. 177 (2016) 286.

[83] J. Massera, P. Gluchowski, M. Lastusaari, L.C.V. Rodrigues, L. Petit, J. Holsa, L. Hupa, M. Hupa, New alternative route for the preparation of phosphate glasses with persistent luminescence properties, J. Eur. Ceram. Soc. (4) (2015) 1225.

[84] W.H. Li, Y.X. Zhuang, P. Zheng, T.L. Zhou, J. Xu, J. Ueda, S. Tanabe, L. Wang, R.J. Xie, Tailoring trap depth and emission wavelength in $Y_3Al_{5-x}Ga_xO_{12}$:Ce^{3+}, V^{3+} phosphor-in-glass films for optical information storage, ACS Appl. Mater. Interfaces 10 (32) (2018), 27150.

[85] D.Q. Chen, Y. Zhou, W. Xu, J.S. Zhong, P. Huang, Persistent and photo-stimulated luminescence in Ce^{3+}/Cr^{3+} activated $Y_3Al_2Ga_3O_{12}$ phosphors and transparent phosphor-in-glass, J. Mater. Chem. C 4 (2016), 11457.

Mechanisms

3

An admitted description of trapping and detrapping progress follows the steps in the succeeding text: excitation, carriers trapping, carriers storing, carriers releasing, and afterglow emission (Fig. 3.1). Charge carriers firstly are excited into an excited state level of activators; then the carriers are captured by "traps," which are some isolated or continuously distributed energy levels inside the forbidden bandgap; subsequently the captured carriers are stored in the traps, until the thermal disturbance is strong enough to release these trapped charge carriers, which finally results in emission of photons. Under the excitation by x-ray, electronic beam, or UV light, the storage of optical energy by trapping electrons takes only a very short time, yet the liberation of electrons due to thermal fluctuation that results in afterglow can last up to several hours under room temperature [1–3].

It should be emphasized that the discussed persistent phosphors here are all inorganic materials, not organic-based macromolecule. In general, inorganic persistent phosphors generate persistent phosphorescence via a trap-controlled progress; thus the afterglow strongly depends on the existence of defects, which play an important role in the mechanism for generating afterglow. In contrast the afterglow from organic molecules (long phosphorescence) is generated based on the excitation of triplet energy level, from which the transition to the spin singlet ground state is not allowed. Controlling surface structure, hydrogen bond, and molecular chain are the main strategies to regulate organic afterglow. Their mechanisms are completely different from that of the persistent phosphorescence observed in inorganic phosphors. In this chapter, we discuss the mechanism of persistent phosphorescence in inorganic persistent phosphors.

3.1 Factors controlling trapping and detrapping progress

Up to now, there is still no widely accepted model to describe the trapping and detrapping progress in persistent phosphors. In general, trap type, concentration, and depth are three essential topics that should be considered.

3.1.1 Trap types

Though traps are common in various persistent phosphors, we focus our discussion here on traps that store and release the charge carriers. Generally speaking, these traps correspond to some lattice defects, that is, intrinsic crystal defects, intentionally introduced defects or both. Usually, there are four ways to create the lattice defects:

(a) heating the material to a high temperature to induced dissociation and disorder;
(b) topochemically reduction or oxidation to induce formation of vacancies or other defects;
(c) irradiation with electron beam, high-energy rays, nuclear particles, or high-power laser;
(d) introduction of impurity ions to induce defect formation.

Persistent Phosphors. https://doi.org/10.1016/B978-0-12-818637-4.00003-3

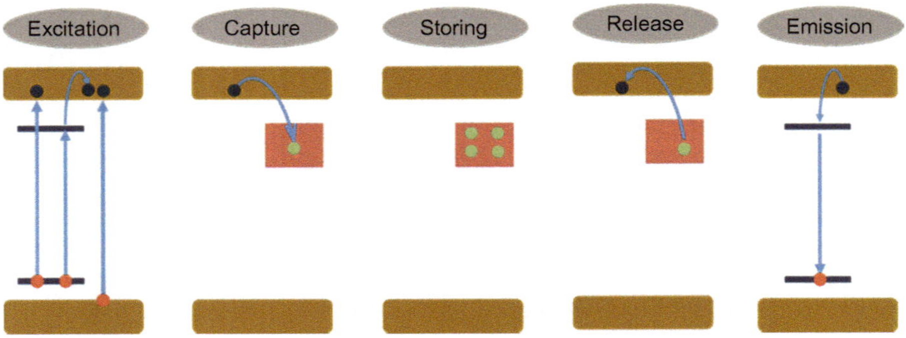

Fig. 3.1 Schematic diagrams of trapping and detrapping progress in persistent luminescence.

Even though the traps can be generated in the aforementioned conditions, it is still difficult to identify the types of traps due to the complex material components and comprehensive operating conditions. For example, the crystal defects in pure Al_2O_3 crystals usually include Al vacancies (V_{Al}), interstitial Al (Al_I), interstitial vacancies (O_I), monovalent oxygen vacancy (V_O^+), and oxygen vacancy (V_O), while in sapphire (including Ti^{4+} and Fe^{3+} impurities in Al_2O_3 crystal), substitutional defects (Ti_{Al}) and (Fe_{Al}) in hexagonal crystal lattices also need to be considered. Actually a lot of work needs to be done to understand which defect is generated during the material synthesis and which the contributing trap is. Moreover, it is impossible to precisely forecast which defect is authentically responsible for the capture, storage, and release of carriers. Until now, no standard database is available for indexing trap types for a specific material composition and synthesis condition.

3.1.2 Trap concentrations

The concentration of contributing traps, that is, the numbers of trap responsible for trapping, storing, and releasing carriers, plays a key role in the performance of persistent phosphors. Generating more contributing traps is beneficial to capture more charge carrier; hence, increasing the capacity of charge carrier reservoir is a considerable issue for prolonging the afterglow time. The trap concentrations can be evaluated via thermoluminescence (TL) measurement, positron annihilation spectra, and photocurrent curves. However, the mechanisms and characterizations of the measurements are different. TL spectra represent the luminescence intensity as a function of temperature, and it is an external method. In contrast, positron annihilation spectra and photocurrent curves are interior characterizations, and it is possible to measure the photoconductivity and the minority carrier lifetime by using these techniques.

3.1.3 Trap depths

The depth of contributed traps is associated with the energy barrier above which the trapped carriers can be released. Usually, shallow traps are quickly emptied at room temperature. Such a rapid release just looks like rushing floodwaters, and it is beneficial for obtaining a high initial intensity. However, a fast release exhausts the stored electrons at full speed and brings about a short afterglow time. Deep traps with the slower release rate, on the other hand, are almost impossible to empty at room temperature. A part of carriers is still stabilized in deep traps; therefore a slow release rate results in a weaker afterglow intensity and long persistent time. If the trap is too deep to release electrons by thermal fluctuation at room temperature, afterglow cannot be observed. Therefore a trap with suitable depth can benefit for the afterglow performance of persistent phosphors.

3.2 Proposed afterglow models

Though different mechanisms have been proposed through the combination of the basic theoretical models and experimental observations, a clear global picture for trapping and detrapping process of charge carriers remains elusive. In this part, we discussed these proposed models, including conduction band (CB)-valence band (VB) recombination model, energy band engineering model, quantum tunneling model, oxygen vacancy and coordination model, electron model, and two-photon absorption.

3.2.1 Conduction band-valence band model

Until the development of aluminate-based materials in the mid-1990s, afterglow mechanism has not been analyzed in very detail. It was generally accepted that the trapping is a fundamental process for energy storage in persistent phosphors. This energy storage is achieved by spatial localization of the excited carriers, in such a way that an electron or a hole is prohibited to freely move unless supplied with plentiful thermal or optical energy. When the trapped electron or hole is released into CB or VB, it can move freely until being trapped by a recombination center or another trap. This model is recommended based on the fact that CB or VB participates into carriers' migration process and that the energy level position of the involved traps should be near to the CB/VB. The CB/VB models include recombination model, hole model, electron model, and two-photon absorption model.

3.2.1.1 Recombination model

The recombination model usually requires simultaneous participation of two traps, that is, electron traps (E-trap), which are located close to the conduction band bottom, and hole traps (H-trap), which are located close to the valence band top (Fig. 3.2). When persistent phosphors are excited by UV light, an electron is pumped into the conduction band, leaving a hole in the valence band. The excited electrons in CB could be captured by an E-trap near the CB, while the holes migrate synchronously and could be captured by a H-trap. After ceasing the irradiation the trapped electrons and holes reenter into the CB/VB. Electrons jump to luminescent center under the excited state through the conduction band, and the holes return to the ground state of the luminescent center via the valence bands, resulting in a recombination of

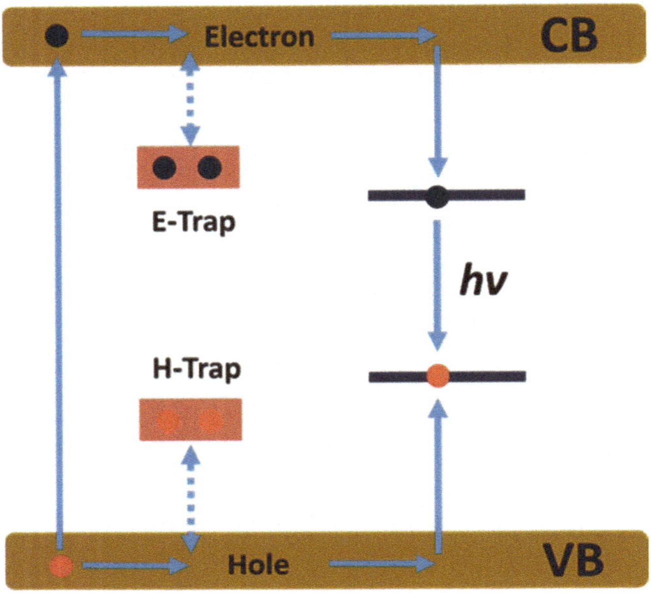

Fig. 3.2 Recombination model.

emitter. Certainly the released electrons/holes also can be retrapped by E-traps and H-traps again and give a retrapping effect [1].

Recombination model is a pioneering model for usual trapping and detrapping process. Later a modified energy-transfer recombination model is proposed and further optimized recently. This modified recombination model takes into consideration of energy-transfer between active centers (Fig. 3.3). *E*-trap and H-trap may capture the carriers, which may combine to give emission afterward. This distinct phenomenon is observed in $CdSiO_3$, $CaSnSiO_5$, and $Ca_3SnSi_2O_9$ PPs. A visible afterglow band at 430 nm is observed in these activator-free $CdSiO_3$, $CaSnSiO_5$, and $Ca_3SnSi_2O_9$ hosts. This blue afterglow is assigned to a trap-mediated recombination, which is unusual in a rare earth-free material designed to achieve the multicolor afterglow via the energy transfer. Based on this model a series of Pr^{3+}, Tb^{3+}, Er^{3+}, Dy^{3+}, Sm^{3+}, Tm^{3+}, and Nd^{3+} doping PPs have been synthesized with typical afterglow performance.

3.2.1.2 *Hole model*

Matsuzawa and coworkers reported the fabrication of $SrAl_2O_4$: Eu^{2+}, Dy^{3+} that showed afterglow duration over 10 h. In their paper, they also elaborated the origin of the extraordinary persistent luminescence (Fig. 3.4) [4]. In this model, holes were assumed to be the majority charge carriers based on the photoconductive results of $SrAl_2O_4$: Eu^{2+}, Dy^{3+}. They further pointed out that when a Eu^{2+} ion was excited by a photon, a hole could possibly escape to the VB, thereby leaving behind an

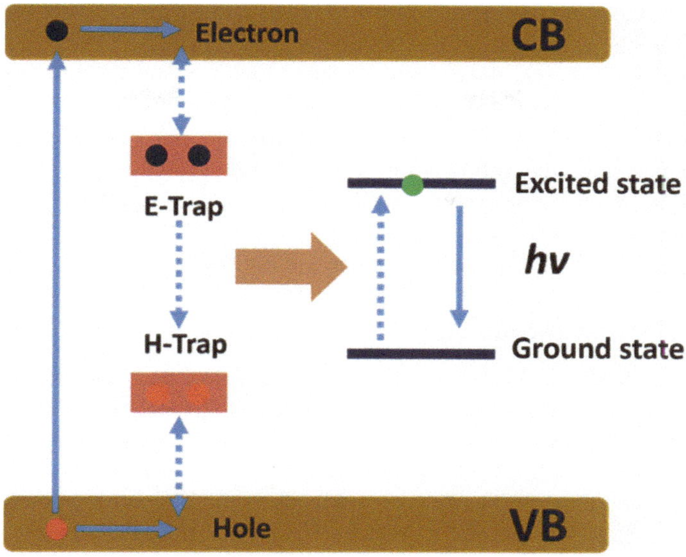

Fig. 3.3 Energy-transfer recombination model.

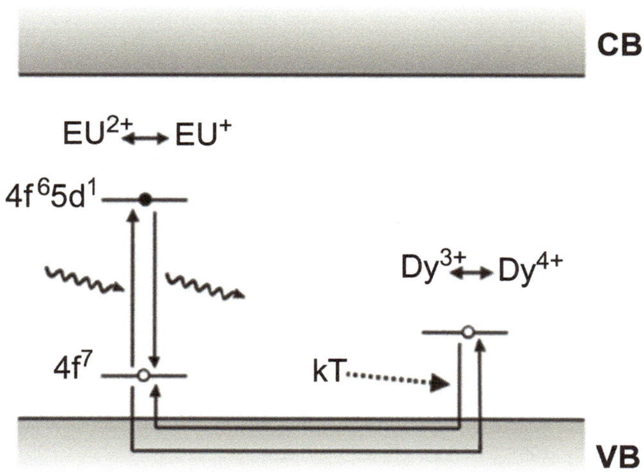

Fig. 3.4 Hole model: afterglow mechanism proposed by Matsuzawa et al. for $SrAl_2O_4$: Eu^{2+}, Dy^{3+}.
From T. Matsuzawa, Y. Aoki, N. Takeuchi, Y. Murayama, A new long phosphorescent phosphor with high brightness, $SrAl_2O_4$: Eu^{2+}, Dy^{3+}, J. Electrochem. Soc. 143 (1996) 2670–2673.

Eu^+ ion. The hole was then captured by a trivalent rare earth ion, such as Dy^{3+}, thus creating a Dy^{4+} ion. After a while, thermal energy caused the trapped hole to be released into the VB again. From there, it moved back to the Eu^+ ion [5], allowing it to return to Eu^{2+} ground state to produce the visible afterglow emission. Their results suggested that Dy^{3+} plays a key role in generating the photoconductivity as hole-trapping center, and the introduction of Dy^{3+} ion produces a new trap level. Abbruscato and coworkers experimentally supported this hypothesis, that is, holes in the valence band had to be the main charge carriers based on earlier measurements of $SrAl_2O_4$: Eu^{2+}, which also showed a weak afterglow.

3.2.1.3 Electron model

In 2003 Aitasalo proposed that electrons were directly excited from the VB into trap levels of unspecified origin [6]. Since the position of CB was much higher than the energy level of the oxygen vacancy trap and a thermally assisted transition to the CB seems to be difficult, they speculated that the energy released on recombination of the electron and the hole was delivered directly to the Eu ions, by means of energy transfer (Fig. 3.5). Electron model also successfully elaborated the doping influence of trivalent rare earth codopants. They opposed the existence of Dy^{4+} and Eu^+ in aluminates or silicates and further pointed out that the reduction of Eu^{2+} to Eu^+ along with the oxidation of Dy^{3+} to Dy^{4+} would lead to a chemically instability of doping ions. They suggested that trivalent RE ions increased the number of lattice defects, because

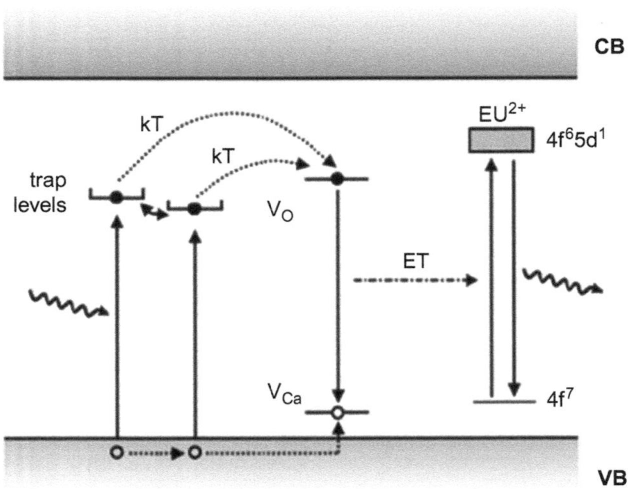

Fig. 3.5 Electron model: afterglow mechanism proposed by Aitasalo et al. for $CaAl_2O_4$: Eu^{2+}, Dy^{3+}.
From T. Aitasalo, P. Deren, J. Hölsä, H. Jungner, J.C. Krupa, M. Lastusaari, J. Legendziewicz, J. Niittykoski, W. Strek, Persistent luminescence phenomena in materials doped with rare earth ions, J. Solid State Chem. 171 (2003) 114–122.

they occupied the divalent alkaline earth sites. In this case spontaneous formation of defect for charge compensation could be a natural consequence [7].

Holsa et al. also recognized that Eu^{+} species were unstable in aluminates [6]. Further, they suggested that Matsuzawa's model virtually ignored the appearance of persistent phosphorescence in Eu^{2+} singly doped $SrAl_2O_4$ phosphors, and R^{3+} might participate in the generation of more traps or improvement of traps properties. Besides, they also synthesized $SrAl_2O_4$: Eu, Sm persistent phosphors, and explained the reason of adding Sm^{3+} into $SrAl_2O_4$: Eu and their detrimental effect for afterglow time. In this host, Sm^{3+} ions were easily reduced to Sm^{2+} ions in reducing atmosphere (mixture of N_2-H_2), leading to a decrease of trap concentrations. Based on this hypothesis, they gave a plausible afterglow mechanism for $CaAl_2O_4$: Eu^{2+}, R^{3+} and ZrO_2: Ti. In their investigation, their employed EPR, XPS, and XANES techniques and observed experimentally $Eu^{2+} \rightarrow Eu^{3+}$ oxidation and no change in the codopants' oxidation state were observed. Although being much more reasonable, the problem with this afterglow mechanisms in general is that several materials—including both $Sr_2MgSi_2O_7$: Eu^{2+}, Dy^{3+} (blue) and $SrAl_2O_4$: Eu^{2+}, Dy^{3+} (green emission)—show rather intense persistent phosphorescence via codoping, albeit weaker and shorter afterglow compared with the same materials without codoping of a secondary dopants. The introduction of a codopant R^{3+} is usually thought to regulate the trap characteristics [8].

Clabau et al. also did not accept the aforementioned Matsuzawa's model. EPR measurement showed that Eu^{2+} concentration decreased during photoexcitation process but increased once the excitation was terminated and continued until the end of persistent luminescence. Thus they suggested Eu^{2+} participated in the trapping and detrapping process, which contradicted the proposal of energy transfer to Eu^{2+} after trapping [9].

3.2.1.4 Two-photon absorption model

An unusual afterglow excitation band at around 530 nm is observed in $CaAl_2O_4$: Eu^{2+}, Dy^{3+} persistent phosphors, implying the presence of a new excitation channel for generating afterglow. Aitasalo et al. suggested an additional energy-transfer model that includes an absorption of two 530-nm photons through excitation of a deep energy level, followed by capturing of two electrons from the shallow energy level. Subsequently, they showed that 430-nm persistent luminescence of $CaAl_2O_4$: Eu^{2+}, Nd^{3+} could be induced by excitation at wavelengths longer than 530 nm. Based on this result, Aitasalo proposed another new afterglow mechanism, the two-photon absorption model (Fig. 3.6). They pointed out that absorbing two photons had to occur in this mechanism, through direct excitation of an electron from the VB into a long-lived intermediate trap level, followed by a so-called excited state absorption process. They believed that the defect served as the electron–hole recombination center in the bandgap after two successive steps of optical excitation. Through this recombination process the energy was eventually transferred to Eu^{2+} to produce the persistent luminescence.

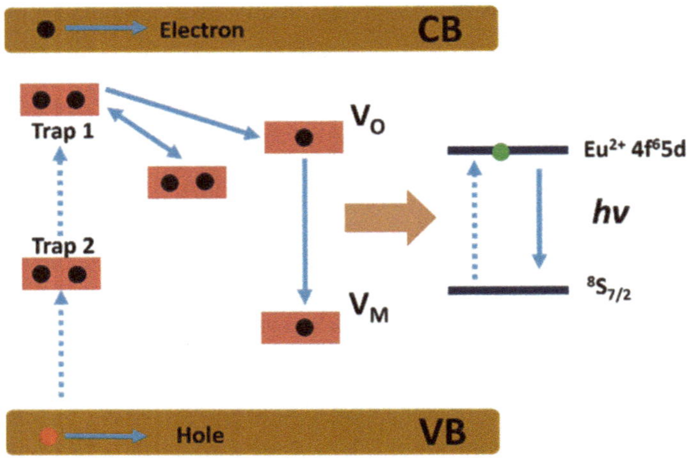

Fig. 3.6 Two-photon absorption model: afterglow mechanism proposed by Aitasalo et al. for $CaAl_2O_4$:Eu^{2+}, Dy^{3+} under the excitation at 530 nm.

By studying the spectral data under recorded different temperatures, Smet et al. observed similar phenomenon from $Ca_2Si_5N_8$: Eu, Tm PPs. They called it "double excitation" process, which could be explained by successive trapping of two electrons by a Tm^{3+}-Eu^{2+} pair.

3.2.2 *Energy band engineering model*

Dorenbos et al. agreed with Aitasalo about his model. They further demonstrated that the suggested carriers on the ground state of Eu^{2+} after excitation were based on faulty reasoning [10]. As he suggested in 2005, the energy levels of lanthanides were localized, in contrast to the delocalized Bloch states of the valence and conduction bands [11]. Therefore the initial electron model should be revised. In his revised model, another obvious distinction was that electrons were excited and assigned to divalent europium ions instead of valence band (Fig. 3.7). Since 5d level of divalent europium is located very close to CB, these excited electrons can be easily released into the CB and subsequently captured by a trivalent RE ion. However, the revised electron model also didn't account for the existence of PP in Eu^{2+} singly doped $SrAl_2O_4$ phosphors.

Besides, other researchers have made great effort to the quantitative calculation of lanthanide energy levels in persistent phosphors [12, 13]. Holsa et al. did plenty of works in this field. Based on the experimental data, they provided a perfect strategy for understanding this afterglow mechanism, including the (i) energy level positions of R^{2+} and R^{3+} ions, (ii) bandgap energy of hosts, and (iii) trap level positions with respect to the bottom of the conduction band of hosts. Diffuse reflection spectra provide information about host bandgap, while the trap level positions could be evaluated by the measurement of TL spectra as will be discussed later. They also provided some methods to assess the ground level and excitation level energies for R^{2+} and R^{3+} ions in the host band structure using $CdSiO_3$: Tb^{3+} as an example [14–16]. They showed

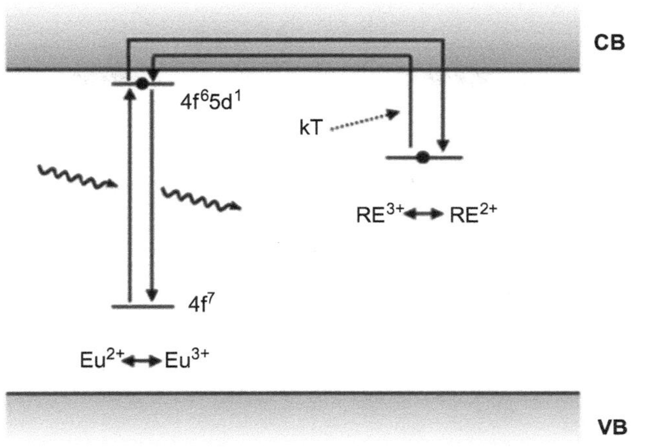

Fig. 3.7 Persistent luminescence mechanism proposed by Dorenbos et al. for aluminates and silicates.
From K. Van den Eeckhout, P.F. Smet, D. Poelman, Persistent luminescence in Eu^{2+}-doped compounds: a review. Materials 3 (2010), 2536–2566, based on P. Dorenbos, Mechanism of persistent luminescence in Eu^{2+} and Dy^{3+} codoped aluminate and silicate compounds, J. Electrochem. Soc. 152 (2005) H107–H110.

that the previous empirical model was based on the fact that the trends in ground state energies for both R^{2+} and R^{3+} ions were independent of the host. This meant that once the energy of ground level for a R^{3+} ion in one host was located, the position of energy levels for any other R^{3+} ion can be determined easily by using the Dorenbos model (Fig. 3.8). For instance, the position of Tb^{3+} ($^7F_6$) ground level could be estimated by referring to the location of Eu^{3+} ($^7F_0$) ground level in $CdSiO_3$, even though we only experimentally identified the energy level of Eu^{3+} ($^7F_0$). Furthermore the transition intensity of different excitation bands, for example, the intensity of $4f^8 \rightarrow 4f^7 5d^1$ (or $4f^1 \rightarrow 5d^1$) transitions in the excitation spectra of Tb^{3+}, was suppressed when the $4f^7 5d^1$ levels were located within the host's conduction band.

The ground state energy level positions for R^{2+} and R^{3+} ions in most wide bandgap materials have been investigated. Based on the aforementioned model, they showed that the best candidates for afterglow centers were (in this order) Eu^{2+}, Ce^{3+}, Tb^{3+}, and Pr^{3+}. First of all, all these centers were known to be strong emitters with weak tendency to lose the excitation energy in cross relaxation or multiphonon deexcitation processes.

As for persistent luminescence, most present applications prefer the use of excitation (daylight or artificial lighting) in the visible spectra region or, if not possible, with near-UV radiation. Eu^{2+}, Ce^{3+}, Tb^{3+}, and Pr^{3+} exhibit the lowest 5d energies in the rare earth series, thus allowing for excitation at relatively longer wavelengths. Furthermore the overlap between the excited 5d levels and the host's CB is large, such that electron

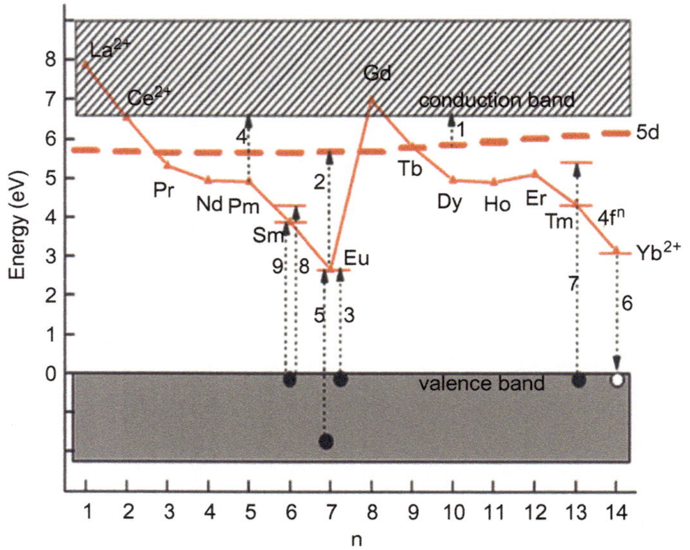

Fig. 3.8 The systematic variation in the energy level positions of divalent lanthanides in wide bandgap ionic crystals; n represents the number of electrons in the $4f^n5d^0$ configuration of the divalent lanthanide.
Reproduced with permission from P. Dorenbos, Energy of the first $^4f_7 \rightarrow {}^4f^5_6d$ transition of Eu+ in inorganic compounds, J. Lumin. 104 (2003), 239–260, Copyright 2003, Elsevier.

transfer from CB to traps could occur with high efficiency. Though the excitation of the trivalent ions (Ce^{3+}, Tb^{3+}, and Pr^{3+}) usually requires the use of UV radiation, Eu^{2+} can be excited in many hosts by the blue light [17–19].

It is generally accepted that a powerful tool for calculating trap level positions is thermoluminescence (TL) measurement, from which trap level positions can be located. Bos et al. and Zych et al. measured TL curves of YPO_4: Ce^{3+}, R^{3+} (R = Nd, Sm, Dy, Ho, Er, and Tm) phosphors and YPO_4: Pr^{3+}, R^{3+} (R = Nd, Dy, Ho, and Er) phosphors [20, 21]. TL peaks of YPO_4: Pr^{3+}, R^{3+} and YPO_4: Ce^{3+}, R^{3+} are located at almost the same temperature. Similar activation energies for electron detrapping from R^{2+} for both Pr^{3+}-doped and Ce^{3+}-doped YPO_4 persistent phosphors have also been confirm by other researchers. Although this was not a typical Eu^{2+}-based compound, these results confirmed that the codopant indeed played the role of traps in some materials.

3.2.3 Quantum tunneling model

Quantum tunneling refers to the quantum mechanical phenomenon where a particle tunnels through a barrier that classically could not be surmounted. If these barriers are thin and low enough, electrons can penetrate potential barriers, as pointed out by Oppenheimer (1928), Fowler and Nordheim (1928), and others. This effect can be interpreted by perpendicular carrier transport in two-dimensional structures. It can easily be understood via the recognition of the wave function of an electron. Instead of immediately stop at a barrier, electrons loss its energies exponentially into the barrier with a slope determined by the barrier height. If the barrier is thin, there is a nonzero amplitude of the wave function, which remains at the end of the barrier, that is, a non-zero probability for the electron to penetrate.

The probability of penetration also depends on the shape of the barrier. In semiconductor devices the potential barrier often can be approximated by a triangular shape, such as the simplified metal semiconductor heterojunction interface and the band-to-band transition case. At sufficiently high field strength, the band edges are tilted, and tunneling through such a barrier is important.

Besides the dependence of the tunneling current on the level-density distribution makes it possible to use this current for obtaining information about carrier distribution. This process involves a similar principle in shifting the Fermi level or quasi-Fermi level on one side of a barrier with respect to the Fermi level on the other side, thus allowing the tunneling happening of carriers through the barrier at suitable energies. The level distribution is profiled near the top of VB or in CB, depending on forward or back bias, and the increase of tunneling current becomes larger when more levels are available. The probing side of the barrier can be either a highly doped semiconductor, a metal electrode, or a superconductor.

This process may occur between the delocalized traps and nearest excited level of emitters. Compared with CB/VB model, this model doesn't require a physical proximity of traps' level and CB/VB. That means this model is mainly utilized to interpret the afterglow mechanism related to deep traps.

Hoogeustraten et al. first investigated the existence of quantum tunneling effect in sulfide [22]. Tanabe and coworkers subsequently used a similar model in studying $SrAl_2O_4:Eu^{2+}$, Dy^{3+} PPs [23]. They found that most of the 5d levels of Eu^{2+} at the Sr site with weak crystal field were located within CB based on the measurement of photocurrent excitation spectra. Based on this discovery, they proposed that the effective footpath for trapping and detrapping is achieved via quantum tunneling process, but not through CB or VB (Fig. 3.9) [24].

Pan et al. reported a series of Cr^{3+}-doped zinc gallogermanate NIR persistent phosphors that demonstrates strong emission at 650–1000 nm with a super-long afterglow of more than 360 h [24]. The efficient activation of afterglow by ultraviolet light implied that the approximate energy of the ionization threshold in $Zn_3Ga_2Ge_2O_{10}$ host was near CB. This was in consistent with the CB/VB model. Nonetheless, the fact is that visible light (400–630 nm) with photon energy below the ionization threshold could still generate weak but long persistent luminescence. They thus suggested another mechanism to explain the electron trapping and detrapping processes, that is, quantum tunneling model. Under the excitation of short-wavelength visible light, the ground state electrons of Cr^{3+} emitters were excited to energy levels of deep traps below the ionization threshold, and the traps were then filled through a tunneling mechanism from accessible Cr^{3+} energy levels. The reverse tunneling recombination generated weak but long persistent luminescence. They also conducted a decay measurement for persistent phosphors at 77 K with different excitation wavelengths

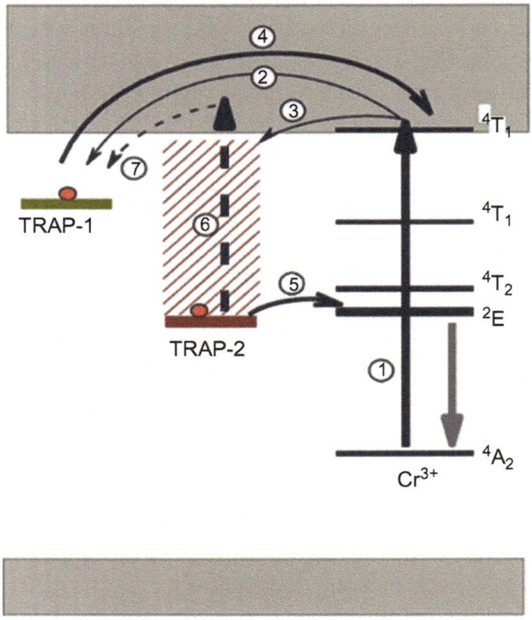

Fig. 3.9 A schematic representation of the persistent NIR luminescence mechanism. The *straight-line arrows* and *curved-line arrows* represent optical transitions and electron transfer processes, respectively.
Reproduced with permission from Z.W. Pan, Y.Y. Lu, F. Liu, Sunlight-activated long-persistent luminescence in the near-infrared from Cr^{3+}-doped zinc gallogermanates, Nat. Mater. 11 (2012), 58–63, Copyright 2012, Nature Publication Group.

between 320 and 600 nm to support this model. Subsequently, they reported a similar phenomenon in a Cr^{3+}-doped $LiGa_5O_8$ NIR persistent phosphor, which shows a long persistent duration beyond 1000 h [25].

3.2.4 Oxygen vacancy model and coordination model

Qiu et al. investigated the photo-stimulated luminescence in Eu^{2+}-doped fluoroaluminate glasses [26], Eu^{2+}-doped strontium aluminosilicate glasses [27], Ce^{3+}-doped alkali borate and silicate glasses, and gold nanoparticle–doped glasses [28, 29]. A new afterglow model, called oxygen vacancy model (oxygen vacancy model has been proposed in 1990s. See Dorenbos' paper on LSO/Ce), was proposed to explain the trapping and detrapping process based on analyzing the EPR spectra. Such model was built depending on the identification and confirmation of trap types, which are universally applicable for understanding the afterglow mechanism of oxide-based persistent phosphors.

In earlier research, Aitasalo et al. suggested an oxygen vacancy as the key trap to capture electrons in their electron model. Clabau et al. also pointed out that oxygen vacancies acted as an efficient electron trap in $SrAl_2O_4$: Eu^{2+}, Dy^{3+} [6]. By comparing

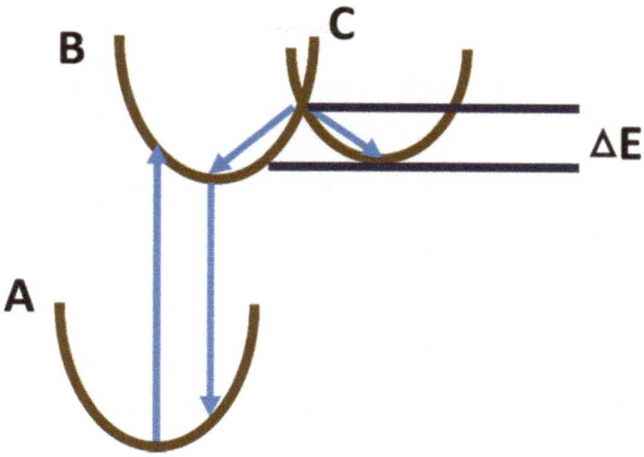

Fig. 3.10 Configurational coordination model for persistent phosphorescence.

afterglow curves of Eu^{2+} singly doped $SrAl_2O_4$ and Eu^{2+}/Dy^{3+}-codoped $SrAl_2O_4$ [9], they found that two TL peak with variable intensity and location but demonstrate similar spectral shape. In addition, they found that the chemical nature of traps was not influenced by codoping, and only oxygen vacancies still worked for storing and releasing electrons. Lim and coworkers studied the afterglow properties of Ti-doped spinel powders. Based on their discovery, it has been demonstrated that 620-nm red emission is associated with Mg^{2+} and O^{2-} vacancies, since a pure spinel was reported to have plenty of intrinsic defects, such as Mg^{2+} and O^{2-} vacancies, and they usually appeared in pair, known as Schottky defects [30].

In addition, Qiu et al. also suggested a coordination model to show the trapping and detrapping process. In Fig. 3.10A and B represent the ground and excited states of Eu^{2+}, respectively; C is the trap level; E is the energy gap between the trap level and the level of Eu^{2+}excited states. This model is used to explain oxygen vacancy model of $SrAl_2O_4$: Eu^{2+}, Dy^{3+} persistent phosphors, which are synthesized under a reduction atmosphere. We propose the trap C as hole trap, and oxygen vacancy is generated during the substitution of Eu^{2+} with Dy^{3+}. A coordination model is demonstrated in the succeeding text. Under the excitation of UV light, the electrons from ground state A are excited and then transferred to the excited state B. Finally the electrons may relax to the trap level C. Because the energy level of trap C is higher than that of excited level B, the electrons in trap level C may return to the excited state B, thus inducing the persistent luminescence. Furthermore, we found that this model allows the visualization of the relationship between afterglow time and trap level. With the increase in the population of the defect level, the absorbed and stored optical energy will rise, which leads to the growth of afterglow time.

3.3 Essential physical processes

Though many afterglow mechanisms are proposed to explain the trapping and detrapping process, the afterglow nature is still controversial. Actually, the essential physical processes are worth being discussed in order to understand the afterglow nature. Since there are few studies on these processes for persistent phosphors, to advance this research, a skillful route from the semiconductor defect which has been widely studied in the last few decades is proposed. When a semiconductor is exposed to an external electromagnetic field, or an electric field, free carriers may be produced, leading to a semiconductivity and photoconductivity. Carriers may also be generated by high energy particles, such as fast electrons or ions. Optical generation of carriers occurs in direct or indirect band-to-band transitions and transitions involving defect levels. Free carriers are also generated by thermal fluctuation, and this process might be enhanced by defect centers. Shallow traps may absorb a phonon with sufficient energy or a few phonons involving several intermediate steps, resulting eventually the population of its excited states; carrier generation from deep traps could occur by multiphonon-induced giant oscillations.

3.3.1 Carrier generation

Due to the Frenkel-Poole effect, generation of carriers by exposing to electric field may occur at low field strength due to field-enhanced thermal disorder between Coulomb-attractive centers. At high field strengths, impact ionization from deep traps or band-to-band impact ionization would appear. At even higher fields beyond 106 V/cm range, tunneling occurs from deep defect centers. In addition to thickness and height of the barrier potential, the tunneling probability strongly depends on the shape of the barrier.

Three types of carrier generation processes are demonstrated in the succeeding text. All these carrier generation processes consume energy, which can be supplied phonons, photons, and electric bias.

The type of generation selected is usually the choice of the experimentalist, whereas the type of predominant recombination is mostly a function of the defect structure of the selected material. It is also influenced by the temperature and sometimes by the field intensity and other parameters.

Optical absorption by band-to-band transition is a basic photoexcitation process, which creates equal electron and hole densities (Fig. 3.11). If photon energies are larger than the bandgap energy, optical absorption becomes stronger; consequently the carrier generation occurs close to and often within 10–100 nm beneath the sample surface. The distorted electron-energy distribution formed immediately after photoexcitation relaxes to achieve a Boltzmann distribution, when most of the electrons relax to the lower CB edge and most of the holes are close to the upper edge of VB. Therefore photoconductivity is primarily contributed from carries with energies near the band edges.

For excitation energy below the band edge, that is, extrinsic photoexcitation, photoconductivity is rather homogeneous throughout the volume of the semiconductor. Optical absorption caused by the transitions of electron-occupied imperfection states to CB (Fig. 3.11B) generates electrons. Another result is the formation of transitions from VB to electron-unoccupied imperfection states (Fig. 3.11C) and the generation of free holes. Free carriers are generated by a ground-to-excited state excitation relating with an imperfection, when a secondary process finishes the excitation into the band triggered by different processes, including impact ionization, thermal ionization, tunneling, and excitation by additional photons.

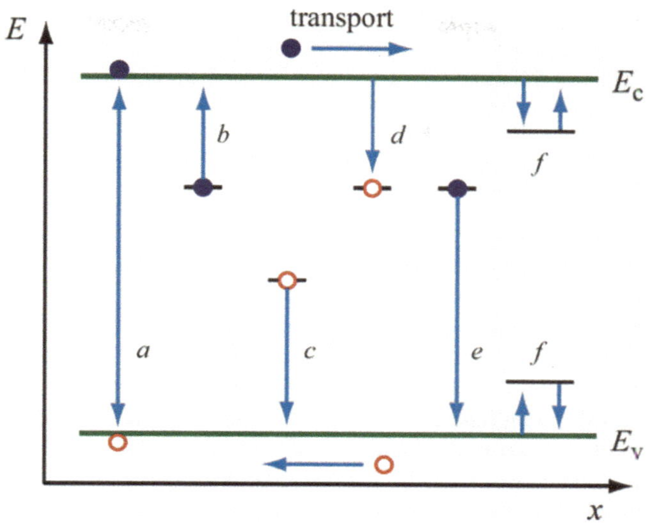

Fig. 3.11 Major transitions and phenomena associated with photoelectronic effects in homogeneous semiconductors: (A) intrinsic absorption, (B) and (C) extrinsic absorption, (D) and (E) capture and recombination, and (F) trapping and detrapping. *Arrows* indicate electronic transitions.

3.3.2 Thermal ionization

Thermal ionization refers to a process by statistical consideration, as it usually requires the presence of several phonons simultaneously (or in a short time interval) to supply plenty of energy. Such a "simultaneous" supplying of phonons is necessary for centers that do not have eigenstates spaced closely enough, into which an intermediate thermal excitation occur. The simultaneous phonon excitation can be interpreted as a process intervening a transient giant oscillation of a lattice atom.

For deep defect centers that possess a phonon ladder, two possibilities appear: a multiphonon absorption or a sequential absorption of phonons, that is, a cascade process. For the latter one the absorption of the next phonon takes place in a temporal threshold of an excited state to achieve the successive excitation, along with the supply of the total ionization energy. In both cases a transient giant oscillation of the defect center results. This leads to the total energy of the center exceed that of the CB bottom. The electron tunnel into the adjacent states of CB is based on this mechanism.

Such a deep defect center may potentially result in transitions from CB to VB. In this case a giant oscillation of a lattice atom, that is, a multiphonon process, will lead to a transfer of an electron from VB into a defect center. If the lifetime of an electron within the center is long enough, a consequent giant oscillation of the center will take this electron into CB. The combination of these two processes is equivalent to a thermally induced band-to-band transition, which is more difficult to understand without this intermediate step.

Every excitation process is followed by an inverted deexcitation step. The lifetime of an electron in any metastable state and the supply of phonons with the appropriate energy for achieving the next excitation process determine that probability of being promoted to a higher excitation state.

3.3.3 Recombination

Carrier recombination is the inverse step of carrier generation. The recombination process must take place after carrier generation. There are many types of recombination, and depending on the type of energy released during recombination, we can differentiate two principle recombination types: (1) thermally induced nonradiative recombination and (2) luminescence by radiative recombination.

The energy of a defect level depends on the relative position of the defect site with respect to its surrounding lattice atoms. Due to vibrations of these atoms, the defect level moves up and down in the bandgap with respect to its equilibrium position.

In one of the mechanisms for the nonradiative process, a sufficiently large lattice vibration shifts the defect level into the conduction band. Here, it is acceptable that an electron is captured recharging for the defect when the lattice is far from the equilibrium position. This leads to a strong vibration, which quickly relaxes with the emission of many phonons. If the relaxation is large enough, a defect level may shift from the upper to the lower half of the bandgap. The defect can now serve as a similar trap for holes, thus completing the process of band-to-band recombination. The thermal activation energies change from a few meV to 0.6 eV for different defects; a higher energy performs a significant fraction of the level depth and requires substantial thermal activation for recombination.

Thermal equilibrium is achieved, while a semiconductor without any external excitation is kept long enough at a specific temperature. Deviation from thermal equilibrium takes place due to nonthermal, additional excitation by light field or an electric field. When such deviations occur and have been stationary, a nonequilibrium steady state resulted.

Once thermal equilibrium of a semiconductor at a specific temperature without external excitation is set up, electrons and holes are then formed by thermal excitation alone. The same number of electrons and holes that are generated within a specific sample volume have to recombine except for statistical fluctuations.

When nonthermal carrier generation appears, the equilibrium density distribution is changed. As a result the Fermi energy splits into two quasi-Fermi levels, and two sets of demarcation lines for each defect center appear. Therefore some levels, which play the role of traps previously, will now act as recombination centers. Much of the content of the following sections deals with this steady-state condition.

On the other hand, recombination requires the energy to be exhausted, which may proceed via the generation of phonons and will result in Auger collision of electrons and meanwhile lead to nonradiative recombination or the emission of photons by radiative recombination. The distinction between nonradiative and radiative processes depends mainly on their relationship with the excitation process and the intermediate processes. One process occurring in optical excitation is the reflection of radiation impinging on the surface of the solid, that is, the coherent reradiation. Reflection of electromagnetic radiation involves the polarization of the semiconductor lattice, which is characterized by its index of refraction and extinction coefficient. The polarization appears in phase with the electric vector of the incoming radiation and produces coherent reradiation, essentially without delay. The reradiation takes place from the undamped fraction of the excitation action. It is indicated from any medium with a refraction index larger than, as seen from the basic reflectance equation, that presented here for normal incidence and interface to vacuum. Certainly, similar discussions are also helpful for understanding the process in the succeeding text: stimulated emission (lasing) and phosphorescence (delayed luminescence).

3.3.4 Potential breakthrough in future

Until now the suggested models still probably failed to satisfactorily account for all the observed phenomena. Modeling about energy level schemes could be a major research direction. At least the following can be predicted: (1) the nature—an electron or hole transfer—of the charging process of persistent luminescence; (2) identification of the most probable centers (what do you mean?) for persistent phosphorescence; (3) the role of (rare earth) codopants, that is, traps capturing electrons in the conduction band; (4) the energy sufficient to charge persistent phosphorescence (i.e., persistent phosphorescence excitation) (what do you mean?). It is now possible to design an efficient and long persistent phosphor via a comprehensive consideration of the bandgap energy of hosts and energy level of luminescent centers. Though some experimental difficulties still exist, the relationship between them now can be partly controlled. The biggest problem of the research in afterglow mechanism we are facing is an all-sided understanding of the role of trap type, trap concentration, and detrapping probability of captured electrons. Physicists and materials scientists need to answer three questions: what are the efficient traps with the abilities of capturing the excited electrons, what is the concentration of such traps, and what is the ratio of stored electrons that can be released.

3.4 Discussions about metastable shallow and deep traps

3.4.1 Shallow and deep levels

Traditional understandings of a variety of effects have used models of isolated point defects with fixed properties, operating independently in the matrix of the solid, and describable in terms of one-electron energy bands and defect levels. More recently, we have come to realize that this classic picture of a simple defect with fixed properties may be the exception—that in real persistent phosphors, there is a wide variety of interactions between defects with deep energy levels and the lattice that leads to variations in local atomic configurations comprising the defect and in some cases to complexes between defects. The central distinguishing feature of most of the defect processes of interest is that they inextricably couple the electronic system and local structural configurations. Consequently, processes involving these defects cannot be fully described by either electronic transitions or defect states alone.

Defects that interact with light in the ways of most interest are generally those whose electronic energy levels are deep enough in the energy gap to permit photoinduced changes to last observably long times. Clearly, this emphasis implies a dependence on temperature, and we shall be concerned with properties of materials generally at temperatures of 20–400 K. We shall see that there are cases for which there is an energy barrier for decay from such a change that is larger than the binding energy of an electron in a donor, so the distinction between shallow and deep must sometimes be qualified.

Typically, deep levels are able to serve as recombination centers for mobile carriers; this is one major evidence of their presence. Shallow levels usually function as traps, as stated earlier. Generally speaking the defects of interest for photochemical interactions are not in the category of traps because their energy levels are normally deeper in the gap, so their electrons are not readily excited into a band by thermal energy. Traps do, however, affect the availability of electrons for capture by deep levels, and it is necessary to account for their presence in analyzing kinetics of electron population changes through recombination centers.

In discussing various dopants and defects, we shall speak generally in terms of electrons, with the extension to holes always implied. Often a donor atom occupies a substitutional lattice site and has one more electron than the host atom that it replaces at that site. It is of central importance that the main binding force for an electron to a shallow donor is the long-range Coulomb force. The electronic states are derived principally from the nearby conduction band; their energies are close to the energies of a Rydberg series, modified from the hydrogen atom in the usual way by the dielectric constant of the material and effective mass of the electron—hence the common term hydrogenic levels.

Such donors produce an effective-mass state with a small (shallow) binding energy for electrons that is nearly fixed in the presence of changes in temperature or hydrostatic pressure or alloying. A major feature of hydrogenic states in semiconductors is that their radii are large compared with the atomic spacing; only in that way does the dielectric constant of the macroscopic material become effective. Therefore shallow levels and large orbits go together. One consequence of the large orbits is that motions of the atoms of the host material have slight effects on the behavior of these electrons.

There are no changes in the atomic positions around shallow donors resulting from optical interaction: The position and coordination of a substitutional donor atom remain unaffected when light changes the large-orbit electronic state. At temperatures high enough for rapid thermal exchange with the adjacent band, optically induced changes in electronic occupation are unimportant since they are promptly overwhelmed by thermal effects. Hence, low temperatures are required for observation of the optical properties of shallow donors.

Deep levels differ from these in essentially all ways. Their states are composed of contributions from both valence and conduction bands and perhaps from other bands as well. The dominant binding force for an electron is not Coulombic, but short-range; thus the lowest levels do not correspond to a hydrogenic series. Since the binding force is not Coulombic, the charge state of a defect is not necessarily the determining property for its capture of charge carriers. There is considerable confusion in the field over the magnitudes of capture cross sections, in part because of this property. Typically the radii of deep-level states are near atomic sizes; that is, these states are localized. This localization is a more important property than the energy level of the defect state, and there are cases of such localized states having energies that are resonant with a crystal band; then terminology becomes a problem because some workers use the term deep to mean localized.

The localized character of the electronic state of a deep-level defect makes it much more sensitive to the positions of the neighboring atoms; in this respect, these defects have properties like those of ionic materials, and the interactions of electronic and configurational changes (i.e., electron-phonon coupling) are of paramount importance. This importance had not been widely recognized before the elucidation of the color center, but it has become the dominant new feature in explaining metastable effects. The higher excited electronic states of deep-level defects, however, may have larger orbits than that of the ground state, and for them, electron-lattice effects become less important, whereas Coulombic effects are larger. An older theory for the capture of a free carrier by a Coulomb-attractive localized center held that it takes place in stages, with the initial capture into an excited large-orbit state, followed by sequential relaxations into the ground state by successive emissions (a cascade) of phonons.

3.4.2 Metastable defects and lattice relaxation

Optical interactions can have major effects on the structural properties of deep-level defects, in contrast to shallow ones. When light causes an electronic excitation of a localized defect, the alteration of the electron wave function induces changes in the surrounding distribution of electronic charge, which cause movement of the neighboring atoms, that is, a configuration change.

This process is generally referred to in abbreviated form as the electron lattice interaction at the defect. Typically, it is local modes of vibration that are most important for this electron-phonon coupling rather than the crystal modes. If, for example, the defect has symmetrically located neighbors, an excitation may alter the nearest-neighbor distance symmetrically, producing a breathing mode of configuration change. This happens in some silicates, but not for the vacancy in Si, where the degeneracy of the tetrahedral bonds causes the symmetry to drop around a vacancy by the Jahn-Teller effect (i.e., any system having a degenerate ground state will spontaneously distort and lower its symmetry).

An essential point is that the term configuration of a defect refers to its position and the positions of all the atoms in its immediate vicinity. If, for example, a hydrostatic stress is applied to the material and all the atoms move closer, that represents a new configuration even if it retains all the initial symmetry (which is not always the case). For these reasons, we shall refer to the defect center, which includes the central defect itself and its immediate neighborhood.

A new configuration induced by an electronic excitation may or may not be relatively stable; if the excited electronic state with the new configuration can promptly (relative to the observation time) relax back to the ground state, it is not stable. The cases of interest here are those for which a new configuration is metastable, so that it will endure long enough to be observable. In these instances understanding a defect requires more than answering the questions: what is it and where it is located? Just as important is, how does it work? That is, what are the rules that control transitions between its ground and metastable states? It is worthy to note that he excited state geometry, even ionized geometry, will be an important factor, needed to be considered in future study of persistent phosphors. At present the all-proposed mechanism model is based on ground state geometry, which can link to the trapping process. However, for the detrapping process, the geometry is not same as ground state ones.

Since these transitions are accompanied by observable changes in material properties, they provide a prime source of information on metastable defects, and kinetics data are often the basis for later identification of the nature of the defects. The transitions of initial interest involve kinetic relations governing capture or release of electrons, and accurate measurements of these are the key to identifying the multiphonon emission mechanism as the controlling one for changes in state.

Metastability implies that there is some property that inhibits relaxation to the ground state. In atomic transitions, such an inhibition arises from selection rules that forbid a transition, and similar suggestions have been advanced for defects in solids, but nearly all recent work in persistent phosphors shows that defect metastability generally implies the existence of an energy barrier between the ground and excited electronic states that must be overcome for relaxation to occur. Moreover, this barrier is associated with a large lattice relaxation (LLR) of the defect center that occurs during an electronic change of state.

All of this is represented simply in a configuration-coordinate (CC) diagram like Fig. 3.12 for a single center. In this context the term relaxation refers to a change in energy and configuration of the center that is associated with a return to the electronic ground state of the defect.

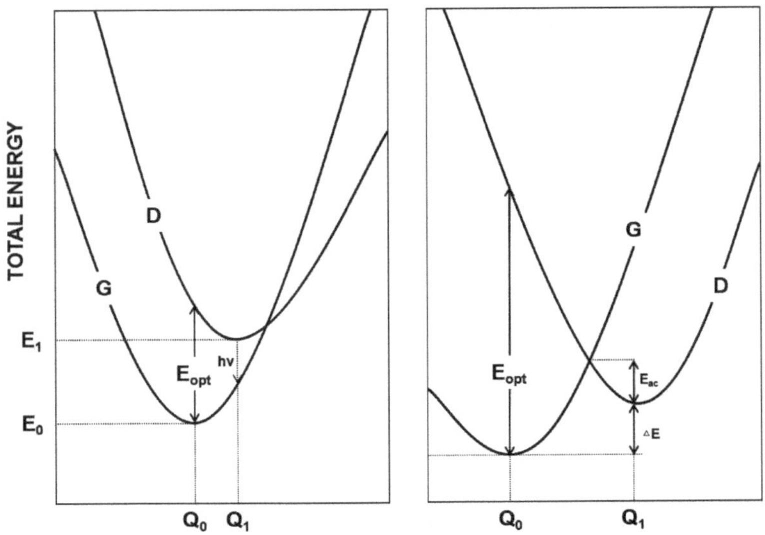

Fig. 3.12 Configuration-coordinate diagrams of a localized center having (A) small or (B) large lattice relaxation. E_{opt} is the minimum excitation energy out of the ground state by a photon, and ΔE is the minimum thermal ionization energy.

In CC diagrams the ordinate is the Gibbs free energy of the solid, but entropic and volumetric changes are always ignored as being small. This total energy explicitly includes both the electronic and strain energies of the center. Although the abscissa is called a configuration coordinate Q, it is not an ordinary Cartesian coordinate. Rather, it is a symbolic representation in one dimension of the entire set of real coordinates that specify the positions of the defect itself and all the nearby atoms that interact with it. Hence a value of Q symbolizes a complete configuration for those atoms, and any change in Q implies a possibly complicated set of changes in atomic positions. The simplest example of this idea is a symmetrical defect in which all the nearest neighbors are equidistant from the central site. Then Q might represent the nearest-neighbor distance, and its change would describe a breathing-mode change in the atom positions around the center. Generally the situation is much more complicated, and each value of Q must be thought of as symbolizing a three-dimensional snapshot of the group of atoms.

A major property of CC diagrams is their assumption of the Born-Oppenheimer (adiabatic) principle that electronic energies can be evaluated as if the atoms are fixed in position, with any change in configuration permitting a reevaluation of the electronic energies in the same way, but using the new positions. This explicitly ignores contributions to the energy from the kinetic energy of the atoms. We shall deal with these diagrams at some length because they are less familiar in discussions of semiconductors than for insulators and because there seems to be no thorough introductory treatment that is designed for semiconductors. It is also necessary to establish the link between the CC descriptions of localized defects and energy bands.

Thus the lower curve labeled G in Fig. 3.12 represents the total system energy for the state of the solid in which the electrons are in their ground state. To take a specific example, this is the system energy when the valence band is full, the conduction band empty, and the defect unoccupied. The configuration in which the atoms have their ideal thermal-equilibrium positions, leading to the minimum energy of the ground electronic state, is the reference configuration Q_0. Any change in configuration, of course, adds strain energy, and for small changes, it is always assumed that the curve has a parabolic shape produced by a fixed restoring force. The width of the parabola is a measure of the restoring force: The stronger the force, the sharper the rise in energy and the narrower the parabola. The normal covalent bonds of tetrahedral semiconductors form such a rigid structure that the force constant is large. Around a defect, this may be altered, and in the case of a-Si/H where a bond is broken, it is likely to be weaker along that axis.

One feature of these localized centers that is often ignored graphically but understood implicitly is their vibrations. If they were to be shown in Fig. 3.12, vibrational energy levels would consist of a series of closely spaced horizontal lines within a parabola; they correspond to the quantum levels of a simple harmonic oscillator. The spacing is close on this scale because quanta of vibrational energy are generally smaller than those of electronic excitations.

These vibronic states permit excursions in Q away from Q_0; these excursions are limited classically by the line of the parabola. Because of their presence a photon with energy higher than the minimum shown in Fig. 3.12 can be absorbed into a final state above a parabola having a higher vibrational state of excitation.

The upper curve labeled D in Fig. 3.12 represents the total system energy for the first excited electronic state; for the example here, we take it to have one electron localized in the defect center and one valence band hole that is mobile and not interacting with the defect (ignoring the state in which hole interaction does occur). The energy minimum occurs at a configuration Q_1 different from that of the ground state for reasons already discussed, but electronic transitions must be vertical in these diagrams (the Franck-Condon principle) because atoms are effectively stationary during the short time of an electronic transition. Hence, optical excitation as shown requires considerably more energy E_{opt} than the difference between the minima $\triangle$E, which is the thermal ionization energy. (There are cases for which this optical ionization energy is even larger than the bandgap of the semiconductor.) Once optical excitation occurs at the original configuration Q_0, the preferred configuration becomes Q_1, so the defect is in a highly excited vibrational state. Hence, there is a rapid decay in vibrational energy by phonon emission as the center seeks Q_1, E_1. These phonons are large-amplitude, predominantly local modes associated with the localized center, and their effect can be thought of as violent rattling of the atoms in the vicinity. This decay process is called multiphonon emission and is central to the initial explanation of the properties of what is now called the color center.

After relaxation to Q_1, E_1 there are two possibilities for decay out of the photoexcited state: radiative and nonradiative. To treat them, we first must distinguish between cases for small and large lattice relaxations shown in Fig. 3.12. In the small-relaxation case, Fig. 3.12A, the minimum of curve D lies within the parabola of G so that there is overlap of the two sets of vibronic states near Q_1. Here, decay from the excited state D can occur by either of two competing processes:

1. nonradiatively by transitions from D to G permitted at any temperature by the overlap, followed by phonon emission to the bottom of G,
2. by luminescent emission of a photon of energy $h\nu$ as shown, at constant Q_1, followed by phonon emission.

This photon always has lower energy than that for optical excitation—a relation called the Stokes shift ($E_{opt} - h\nu$). If the initial excitation is to an energy on D that is above the crossing of G and D (as in the case shown), there will be little luminescence.

For the case of large lattice relaxation shown in Fig. 3.12B, the minimum of D lies outside the parabola of G, so there can be no radiative decay from D to G at the configuration Q_1. Since there is no overlap of vibronic states of D and G near Q_1, the only path for decay is the nonradiative one that requires thermal energy in the amount E_{ac} as shown. It is clear that in this way E_{ac} becomes an energy barrier for thermally activated relaxation to occur from the excited electronic state, so that once the defect is in this state, it will persist if thermal energy is low; this is the metastable condition that is central to the photoinduced effects of interest in this book. (The term bistable is sometimes used instead of metastable for such a center.) Such metastability is the accepted explanation for persistent photoconductivity in homogeneous materials.

References

[1] K. Van den Eeckhout, P.F. Smet, D. Poelman, Persistent luminescence in Eu^{2+}-doped compounds: a review, Materials 3 (2010) 2536–2566.

[2] K. Van den Eeckhout, D. Poelman, P.F. Smet, Persistent luminescence in non-Eu^{2+}-doped compounds: a review, Materials 6 (2013) 2789–2818.

[3] J. Botterman, P.F. Smet, Persistent phosphor $SrAl_2O_4$: Eu,Dy in outdoor conditions: saved by the trap distribution, Opt. Express 23 (2015) A868–A881.

[4] H. Yamamoto, T. Matsuzawa, Mechanism of long phosphorescence of $SrAl_2O_4$: Eu^{2+}, Dy^{3+} and $CaAl_2O_4$: Eu^{2+}, Nd^{3+}, J. Lumin. 72 (1997) 287–289.

[5] T. Nakamura, K. Kaiya, N. Takahashi, T. Matsuzawa, C.C. Rowlands, V. Beltran-Lopez, G.M. Smith, P.C. Riedi, High frequency EPR of europium (II)-doped strontium aluminate phosphors, J. Mater. Chem. 10 (2000) 2566–2569.

[6] T. Aitasalo, P. Deren, J. Holsa, H. Jungner, J.C. Krupa, M. Lastusaari, J. Legendziewicz, J. Niittykoski, W. Strek, Persistent luminescence phenomena in materials doped with rare earth ions, J. Solid State Chem. 171 (2003) 114–122.

[7] J. Holsa, T. Aitasalo, H. Jungner, M. Lastusaari, J. Niittykoski, G. Spano, Role of defect states in persistent luminescence materials, J. Alloys Compd. 374 (2004) 56–59.

[8] P. Gluchowski, W. Strek, M. Lastusaari, J. Holsa, Optically stimulated persistent luminescence of europium-doped $LaAlO_3$ nanocrystals, Phys. Chem. Chem. Phys. 17 (2015) 17246–17252.

[9] F. Clabau, X. Rocquefelte, S. Jobic, P. Deniard, M.H. Whangbo, A. Garcia, T. Le Mercier, Mechanism of phosphorescence appropriate for the long-lasting phosphors Eu^{2+}-doped $SrAl_2O_4$ with codopants Dy^{3+} and B^{3+}, Chem. Mater. 17 (2005) 3904–3912.

[10] P. Dorenbos, Energy of the first $^4f_7 \rightarrow {}^4f_6^5d$ transition of Eu+ in inorganic compounds, J. Lumin. 104 (2003) 239–260.

[11] P. Dorenbos, The Eu^{3+} charge transfer energy and the relation with the band gap of compounds, J. Lumin. 111 (2005) 89–104.

[12] D.C. Rodriguez Burbano, S.K. Sharma, P. Dorenbos, B. Viana, J.A. Capobianco, Persistent and photostimulated red emission in CaS: Eu^{2+}, Dy^{3+} nanophosphors, Adv. Opt. Mater. 3 (2015) 551–557.

[13] F.T. You, A.J.J. Bos, Q.F. Shi, S.H. Huang, P. Dorenbos, Thermoluminescence investigation of donor (Ce^{3+}, Pr^{3+}, Tb^{3+}) acceptor (Eu^{3+}, Yb^{3+}) pairs in $Y_3Al_5O_{12}$, Phys. Rev. B: Condens. Matter Mater. Phys. 85 (2012) 115101.

[14] L.C.V. Rodrigues, J. Holsa, M. Lastusaari, M.C.F.C. Felinto, H.F. Brito, Defect to R^{3+} energy transfer: colour tuning of persistent luminescence in $CdSiO_3$, J. Mater. Chem. C 2 (2014) 1612–1618.

[15] L.C.V. Rodrigues, H.F. Brito, J. Holsa, R. Stefani, M.C.F.C. Felinto, M. Lastusaari, T. Laamanen, L.A.O. Nunes, Discovery of the persistent luminescence mechanism of $CdSiO_3$:Tb^{3+}, J. Phys. Chem. C 116 (2012) 11232–11240.

[16] L.C.V. Rodrigues, H.F. Brito, J. Holsa, M. Lastusaari, Persistent luminescence behavior of materials doped with Eu^{2+} and Tb^{3+}, Opt. Mater. Express 2 (2012) 382–390.

[17] M. Lastusaari, A.J.J. Bos, P. Dorenbos, T. Laamanen, M. Malkamaki, L.C.V. Rodrigues, J. Holsa, Wavelength-sensitive energy storage in $Sr_3MgSi_2O_8$: Eu^{2+},Dy^{3+}, J. Therm. Anal. Calorim. 121 (2015) 29–35.

[18] V. Kale, M. Lastusaari, J. Holsa, T. Soukka, Intense UV upconversion through highly sensitized $NaRF_4$: tm (R: Y, Yb) crystals, RSC Adv. 5 (2015) 35858–35865.

[19] L.U. Khan, H.F. Brito, J. Holsa, K.R. Pirota, D. Muraca, M.C.F.C. Felinto, E.E.S. Teotonio, O.L. Malta, Red-green emitting and superparamagnetic nanomarkers containing Fe_3O_4 functionalized with calixarene and rare earth complexes, Inorg. Chem. 53 (2014) 12902–12910.

[20] A. Lecointre, A. Bessiere, A.J.J. Bos, P. Dorenbos, B. Viana, S. Jacquart, Designing a red persistent luminescence phosphor: the example of YPO_4: Pr^{3+}, Ln^{3+} (Ln = Nd, Er, Ho, Dy), J. Phys. Chem. C 115 (2011) 4217–4227.

[21] A.J.J. Bos, N.R.J. Poolton, J. Wallinga, A. Bessiere, P. Dorenbos, Energy levels in YPO_4: Ce^{3+}, Sm^{3+} studied by thermally and optically stimulated luminescence, Radiat. Meas. 45 (2010) 343–346.

[22] W. Hoogeustraten, Electron traps in zinc sulphide phosphors, Philips Res. Rep. 13 (1958) 515–693.

[23] S. Tanabe, S. Watanabe, F. Saif, M. Matsuzawa, Survival probability of a truncated radial oscillator subject to periodic kicks, Phys. Rev. A: At. Mol. Opt. Phys. 65 (2002) 033420.

[24] Z.W. Pan, Y.Y. Lu, F. Liu, Sunlight-activated long-persistent luminescence in the near-infrared from Cr^{3+}-doped zinc gallogermanates, Nat. Mater. 11 (2012) 58–63.

[25] F. Liu, W.Z. Yan, Y.J. Chuang, Z.P. Zhen, J. Xie, Z.W. Pan, Photostimulated near-infrared persistent luminescence as a new optical read-out from Cr^{3+}-doped $LiGa_5O_8$, Sci. Rep. 3 (2013), 01554.

[26] J.H. Si, K. Kitaoka, J.R. Qiu, T. Mitsuyu, K. Hirao, Optically encoded second-harmonic generation in germanosilicate glass by a femtosecond laser, Opt. Lett. 24 (1999) 911–913.

[27] J.R. Qiu, K. Kojima, K. Miura, T. Mitsuyu, K. Hirao, Infrared femtosecond laser pulse-induced permanent reduction of Eu^{3+} to Eu^{2+} in a fluorozirconate glass, Opt. Lett. 24 (1999) 786–788.

[28] Y. Shimotsuma, P.G. Kazansky, J.R. Qiu, K. Hirao, Self-organized nanogratings in glass irradiated by ultrashort light pulses, Phys. Rev. Lett. 91 (2003) 247405.

[29] J.R. Qiu, X.W. Jiang, C.S. Zhu, M. Shirai, J. Si, N. Jiang, K. Hirao, Manipulation of gold nanoparticles inside transparent materials, Angew. Chem. Int. Ed. 43 (2004) 2230–2234.

[30] J.H. Lim, B.N. Kim, Y. Kim, S. Kang, R.J. Xie, I.S. Chong, K. Morita, H. Yoshida, K. Hiraga, Non-rare earth white emission phosphor: Ti-doped $MgAl_2O_4$, Appl. Phys. Lett. 102 (2013), 031104.

Characterization and simulation 4

4.1 Instrument measurement

In the applications of persistent phosphors, interactions between the luminescence ion and the trap's electronic states determine the afterglow performance and provide mechanisms for manipulating the optical properties. Continued advances in these technologies require knowledge of the relative energies of luminescence center and trap state in crystal band states so that properties of available materials may be fully understood and new materials may be logically developed.

Many different experimental techniques have been used to study the atomic origin of luminescence center and trap state and explore the relationships between the luminescence ion and the trap's electronic states and host electronic states by measuring either ΔE_{VB}, ΔE_{CB}, or ΔE_{Trap} (see Fig. 4.1). In this section, we list the experimental techniques in the following order:

I. characterization for atomic state of luminescent center and trap
II. characterization for the ΔE_{VB} and ΔE_{CB} of the luminescent center
III. characterization for the ΔE_{Trap} of the trap

For the instrument measurement [I], we have listed the experimental approaches of steady-state and time-resolved spectrum, extended x-ray absorption fine structure (EXAFS) and x-ray absorption near edge structure (XANES), and electron spin resonance (EPR). These techniques can provide the basic information of photoluminescence spectra and decay, valence state, and local coordination of luminescent center and trap sites. For the instrument measurement [II] and [III], the experimental techniques can be divided into three different categories based on how the electron transfer processes are detected. The first category contains methods that directly measure the absorption due to electron transfer transitions. These absorption spectroscopy approaches are the most direct methods for measuring interactions between luminescent center/trap and the host crystal and also allow the study of exciton and other localized electron transfer processes that do not directly involve the host band states. While these methods provide important information for understanding the material properties, they are often difficult to interpret by themselves since the observed spectra depend on both the transition probabilities and the joint density of initial and final states. Furthermore, assignment of initial and final states for each observed transition requires independent theoretical or experimental information, introducing additional uncertainty. For the rare earth activated phosphors, a fact responsible for the 4f electrons' atomic-like properties and nonbonding character, direct excitation of 4f electrons into crystal band states is a weak process difficult to observe, due to the extremely weak spatial overlap between 4f electron states and host band states.

The second category contains a diverse assortment of techniques that indirectly detect electron transfer by monitoring changes in the material properties or relaxation

Persistent Phosphors. https://doi.org/10.1016/B978-0-12-818637-4.00004-5

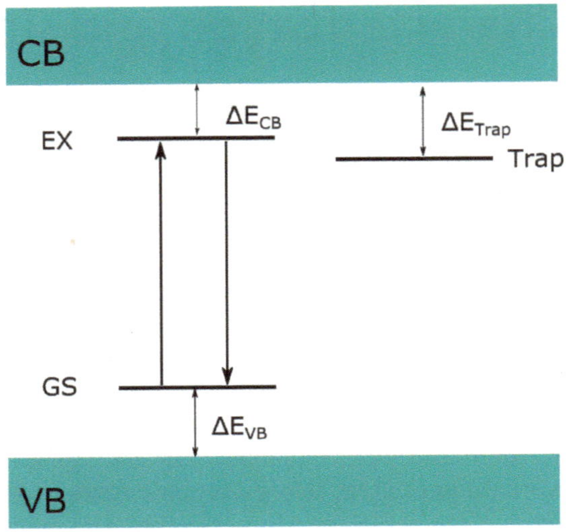

Fig. 4.1 Typical energy level scheme for the persistent phosphors. "GS" and "EX" stand for the ground and excited state energy level of luminescent center, and the trap state is shown from electron's viewpoint briefly. The ΔE_{VB}, ΔE_{CB}, and ΔE_{Trap} are the binding energies difference of involved states.

of the system. These excitation techniques are very sensitive methods for detecting electron transfer and often provide additional related information such as the branching ratios for different relaxation pathways or the thermal stability of trap states. A prominent example of this approach is photoconductivity spectroscopy, where the change in the electrical resistance of an insulating material due to the excitation of charge carriers into the host band states can be used to unambiguously assign the final state of a transition as involving either an electron in the conduction band or a hole in the valence band. Two-step photoconductivity methods are particularly useful for interpreting the spectrum and avoiding complications due to overlapping transitions. Several approaches that rely on the capture of photoinduced charge carriers by electron or hole trap states in the crystal are also included in the second category, including thermoluminescence, thermally stimulated luminescence excitation, and afterglow measurements. By determining the photon energy at which electrons begin to be transferred to or from trap states, the electron transfer energies may be established. Afterglow measurements detect the population transferred to trap states by monitoring the very slow components of the luminescence that result from gradual depopulation of trap states; hence the electron transfer energies may be determined by measuring the threshold energy at which afterglow begins to appear. In thermoluminescence techniques, population in the trap states is monitored by varying the crystal temperature and detecting the luminescence resulting from the thermal depopulation of shallow traps. For thermally stimulated luminescence excitation spectroscopy, the luminescence of a specific ion is monitored, while the trap states are thermally depopulated. Since the luminescence must arise from an ion recapturing a charge carrier, this approach clearly identifies the ions originally responsible for producing the charge carriers, allowing the electron transfer thresholds to be detected and assigned. Photoluminescence techniques can also be used to study electron transfer by

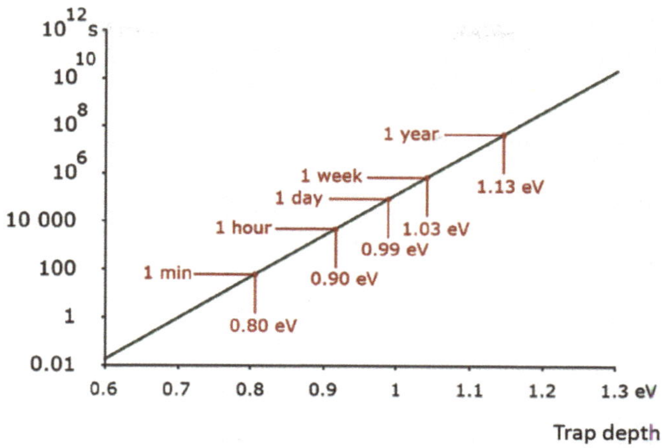

Fig. 4.2 Average lifetime of an electron caught by a trap level, at room temperature and with a normal escape frequency.
Reproduced with permission from Clabau, F., Phosphorescence: Mécanismes et nouveaux matériaux, PhD thesis, Institut des Matériaux Jean Rouxel, Nantes, 2005.

monitoring their effect on the nonradiative relaxation pathways responsible for luminescence quenching.

The final category contains electron spectroscopy methods that directly measure the electron binding energies. Electron photoemission spectroscopy measures the binding energies for occupied electronic states relative to a common reference level, complementing optical methods from the previous two categories by providing fundamental information that may be used to interpret the optical spectra. Combined with the methods of inverse photoemission and bremsstrahlung isochromatic spectroscopy, which measure the binding energies of unoccupied electronic states, photoemission can be used to provide a clear picture of the broad electronic structure in a material.

To achieve efficient persistent luminescence at room temperature, the trap should have a specific depth, which is neither too shallow nor too deep. If the trap is too shallow, the electrons will easily escape from it, and the afterglow will cease in a matter of seconds or less. If the trap is too deep, the thermal energy available at room temperature is not sufficient to release any trapped electron. A trap depth of around 0.6–0.7 eV is often stated as ideal to achieve efficient persistent luminescence at room temperature (Fig. 4.2) [1].

4.1.1 Steady-state and time-resolved spectra

Steady-state and time-resolved spectra are most important and basic characterization techniques of optical material. Steady-state spectra are mainly used for research of optical properties (excitation and emission spectra) through the continuous wave. Time-resolved spectra, which are usually employed to research the dynamic processes (electronic states and energy transfer processes), reach more than picosecond to femtosecond by ultrafast laser pulse. The schematic illustration of the prototype setup of steady-state and time-resolved spectra are shown in Figs. 4.3 and 4.4, respectively [1–3]. For the measurement of emission spectra, the targeted excitation photon is created after the xenon light source entering the monochromator. Under this special excitation energy, the sample is excited to some excited states, and emission processes

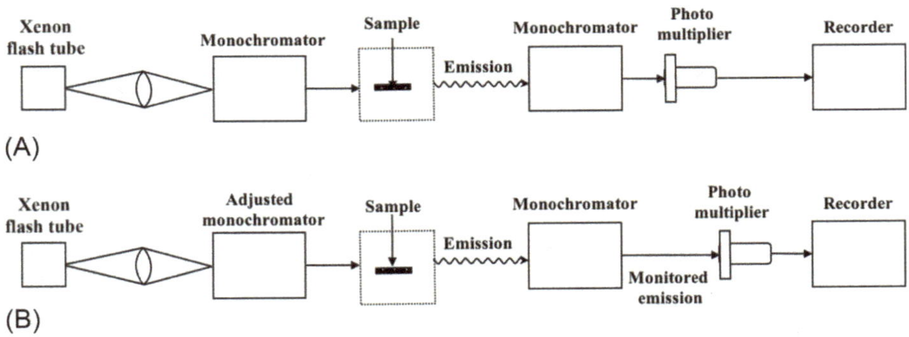

Fig. 4.3 A detailed schematic diagram of the setup used for the steady state spectroscopy: (A) emission spectrum and (B) excitation spectrum. The emission beam is indicated by the wave lines. In the measurement of emission spectrum, the recorder writes the intensity of emission photons with lower energy than the excitation energy, while the recorder writes the intensity of monitored emission photon with adjusted excitation energy in the measurement of excitation spectrum.

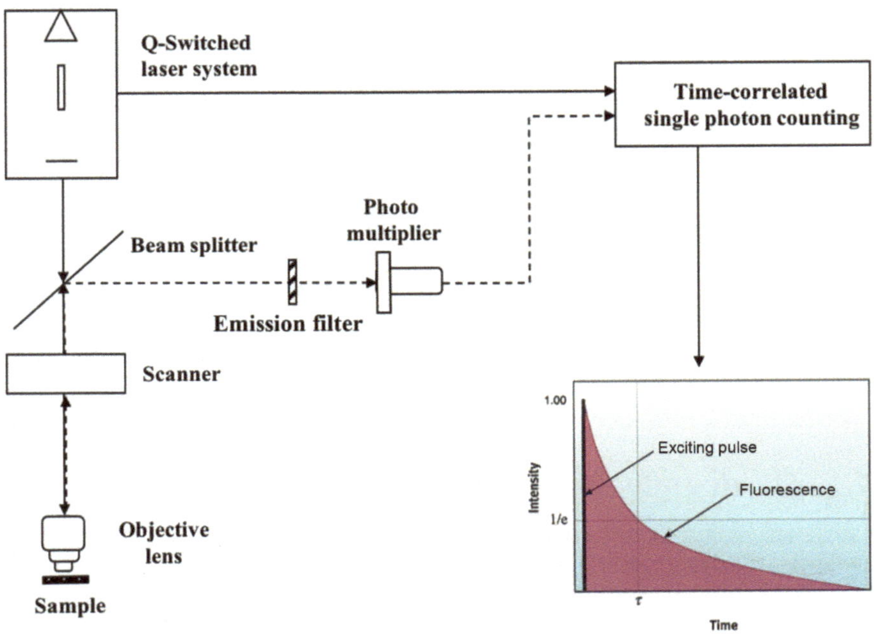

Fig. 4.4 A detailed schematic diagram of the setup used for the time-resolved spectroscopy.

from these excited states are recorded. The excitation spectrum measurement is similar with the case of emission spectrum. However, the recorded information is the intensity of special emission energy, under the change of excitation photon energy. Here, it is worthy to emphasize that the intensity of emission photon is recorded in both excitation and emission spectra measurements. It will be helpful to analyze the relationship of excited state and conduction band, with keeping this information in mind. Also the use of temperature-dependent PL spectra can be reponsible for the determination of ΔE_{CB} (defined in Fig. 4.1), which is the thermal quenching barrier of phosphors. In principle, there are two independent models for the analysis of such thermal quenching: autoionization model and 4f-5d crossover. In our previous work, we have shown that the 4f-5d crossover cannot be the main mechanism for the rare earth activated phosphors, even with the consideration of multidimensional vibrational coordinates. Therefore the energies barrier from temperature-dependent PL spectra can be directly linked to the ΔE_{CB} [4].

The measurement of time-resolved spectrum for persistent phosphor is routine compared with the procedure for conventional phosphors. But the analysis of decay curve for persistent phosphors is much more complex than the latter case. When the optical excitation of the persistent luminescent phosphor is ended, the luminescence of persistent phosphor does not vanish immediately since traps are still being emptied, leading to recombination at the luminescent centers. As the number of filled traps continues to decrease, also the rate at which the remaining filled traps are emptied lowers. And thus the "residual" luminescence–or afterglow–decreases, leading to a specific shape of the afterglow decay, which can be exponential, multiexponential, hyperbolic, or even more complex. The exact shape of the charging profile and the afterglow decay is difficult to predict or to model as it depends on many parameters: the trap depth, the temperature, the excitation intensity, the concentration of traps, and the trapping and detrapping mechanisms.

4.1.2 EXAFS and XANES

x-Ray absorption spectroscopy (XAS) is a common used characterization technology for analyzing partial fine structure (including valence state, coordination condition and distances, coordination number, and category of the atoms located around the checked element) of matter (can be gases, liquids, or solids). The operational principle of XAS is that the x-ray absorption coefficient of sample as a function of energy is set up by using x-ray radiation, that is, incident and transmitted x-ray intensity is recorded. While the incident x-ray energy matches the electron binding energy of sample, an increasing number of x-ray can be absorbed, which leads to the number of transmitted x-ray to fall off. This is called an absorption edge. The reason XAS has good selectivity is that each element has a cluster of specific absorption edge corresponding to their binding energies. XAS can be divided into two parts according to the range of absorption edge: one is x-ray absorption near edge structure (XANES) typically in the 15–150 eV range, and the other is extended x-ray absorption fine structure (EXAFS), which enlarged the scope from 150 to 1000 eV. Generally speaking the XANES is normally used to determine the valence state and coordination condition of the element in the sample; meanwhile, the EXAFS is used to show the distances, coordination number, and category of the atoms located around the selected element.

Brito et al. supervised the synthesis progression by recording ex situ XANES [5]. According to the experimental results (Fig. 4.5A), before the experiments can even begin (0 min), the precursor is made up of elemental sulfur and Gd_2O_3; then, it is translated into Gd_2O_2S within 2 min, and with the increasing irradiation time, the band intensity at 2482 eV exhibits a decrease first and then an increase in the variation of oxidation. Nunes and his coworkers adopted the EXAFS to analyze the information on the distance between the Tb^{3+} ions and surrounding atoms [6]. As Fig. 4.5B shows, the distances of Tb-O at 10 K are about 2.3 Å, which similar to the value at 300 K. It reveals that no transformations happened when a change in the coordination number occurred. In addition, the similarities of bond distances between Tb-O and Cd-O are indicating that Tb^{3+} ion in place of Cd^{2+} ion.

In the vast majority of situations, in effect, the combination of XANES with EXAFS is a powerful means for analyzing partial fine structure of LPPs. Brgoch et al. prepared a $Zn(Ga_{1-x}\ Al_x)_2O_4{:}Cr^{3+}$ (x = 0–1) phosphor. In this research, firstly, they investigated the local structural environment of Cr^{3+}, and the data indicated that

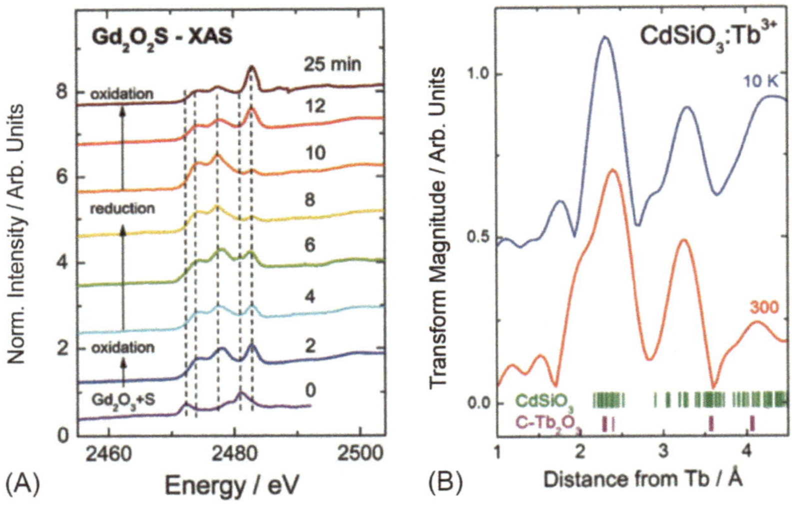

Fig. 4.5 (A) SR-XANES of the sulfur K-edge of the Gd_2O_2S materials obtained by the MASS method for different durations of microwave exposure. (B) EXAFS results for the $CdSiO_3{:}Tb^{3+}$(1%Tb) material at 10 and 300 K.

A: Reproduced with permission from J.M. Carvalho, C.C.S. Pedroso, I.P. Machado, J.H. Rodrigues, P. Głuchowski, M. Lastusaari, H.F. Brito, Persistent luminescence warm-light LEDs based on Ti-doped RE_2O_2S materials prepared by rapid and energy-saving microwave-assisted synthesis, J. Mater. Chem. C. 6 (2018), 8897–8905, Copyright 2018, Royal Society of Chemistry; B: Reproduced with permission from L. Rodrigues, H. Brito, J. Holsa, R. Stefani, M. Felinto, M. Lastusaari, T. Laamanen, L. Nunes, Discovery of the persistent luminescence mechanism of $CdSiO_3{:}Tb^{3+}$, J. Phys. Chem. C. 116 (2012), 11232–11240, Copyright 2012, American Chemical Society.

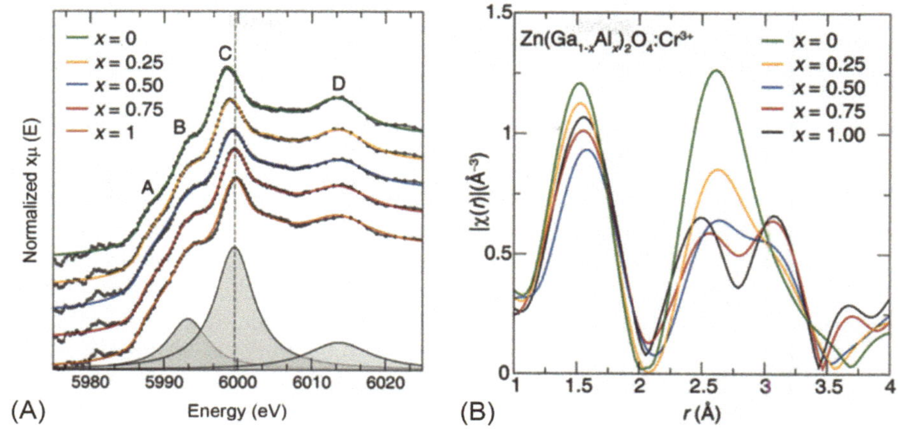

Fig. 4.6 (A) Cr K edge XANES for $Zn(Ga_{1-x}\ Al_x)_2O_4:Cr^{3+}$ ($x = 0$–1). The four prominent features: A–B is the 1 s → 3d transition, C is 1 s to continuum, and D is a scattering peak resulting from the crystal structure (rather than the electronic structure); Lorentzian peaks that fit to the data are *gray*. The *dashed line* is a guide for the eye to show the shift in the energy of peak C. (B) Real-space k^2-weighted magnitude of the Cr K EXAFS across the solid solution $Zn(Ga_{1-x}\ Al_x)_2O_4:Cr^{3+}$ ($x = 0$–1). First coordination shell ($[CrO_6]$) at 1.5 Å and the second ($[Ga/AlO_6]$) and third ($[ZnO_4]$) coordination shells at 2.5 Å and 3.0 Å, respectively. Data are not phase corrected, to be consistent with previous literature reports.
Reproduced with permission from E. Finley, M.W. Gaultois, J. Brgoch, Unlocking the key to persistent luminescence with X-ray absorption spectroscopy: a local structure investigation of Cr-substituted spinel-type phosphors, Phys. Chem. Chem. Phys. 21 (2019), 19349–19358, Copyright 2019, Royal Society of Chemistry.

the C feature band was shift to higher energy with the increasing of Al^{3+} concentration due to the increased electronegativity of Al^{3+} than Gd^{3+} ions and the changes in bond length of Cr-O (Fig. 4.6). In addition, as shown in Fig. 4.6, EXAFS region has shown more details about how local symmetry distortions vary with a function of x that further proved the structural changes of phosphor with the increase of Al^{3+} concentration [7].

4.1.3 Electron spin resonance

Electron paramagnetic resonance (EPR), also called electron spin resonance (ESR), deals with the interaction between electromagnetic radiation and magnetic moments of electrons [8]. It is a technique applicable to systems in a paramagnetic state. The species exist either in a paramagnetic ground state or may be temporarily excited into a paramagnetic excited state, for instance, by irradiation. Typical systems that have been studied include the following:

(1) Free radicals in the solid, liquid, or gaseous phases. A free radical is herein defined as an atom, molecule, or ion containing one unpaired electron.
(2) Transition ions including actinide ions.

(3) Various point defects (localized imperfections, with electron spin distributed over relatively few atoms) in solids. Best known in this class is the F center, an electron trapped at a negative-ion vacancy in crystals and glasses. Deficiency of an electron (a "positive hole") may also give rise to a paramagnetic entity.
(4) Systems with more than one unpaired electron. Excluding ions in category 2, these include triplet-state systems, biradical and multiradical.

An EPR spectrum is observed, making the magnetic field vary (abscissa of the spectrum) and recording the absorption of the microwave irradiation. This EPR signal appears at the resonant field where the energy between the two electronic spin sublevels created by the magnetic field (gβH, where g is the land factor, β the Bohr magneton, and H the magnetic field applied to the sample containing radicals with spin $S = 1/2$ for instance) is equal to the energy of the fixed microwave (hν, where h is the Planck constant and ν is the microwave frequency). This equality is the fundamental equation in EPR: $h\nu = g\beta H$ (Fig. 4.7).

EPR has proved to be an important tool in the study of semiconductors, particularly in identifying and elucidating the structures of point defects and impurity ions. Now, many researchers are trying to explore the species of traps in LPPs by using EPR. Here, we focus primarily on solids containing independent unpaired-electron species, limited to relatively few atoms in each center. Depending on the anisotropy, some important classes of paramagnetic systems mainly include free radicals, transition ions surrounded by ligands, and point defects.

Up to now a large number of researchers are trying to investigate the type of traps in LPPs by using EPR. Our group had demonstrated the trap type of $CaTiO_3$:Cr phosphor by EPR. The EPR spectra of CTC1 (oxidized phosphor) before the irradiation

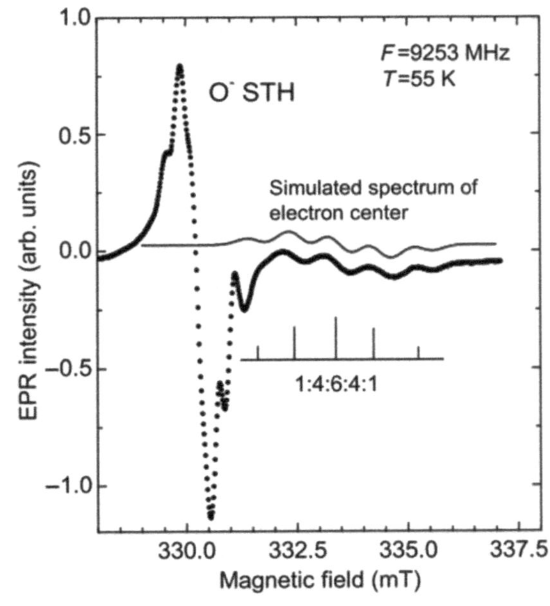

Fig. 4.7 EPR spectrum created by x-ray irradiation of YSO at 20 K. The strong line at 330 mT belongs to a trapped hole center; the second line at 333.5 mT with HF structure belongs to a trapped electron center. The spectrum is measured at B||c. The simulated spectrum of a trapped electron center is shown as well. From X. Qin, Y. Li, Y. Li, Y. Wu, R. Chen, K. Sharafudeen, et al., Inducing NIR long persistent phosphorescence in Cr-doped perovskite titanate via redox, J. Alloys Compd. 666 (2016), 387–391.

reflected no signal, while CTC5 (reduced phosphor) shows a prominent signal at g-value of 1.97, and with the irradiation time changes from 1 to 10 min, the EPR signal at 1.97 was decreased after sudden increase. The intense EPR signal at $g = 1.97$ could be due to the newly trap [CrTi], namely, Cr^{3+} precipitated in the octahedral Ti^{4+} [9]. Another paradigmatic example for the usage of EPR measurement was the study of nature of trap state in Y_2SiO_5:Ce (YSO:Ce) scintillator material, which suffer from the "afterglow problem" for degrading the scintillating performance. The EPR investigation of the YSO single-crystal x-ray irradiated at different temperatures has shown that the holes created by irradiation are first self-trapped at Si-unbound oxygen ions forming O^- paramagnetic centers [10].

4.1.4 Photoelectron emission

Electron photoemission spectroscopy (PES) is a powerful tool in the study of electronic structure that has found applications throughout nearly all of the fundamental and applied sciences [11, 12]. The technique of photoemission spectroscopy (also known as photoelectron spectroscopy) is based on the photoelectric effect, in which electrons are ejected when sufficiently energetic photons are incident on a material. The photoelectric effect was first observed more than a century ago by Hertz in 1887 and was later explained by Einstein in his famous 1905 paper that first proposed the concept of the wave/particle duality of light. Einstein described the photoelectric effect in terms of a "photoelectric law" given by

$$E_{KB} = \hbar\omega - E_{BE} \tag{4.1}$$

where E_{KE} is the kinetic energy of the ejected electron, $\hbar\omega$ is the energy of the incident photon, and E_{BE} is a constant that Einstein called the escape energy. From the photoelectric law, we see that when the absorption of a photon ejects an electron from the material, a portion EBE of the photon energy is used to remove the electron from the material and the remaining energy appears as the electron's kinetic energy. As a result the constant E_{BE} is generally referred to as the electron binding energy (for the specific case of emission of the least bound valence electrons from metals, E_{BE} is equal to the work function). By using the photoelectric law, the electron binding energy may be determined if both the photon energy and electron kinetic energy are known. This basic relationship provides the foundation for photoemission spectroscopy, as demonstrated in Fig. 4.8.

The natural energy reference for the photoemission process is the vacuum level, which corresponds to the energy of a free electron in vacuum with no kinetic energy, as is implicit in the form of Eq. (4.1). Nevertheless, difficulty in precisely determining the vacuum level and the additional complications due to the photoelectron's interactions with the surface during escape from the solid have led to the common practice of referencing binding energies in metals relative to the Fermi level of the conductively coupled sample-analyzer system. The Fermi level is the highest energy—and lowest binding energy—occupied state in the metal (formally equivalent to the chemical potential at $T = 0$) and provides several significant advantages as a reference.

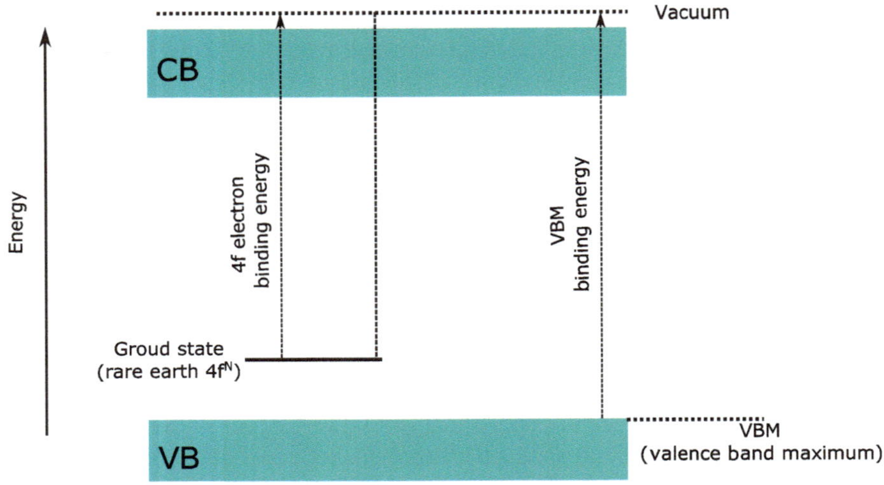

Fig. 4.8 Energy level diagram illustrating the principle of photoemission spectroscopy.

The most important advantage is that the Fermi level must have the same energy throughout the entire conductively coupled system; consequently the Fermi level provides a reference that is independent of the material's work function, which is a property that is difficult to accurately measure and that can vary appreciably for different surface conditions and orientations. Furthermore, the sharp Fermi edge in metals is a convenient reference since it can be easily determined from the metal photoemission spectrum. One disadvantage of this approach is that binding energies referenced to the Fermi level cannot be directly related to the results of theoretical models, which are generally referenced relative to the vacuum level. The merits and widespread use of the Fermi level reference in metals have also led to its application in semiconducting and insulating materials where there is no Fermi edge in the spectrum. Unfortunately, that approach can produce unreliable and inconsistent results for nonmetals since the Fermi level of these materials can be strongly influenced by low levels of defects or impurities.

4.1.5 Photocurrent and thermocurrent

Photoconductivity is an optical and electrical phenomenon in which a material becomes more electrically conductive owing to the absorption of electromagnetic radiation, such as visible light, ultraviolet light, infrared light, or gamma radiation [13]. When light is absorbed by a material, the number of free electrons and holes increases and raises its electrical conductivity. To cause excitation the light that strikes the semiconductor must have enough energy to raise electrons across the band gap or to excite the impurities within the band gap. Photocurrent is the electric current that flows as a result of a photoconductivity or a photovoltaic effect. The photocurrent discussed here is not related to the photovoltaic effect, but concerns the generation, migration, and recombination of free carriers in the conduction and valence bands, that is, photoconductivity.

Photoconducting materials can be divided into two classes: idiochromatic and allochromatic (Fig. 4.9). Idiochromatic materials are those that show photoconductivity when in a pure condition, the photoconductivity being associated directly with the intrinsic properties of the material itself. Such materials exhibit photoconductivity when energy is absorbed by the atoms of the crystal but in general show a maximum photosensitivity at the wavelength corresponding to the minimum energy required producing a free electron (the intrinsic absorption edge). Allochromatic materials are those in which the photoconductivity is associated directly with impurities or with crystal imperfections, such as vacancies. The photoconductivity associated with absorption in the characteristic band of the F center in alkali halides is typical of allochromatic behavior. The absorption band and the corresponding photoconductivity excitation spectra are very similar to each other. The incorporation of defects in a photoconductor crystal may have one or more of the following effects: (1) change the conductivity, (2) change the photosensitivity (imperfections, which act as efficient recombination centers, decrease the sensitivity), (3) change the speed of response (imperfections, which act as trapping centers or which increase the sensitivity, decrease the speed of response), and (4) extend the spectral response. Quantitatively the location and density of defect levels may be determined by one or more of the following experimental measurements: (1) the variation of free-carrier density with temperature; (2) the appearance of new absorption, associated with direct absorption by the imperfection centers; (3) the appearance of new photoconductivity excitation bands, associated with direct ionization of imperfection centers; (4) the appearance of new luminescence emission bands, corresponding to radiative recombination transitions involving the imperfection centers; (5) the appearance of new luminescence excitation bands, associated with direct excitation of imperfection centers; (6) the dependence of the measured carrier lifetime on carrier density, when the imperfection centers act as recombination centers; and (7) the appearance of new peaks in curves of thermally stimulated current (TSC) or TL emission. The increased conductivity or luminescence emission obtained when a preexcited crystal is heated to release trapped carriers can be used to determine the ionization energies of the trapping centers.

Photoconductivity curves (or photoconductivity excitation curves) could be used to identify the trap species and the possible trapping and detrapping processes. Prof. S. Tanabe and coworkers from Kyoto University developed an automatic measurement system of photoconductivity in inorganic phosphors. Through this system the wavelength and temperature dependence of photocurrents could be measured automatically (Fig. 4.9B). The numerical values of photocurrents were measured and recorded as a function of temperature and excitation wavelength automatically. They investigated the photocurrent features of $SrAl_2O_4$:Eu^{2+}, $SrAl_2O_4$:Eu^{2+},Dy^{3+}, and $CaTiO_3$:Pr^{3+} [14, 15]. The excitation wavelength dependence of photocurrent showed a similar profile to the PLE spectrum of Eu^{2+} ions. The result clearly indicated the interaction between the 5d level of Eu^{2+} with the conduction band. The activation energy of the photocurrent was evaluated from its temperature dependence for each wavelength. While most 5d levels of Eu^{2+} at the Sr site with a weak crystal field were

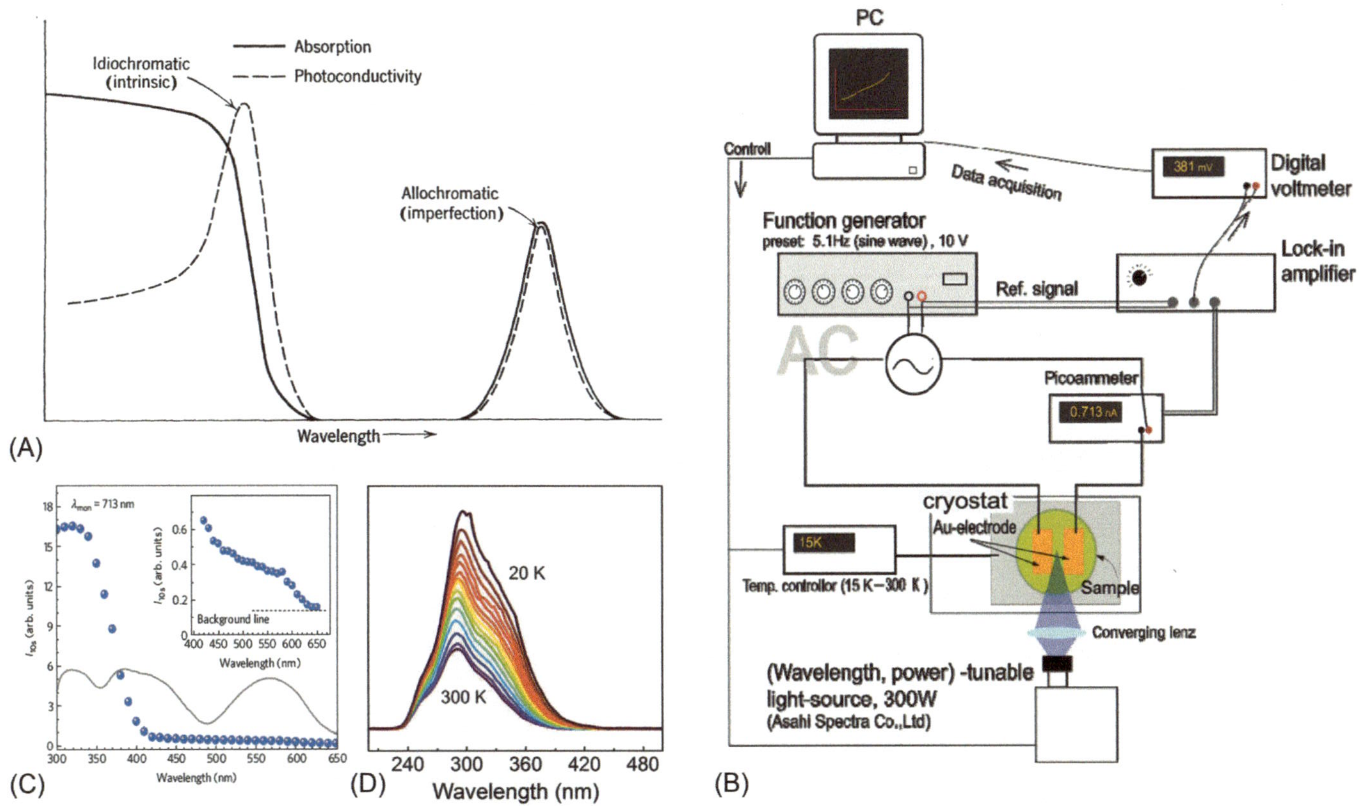

Fig. 4.9 (A) Wavelength dependence of absorption and photoconductivity for an idiochromatic effect (intrinsic) or an allochromatic effect (imperfection). (B) Schematic diagram of photocurrent measurement system as function of temperature and excitation wavelength. All the equipment was automatically controlled by a PC program. (C) Afterglow intensity I10s monitored at 713 nm for $Zn_3Ga_2Ge_2O_{10}$:Cr as a function of the excitation wavelengths over the 300- to 650-nm spectral range. The inset shows an enlargement of the 400–650 nm region. The *gray line* shows the excitation spectrum monitored at 713 nm. (D) Temperature-dependent photocurrent excitation spectra of $Zn_3Ga_2Ge_2O_{10}$:Cr for the range of 20–300 K.
Reproduced with permission from Y. Li, S.F. Zhou, G.P. Dong, M.Y. Peng, L. Wondraczek, J.R. Qiu, Anti-stokes fluorescent probe with incoherent excitation, Sci. Rep. 4 (2014), 04059, Copyright 2012, Nature Publishing Group.

located within the conduction band, only the 5d levels of Eu^{2+} above 3.1 eV (400 nm) at the Sr site with a strong crystal field were located within the conduction band [16].

The long NIR phosphorescent phosphor $Zn_3Ga_2Ge_2O_{10}$:Cr could be excited by ultraviolet and visible light, revealing that some shallow traps are near the CB while others are far away from the CB (Fig. 4.9C) [17]. The corresponding photocurrent excitation spectra are provided in Fig. 4.9D. Depending on the distance between the traps and the conduction band, only one excitation peak responsible for the electrons released from the shallow traps near the conduction band could be observed in the photocurrent excitation spectra. The photocurrent single may correspond to the intrinsic defects in $Zn_3Ga_2Ge_2O_{10}$:Cr [18].

4.1.6 Thermoluminescence spectra and TL excitation spectra

Thermoluminescence (TL) techniques are usually employed to give an estimate of the trap depth [19]. TL is defined as the emission of light during heating of a solid following the previous absorption of energy during irradiation. The alternative radiation source is high-energy ray (x-ray and gamma-ray), ultraviolet-visible-infrared light, and electron beam. In fact, this radiation is the energy store, whereas the heating is just a trigger to help release the absorbed energy. Irradiation source, irradiation duration, interval after ceasing the irradiation, and environmental temperature all influence the electrons distribution in different traps (from shallow traps to deep traps) [20, 21]. The experimental setup for the thermoluminescence measurement is shown in Fig. 4.10.

By performing a series of TL experiments with varying excitation durations and varying excitation temperatures and employing the initial rise analysis method, the depth and shape of trap distribution can be estimated. Smet et al. pointed out the main steps that should be taken for such a trap depth probing:

(a) Measuring the thermal quenching of the sample to properly correct the TL data.
(b) Measuring the dose dependence of the TL glow curves to verify the presence of higher order kinetics.
(c) Measuring the TL glow curves after exciting at various temperatures (T). The initial rise analysis can be performed to estimate the trap depth. An estimated trap depth that varies continuously as a function of T is a strong indication for the presence of a continuous trap depth distribution.
(d) If the trapping process is not thermally activated, the area under the glow curves obtained in (c) can be used to estimate the shape of the continuous trap depth distribution. If thermal activation of the trapping is seen, the areas should first be corrected for this.

Apart from TL spectra, if we plot the integrated TL signal intensity as a function of excitation wavelength, we will get a TL excitation (TLE) spectrum (Fig. 4.11D). Bos et al. presented their TLE facility (Fig. 4.11A) [22]. In detail the tunable monochromatic UV-Vis light was used to excite the electrons into the conduction band. The created free electrons would be captured in the intrinsic or introduced trapping centers of the sample. Traps were subsequently read out by heating the sample while the TL was recorded. By varying the monochromatic wavelength, repeating the TL readout,

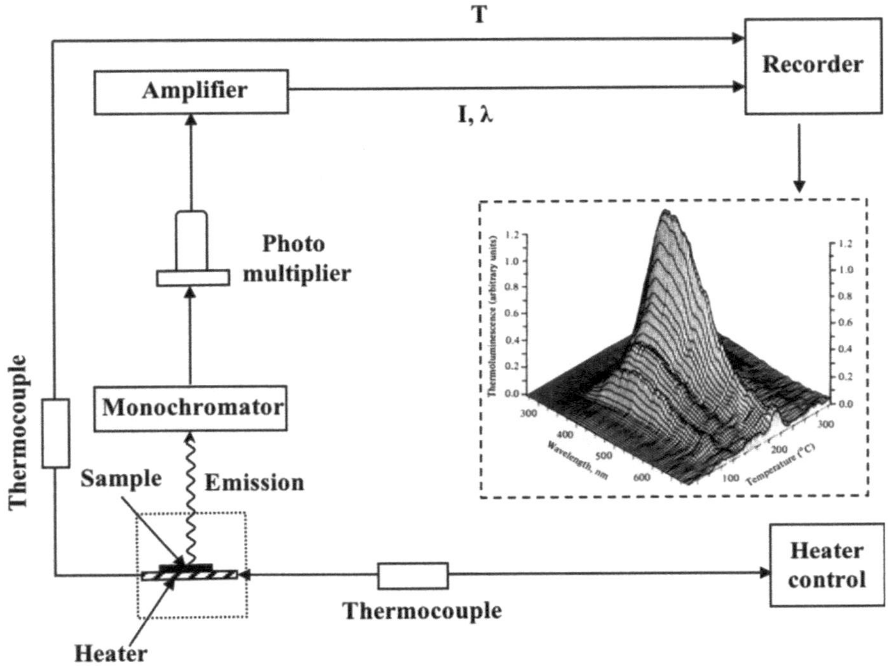

Fig. 4.10 A detailed schematic diagram of the setup used for the thermoluminescence spectroscopy and a typical 3-D profile result.

and displaying the temperature integrated TL yielded as a function of the subsequent wavelengths, the TLE spectrum was obtained. In subsequent wavelengths the TLE spectrum was obtained.

Bessiere et al. showed a distinctive application of TLE spectra, that is, analyzing the attribution of the PLE band by comparison of TLE spectra and PLE spectra. The total area of each glow curve was reported as well as the TSL excitation intensity of Cr^{3+}-doped $ZnGa_2O_4$ phosphors (red open circles in Fig. 4.11C). The TSL excitation spectrum was divided into five wavelength ranges. By comparing the PLE and TLE spectra, the former four ranges could be assigned to the $^4A_2 \rightarrow {}^4T_2$ (4F) of Cr^{3+}, the $^4A_2 \rightarrow {}^4T_2$(4F) transitions of Cr^{3+}, the $^4A_2 \rightarrow {}^4T_1$ (4P) transitions of Cr^{3+}, and band-to-band transition, respectively [23].

4.1.7 Positron annihilation lifetime spectroscopy

Up to now, there have been few reports giving an accurate description for determining trap concentrations. It is shown that positron annihilation lifetime spectroscopy has tremendous potential as a powerful tool for quantifying the types and densities of defects in solids. Positron lifetime, t, is a function of the electron density at the annihilation site. Positron lifetime, t, is inversely proportional to the local electron density at the site of positron annihilation. Positrons are trapped in open-volume defects, such

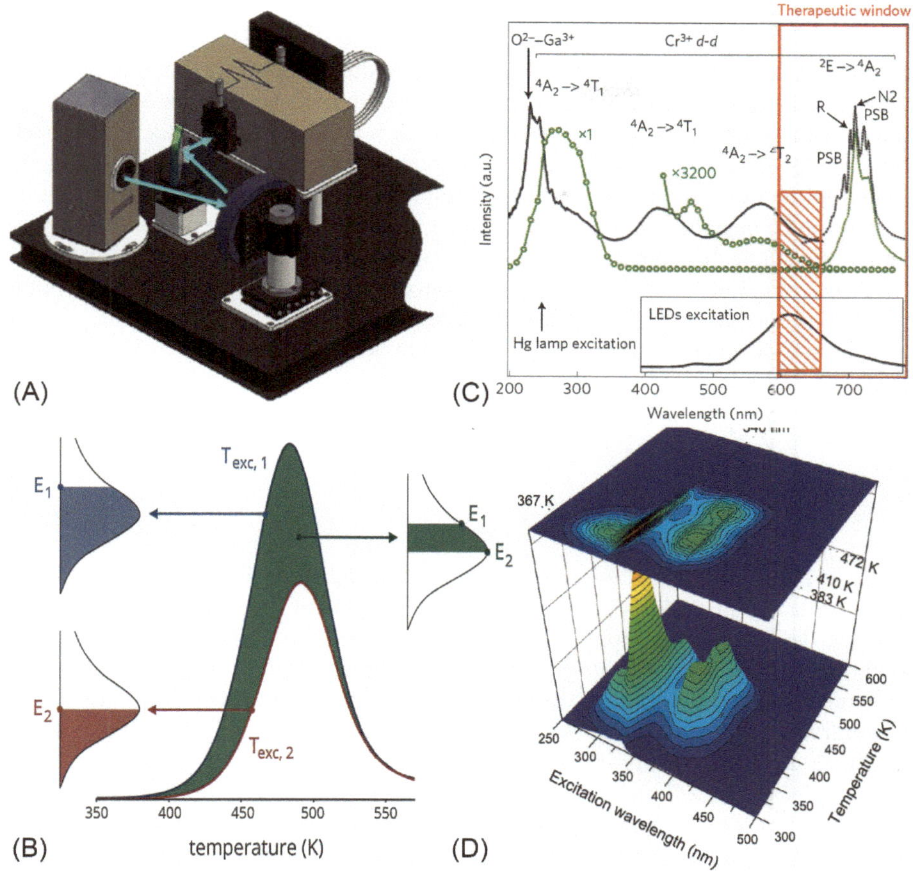

Fig. 4.11 (A) Three-dimensional drawing of the TL excitation part of the setup with the Xe lamp, focusing mirrors, monochromator, shutter, and optical fiber. (B) The density of the continuous trap depth distribution at a certain depth can be estimated from the difference between the integrated intensities at two different excitation temperatures. (C) PL excitation (*black solid line*) and emission (*black dotted line*) spectra compared with TSL excitation (*green open points*) and emission (*green dotted lines*) spectra of Cr^{3+}-doped n-$ZnGa_2O_4$. (D) TL intensity of $ZnS{:}Cu^+$ as a function of excitation wavelength and temperature as contour plot (top) and in 3-D representation (bottom).

A: Reproduced with permission from A.J.J. Bos, R.M. van Duijvenvoorde, E. van der Kolk, W. Drozdowski, P. Dorenbos, Thermoluminescence excitation spectroscopy: a versatile technique to study persistent luminescence phosphors, J. Lumin. 131 (2011), 1465–1471, Copyright 2011, Elsevier; B: Reproduced with permission from K. Van den Eeckhout, A.J.J. Bos, D. Poelman, P.F. Smet, Revealing trap depth distributions in persistent phosphors, Phys. Rev. B 87 (2013), 045126, Copyright 2013, American Physical Society; C: Reproduced with permission from T. Maldiney, A. Bessiere, J. Seguin, E. Teston, S.K. Sharma, B. Viana, A.J.J. Bos, P. Dorenbos, M. Bessodes, D. Gourier, D. Scherman, C. Richard, The in vivo activation of persistent nanophosphors for optical imaging of vascularization, tumours and grafted cells, Nat. Mater. 13 (2014), 418–426, Copyright 2014, Nature Publishing Group; D: Reproduced with permission from A.J.J. Bos, R.M. van Duijvenvoorde, E. van der Kolk, W. Drozdowski, P. Dorenbos, Thermoluminescence excitation spectroscopy: a versatile technique to study persistent luminescence phosphors, J. Lumin. 131 (2011), 1465–1471, Copyright 2011, Elsevier.

as in vacancies and their agglomerates, and the positron lifetime, t, can increase with respect to the defect-free sample owing to the locally reduced electron density of the defect. Thus a longer lifetime component, which is a measure of the size of the open volume, appears. The strength of this component, that is, its intensity, is directly related to the defect concentration. In principle, both pieces of information, that is, the kind and concentration of the defect under investigation, can be obtained independently by a single measurement. This is the major advantage of positron lifetime spectroscopy compared with angular correlation of annihilation radiation or Doppler broadening spectroscopy with respect to defect issues.

The time-dependent positron decay spectrum, $M(t)$, is given by

$$M(t) = \sum_{i=1}^{k+1} I_i \exp\left(-\frac{t}{\tau_i}\right) \tag{4.2}$$

$$\kappa = \mu C \tag{4.3}$$

where τ_i is the lifetime of every component after decomposing the positron decay spectrum using an analysis program, I_i is the strengths of the different components, C is the defect concentration, and κ is the annihilation probability of the positron in one defect. Thus τ_i gives information on the types of defect. The strength of this component, I_i, and the captured probability, κ, are in direct proportion to the defect concentration.

Qiu et al. systematically investigated the trap concentrations of Cr-doped gallates ($Zn_3Ga_2GeO_8$:Cr, SZG and $Zn_3Ga_2SnO_8$:Cr, GZG) by positron annihilation lifetime spectroscopy [24]. The short lifetime component, τ_1, reflected the characteristics of free-state annihilation for the perfect crystal lattice. The long lifetime component, τ_2, was attributed to the combined effect of positrons annihilating in vacancy-type defects, mainly including the cationic deficiency, oxygen vacancy, and interfaces. Inconspicuous variations in the lifetime components τ_1 and τ_2 between two phosphors further provided proof of similarity in traps. However, the intensity, I_2, and the captured probability, κ, of the SZG powders were both superior to those of the GZG powders, showing the total concentration of vacancies in the SZG powders is higher than those in the GZG powders.

4.1.8 Excited state spectroscopy

The trapping and detrapping process in persistent phosphors involved the interaction of excited state and trap in the condition of relaxed excited state geometry. In the aforementioned characterization measurement, the information obtained mostly is from the ground state geometry. Therefore the effect of relaxed geometry is neglected. To meet such void, we briefly list the measurement of excited state spectroscopy from laser pulse. The schematic energy level diagram behind the excited state spectroscopy is depicted in Fig. 4.12.

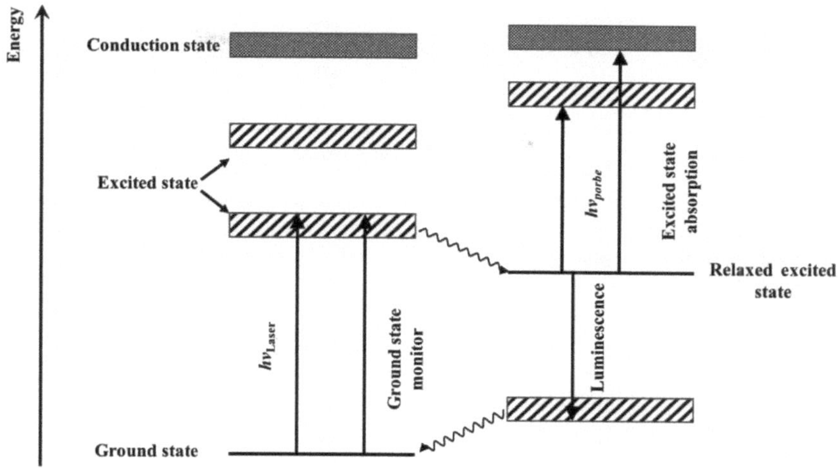

Fig. 4.12 Schematic energy level diagram of a strongly coupled luminescent center in solids. The absorption of a laser photon ($h\upsilon_L$) and subsequent lattice relaxation leave the luminescent center in a relaxed excited state. During the lifetime τ of this excitation, absorption and emission from this excited state can be observed from an intense monochromatized conventional source excitation ($h\upsilon_{probe}$).

The excited state spectroscopy with laser pulse combines the advantages of high photon densities available from laser source and the wavelength flexibility of conventional monochromators. Indeed a typical laser pulse of 10-MW power, 30-ns duration, and of cross-sectional area $A = 0.25$ cm^2 contains about 10^{18} photon/s, while a typical sample with luminescent center concentration of $c = 10^{17}$/cm^3 and a thickness $d = 0.2$ cm contains about 10^{15} luminescent center in a cylinder of cross-sectional area $A = 0.25$ cm^2. Therefore it is possible to populate fully the relaxed excited state of luminescent center system with laser source. As a result the "new" excited state sample can be created. In principle the measurement for the ground state sample can also be performed on this excited state sample.

The physical background for the analysis of excited state spectroscopy measurement is shown Fig. 4.13. Two beams of light simultaneously penetrate a sample. One beam, the probe beam with photon energy hυ, originated from a xenon flash tube and passes a monochromator before it enters the sample. Another monochromator behind the sample sets the required wavelength resolution. The second beam comes from laser source. Depending on the properties of the sample and the setting monochromator, a fraction I of the probe light intensity I_0 incident on sample passes the sample and is recorded by photomultiplier. Any changes ΔI of transmission induced by the passage of the laser beam are also recorded. It can be expected that if the spectrometer happens to be set at an absorption band associated with the luminescent center in the ground state, then ΔI is an increase for the transmission and if the spectrometer happens to be set at an absorption band associated with the luminescent center in the excited state, then ΔI is a decrease for the transmission. Fig. 4.14 shows

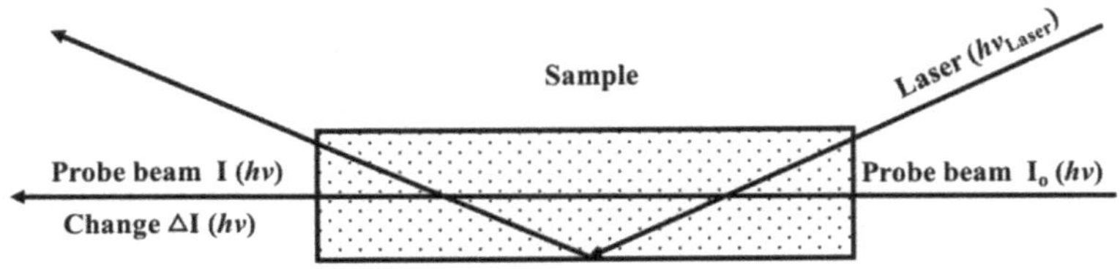

Fig. 4.13 Schematic arrangement used for the simultaneous passage of the laser beam ($h\upsilon_L$) and probe beam (hυ). Intensity I prior to the passage of the laser beam and the changes ΔI induced by the laser are received by a detector tuned to the probe beam photon energy hυ.

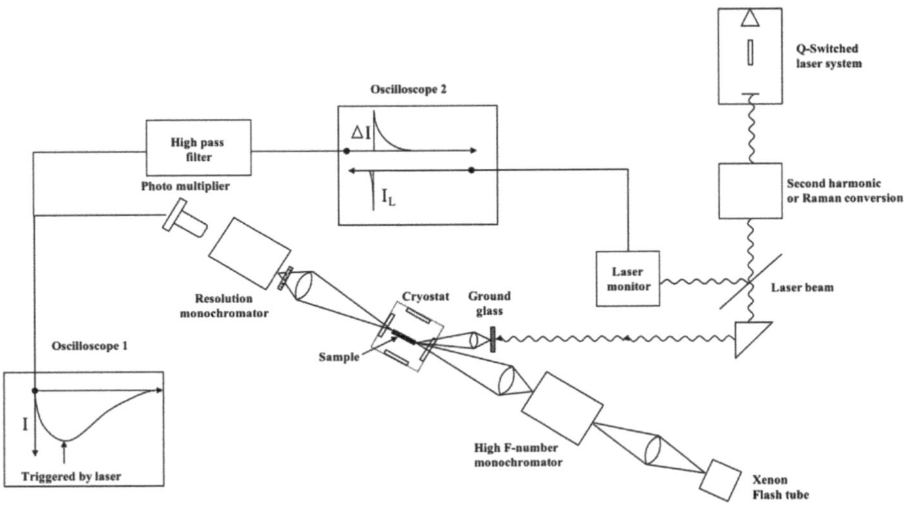

Fig. 4.14 A detailed schematic diagram of the setup used for the excited state spectroscopy with laser pulse. The path of the laser beam is indicated by the wave lines. Oscilloscope 1 records the (slowly varying) probe beam intensity I. Oscilloscope 2 handles the fast signals ΔI and I_L.

the detailed schematic diagram of the experimental setup used for the excited state spectroscopy.

4.2 Theoretical simulation

Aforementioned experimental techniques have provided very valuable information of the atomic nature and energy levels about luminescent center and trap sites. Still, there are missing blocks for the built of full physical picture in the persistent phosphors. Three main disadvantages of experimental results exist: (1) instrument measurement [I] shed light on the ground state properties of luminescent center, (2) instrument measurements [II] and [III] cannot distinguish the role of electron and hole, and (3) the excited state geometry is lacking from experiment. As an example the results from EPR measurement can provide the promising "activator" crystalline sites in the host matrix. These promising "activator" crystalline sites, however, cannot directly link the exact "luminescent center," due to the possible occurrence of autoionization process for the promising "activator" crystalline sites, which will result in full thermal quenching, that is, no luminescence at all.

As a parallel branch for materials science, theoretical simulation can provide the advantages for the problems existing from experiment. In this section, we will briefly introduce the theoretical simulation from the viewpoint of first principles calculation and semiempirical idea. Full physical picture in the persistent phosphors can be expected to build from the simulation results, especially the information of luminescent center identification and trap state calculation.

4.2.1 Physics of luminescent center in solids

To identify the luminescence center in persistent phosphors, here, we first want to introduce the physical background for the photoluminescence of impurities in crystals. Indeed the present quantum-mechanical theory of the shapes of absorption and luminescence bands can date back to the pioneering study about dimolecular spectra by Frank and Condon in 1920s [25], where the well-known Franck-Condon principle is established, on basis of adiabatic approximation [26].

The Schrodinger equation for the stationary states of the system, as we know from quantum mechanics, can be constructed as

$$-\frac{\hbar^2}{2m}\sum_i \Delta_i\Psi - \frac{\hbar^2}{2}\sum_\alpha \frac{\Delta_\alpha}{M_\alpha}\Psi + V(r,R)\Psi = E\Psi \tag{4.4}$$

with

$$V(r,R) = \frac{1}{2}\sum_i\sum_I \frac{e^2}{|r_i - r_I|} + \frac{1}{2}\sum_\alpha\sum_\beta \frac{e^2 Z_\alpha Z_\beta}{|R_\alpha - R_\beta|} - \sum_\alpha\sum_i \frac{e^2 Z_\alpha}{|r_i - R_\alpha|} \tag{4.5}$$

We will solve Eq.(4.4) in the adiabatic approximation that the solution of Eq. (4.4) as a product of two functions:

$$\Psi(r,R) = \Phi(r,R)\varphi(R) \tag{4.6}$$

We will see in the succeeding text that the function $\Phi(r,R)$ is an electronic wave function that depends parametrically on the nuclear coordinates and $\varphi(R)$ is a wave function that.

describes the nuclear motion. Substituting (4.6) into (4.4), we find

$$\begin{aligned}&-\frac{\hbar^2}{2m}\varphi(R)\sum_i \Delta_i\Phi(r,R) - \frac{\hbar^2}{2}\sum_\alpha \frac{\Delta_\alpha}{M_\alpha}\Phi(r,R)\varphi(R)\\ &+V(r,R)\Phi(r,R)\varphi(R) = E\Phi(r,R)\varphi(R)\end{aligned} \tag{4.7}$$

The solution of the Schrodinger equation (4.7) now reduces to the solution of two equations. We can separate each term in this equation into the product of the wave functions,$\Phi(r,R)\varphi(R)$. Then, we find an equation in which we have terms containing only nuclear or only electronic wave functions. This equation can be written as a system of two equations:

$$-\frac{\hbar^2}{2m}\sum_i \Delta_i \Phi(r,R) + V(r,R)\Phi(r,R) = W(R)\,\Phi(r,R) \tag{4.8}$$

$$-\frac{\hbar^2}{2}\sum_i \frac{\Delta_\alpha}{M_\alpha}\varphi(R) + W(R)\varphi(R) = \mathrm{E}\varphi(R) \tag{4.9}$$

Each equation has the form of a Schrodinger equation. The first equation of the system describes the stationary states of the electrons in the adiabatic approximation. It is a differential equation in the electronic coordinates. There is no differentiation with respect to nuclear coordinates; therefore Eq. (4.8) can be regarded as a set of Schrodinger equations for the electrons in the crystal with nuclei fixed at positions R. For each nuclear position R, we have an electronic equation of motion. The energy eigenvalue $W(R)$ of the electronic state (it is evident from our definition of the operator $V(r, R)$ that $W(R)$ also includes the potential energy of the interaction between nuclei) depends on the nuclear positions. We say that the electronic Schrodinger equation (4.8) depends parametrically on the nuclear coordinates' positions.

Eq. (4.9) is a Schrodinger equation that governs the motion of the nuclei in the molecule. The function $W(R)$, the eigenvalue of the electronic equation that depends parametrically on the nuclear coordinate, plays a large role in this equation. In Eq. (4.9), which we will also call the vibrational equation, it plays the role of the potential energy for nuclear motion. Each eigenvalue of Eq. (4.9) is also an eigenvalue of the total energy of the molecule in the adiabatic approximation.

The degree of complexity of the nuclear equation of motion is determined by the complexity of W(R). Its exact form is not known, but in one exceedingly important case—that of small oscillations of the nuclei—we know it approximately:

$$W_l(R) = W_{l0}(R_{l0}) + \frac{1}{2}\left\{\frac{\partial^2 W_l}{\partial R_{1x}^2}(R_{1x} - R_{1x,l0})^2 + \frac{\partial^2 W_l}{\partial R_{1y}^2}(R_{1y} - R_{1y,l0})^2\right.$$

$$\left. + \frac{\partial^2 W_l}{\partial R_{1z}^2}(R_{1z} - R_{1z,l0})^2 + \frac{\partial^2 W_l}{\partial R_{2x}^2}(R_{2x} - R_{2x,l0})^2 + \ldots\right\} \tag{4.10}$$

Here, R_{ax}, R_{ay}, and R_{az} are the Cartesian coordinates of the ath atom in the center-of-mass system of the molecule, and the derivatives are taken at the equilibrium configuration. By introducing appropriate new variables q_{ls} (normal coordinates), the expression for W(R) simplifies substantially; it takes the form of a diagonal quadratic form, that is, in this coordinate system the cross terms vanish. When we write the variables in Eq. (4.9) in normal coordinates, they are separated, and the problem reduces to finding the energy levels and wave functions for the stationary states of a set of N independent harmonic oscillators. We find for the total vibrational energy

$$E_{l;i_1,i_2,\ldots,i_N}=\sum_{s=1}^{N}\hbar\omega_{ls}\left(i_s+\frac{1}{2}\right) \tag{4.11}$$

Following the aforementioned procedure the photoluminescence of luminescent center in solids can be analyzed from the well-known configuration coordinate diagram. In practice, one typical effective vibration mode has been chosen, breathing mode, to stand for the full phonon band spectra. Therefore the multidimensional problem can be further simply to one-dimensional one. The typical one-dimensional configuration coordinate diagram has been shown in Fig. 4.15 [26].

Based on this idea the absorption/emission energy, Franck-Condon shifts, and the Stokes shift of luminescent center in solids can be determined semiclassically as follows. The absorption process with energy

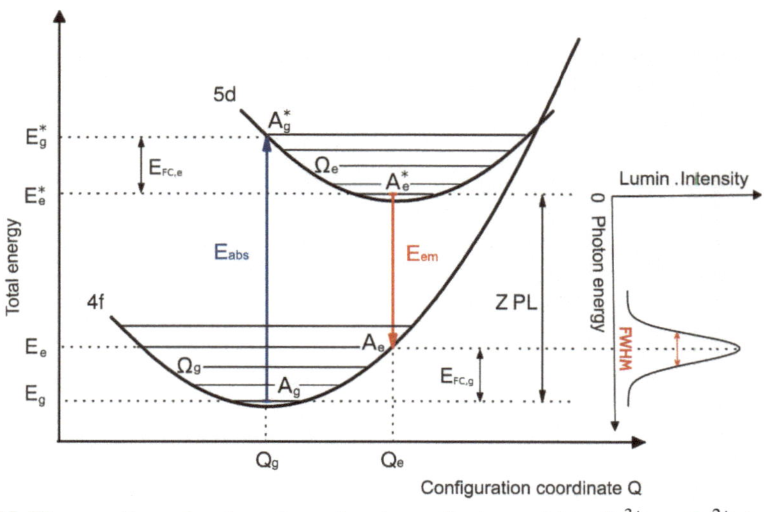

Fig. 4.15 The one-dimensional configurational coordinate model in Ce^{3+}- or Eu^{2+}-doped phosphors that share 4f-5d excitation. The generalized configuration coordinate Q results from the linear combination of ground state atomic positions (Q_g) and excited state ones (Q_e). Total energies as a function of Q are reported for the electronic ground state (labeled 4f) and for the electronic excited state (labeled 5d); the absorption and emission occur between these energies at E_g and E_e, respectively. The energy lost to the lattice after light absorption defines the Frank-Condon shift $E_{FC;e}$, and similarly, $E_{FC;g}$ is the energy lost to the lattice after light emission. Their sum gives the Stokes shift. One can associate vibrational frequencies to the ground and excited state curves, Ω_g and Ω_e. They are linked to each spectrum shape through the so-called Huang-Rhys factor S, the ratio between a Frank-Condon shift, and the related vibrational frequency. ZPL denotes the zero-phonon line, the direct transition energy when no phonon is involved. FWHM is the full width at half maximum.

$$E_{abs} = E_g^* - E_g \tag{4.12}$$

is followed by multiphonon emission, with Franck-Condon shift in the excited state.

$$E_{FC,e} = E_g^* - E_e^*; \tag{4.13}$$

then the photon emission proceeds, with energy.

$$E_{em} = E_e^* - E_e; \tag{4.14}$$

and the system relaxes into the electronic ground state, with a release of energy given by the Franck-Condon shift in the ground state:

$$E_{FC,g} = E_e - E_g \tag{4.15}$$

The two Franck-Condon shifts combine to give the observable Stokes shift

$$\Delta S = E_{FC,g} + E_{FC,e} \tag{4.16}$$

Sometimes, experiments yield also the zero-phonon line that corresponds to

$$E_{ZPL} = E_e^* - E_g \tag{4.17}$$

For this purpose, one supposes that the large number of vibrational modes contributing to the line shape can be simplified into one effective vibrational mode. The parameters entering the 1-D model are the normal coordinate ΔQ (connecting Q_g and Q_e), the displacement of the nuclei at the potential energy minimum R, the modal mass of the effective vibration, and the effective vibrational frequencies Ω_g and Ω_e. These parameters can be calculated as follows [27]:

$$\Delta Q^2 = \sum_{\alpha,i} m_\alpha \left(R_{\alpha i,e} - R_{\alpha i,g}\right)^2 \tag{4.18}$$

$$\Delta R^2 = \sum_{\alpha,i} \left(R_{\alpha i,e} - R_{\alpha i,g}\right)^2 \tag{4.19}$$

$$M = \Delta Q^2 / \Delta R^2 \tag{4.20}$$

where α denotes atoms in the supercell calculation, i denotes the Cartesian directions, m_α is the mass of atom m_α, and the $R_{\alpha i,\ g(e)}$ are the atomic coordinates in the ground and excited states, respectively. The modal mass M is an average of the masses of the ions involved in the displacement, weighted by the square of the nuclei displacements.

Based on the effective vibrational model and the total energy obtained through the ΔSCF method, the effective vibration frequencies are

$$\Omega_g^2 = 2E_{FC,g}/\Delta Q^2 \tag{4.21}$$

$$\Omega_e^2 = 2E_{FC,e}/\Delta Q^2 \tag{4.22}$$

Then the Huang-Rhys factors that denote the average numbers of phonons emitted in the ground and excited state geometry can be obtained as

$$S_{abs} = E_{FC,e}/(\hbar\Omega_e) \tag{4.23}$$

$$S_{em} = E_{FC,g}/(\hbar\Omega_g) \tag{4.24}$$

Treated quantum mechanically, the harmonic oscillator in the excited state will yield a series of quantized energy levels and their corresponding wave functions. The luminescence emission spectrum line shape at 0 K can be expressed using matrix elements of the transition as [28].

$$I(\hbar\Omega) = I_0 e^{-S_{em}} S_{em}^n n! \delta(E_{ZPL} - n\hbar\Omega_g - \hbar\Omega) \tag{4.25}$$

where I_0 is a normalization factor.

A semiclassical approximation can be used to describe the expectation value of Q and compute the density of transition as a function of the emitted energy. From there the FWHM of the emission peak W at 0 K can be calculated as

$$W(0K) = S_{em}\hbar\Omega_g\sqrt{8\ \ln 2}/\sqrt{S_{abs}} \tag{4.26}$$

$$W(T) = W(0)\sqrt{\coth\left[\hbar\Omega_g/(2k_BT)\right]} \tag{4.27}$$

where k_B is the Boltzmann constant.

4.2.2 Luminescent center identification

After the brief introduction of the theory of luminescent center in solid, we return back the quantitative solution from modern quantum mechanical calculation. Nowadays, density functional theory (DFT) is the most popular method used in materials sciences. The basic idea of DFT methodology is that the electronic density n(r) is considered as the fundamental variable of the problem and hence the space in which the problem is solved is reduced by a factor N. This section presents the two main theorems that come with this idea: [29].

$$E[n(r)] = T_s[n(r)] + E_H[n(r)] + E_{XC}[n(r)] + \int n(r)V_{ext}(r)dr \tag{4.28}$$

where

$$T_s[n(r)] = \frac{1}{2}\sum_{i=1}^{N}\int \Phi_i^*(\mathrm{r})\nabla^2\Phi_i(r)dr \qquad (4.29)$$

$$E_H[n(r)] = \frac{1}{2}\int n(r)V_H(r)dr \qquad (4.30)$$

Here, it is worth to mention that the DFT is a ground state theory, which means that the solution of excited state eigenstate cannot directly link to the optical spectra. More advanced theories like many-body perturbation theory (GW + BSE) are needed to describe the excited state of luminescent center. Fig. 4.16 shows the different capability of DFT and many-body perturbation theory [30]. At present the Bethe-Salpeter equation (BSE) is the best approach to study the optical properties of solids. It describes neutral excitations as coherent superpositions of electron-hole pairs. However, the computational burden of such approach is quite heavy and not feasible with supercells of nearly 100 atoms.

Instead of BSE, the neutral excitation of luminescence ion can be simulated on the basis of the constrained DFT method (CDFT) [31–33]. The electron-hole interaction, an essential contribution in the BSE, is mimicked by promoting the ground state electron to the excited state by constraining the seven ground bands to be unoccupied while occupying the lowest excited state lying higher in energy. Fig. 4.17 shows the description of CDFT approach for the Eu^{2+}-doped phosphors.

The CDFT method has shown promising capability for the luminescence center identification. Fig. 4.18 shows the comparison between theory and experiment leading

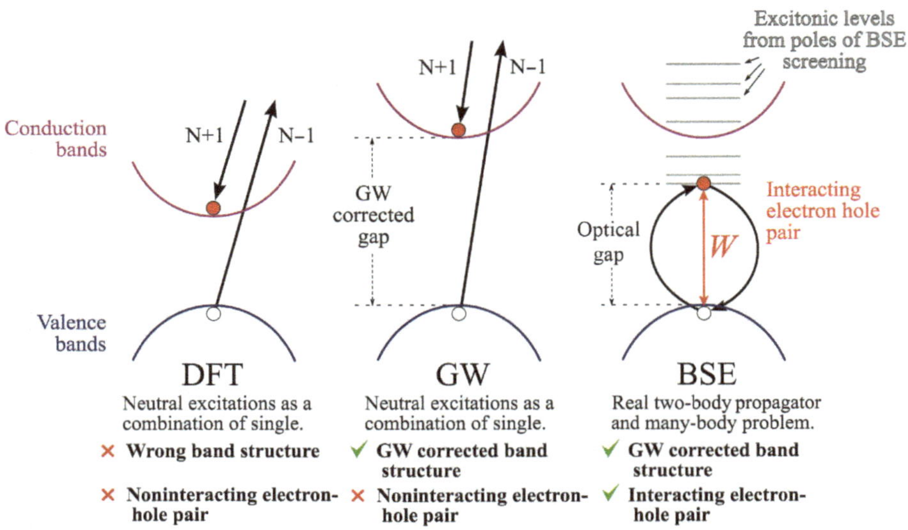

Fig. 4.16 Brief introduction about the capability of DFT and many-body perturbation theory.

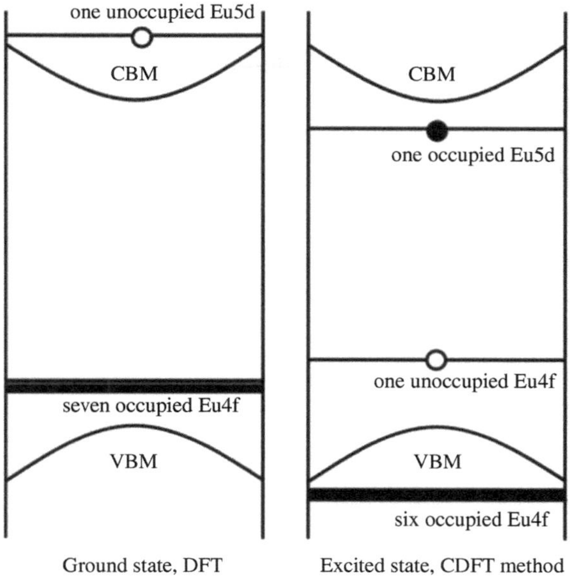

Fig. 4.17 CDFT description of photoluminescence of Eu^{2+}-doped phosphors. Schematic electron occupancies are shown for the ground and excited state calculations.
Reproduced with permission from Y. Jia, A. Miglio, S. Ponce, M. Mikami, X. Gonze, First-principles study of the luminescence of Eu^{2+}-doped phosphors, Phys. Rev. B 96 (2017), 125132, Copyright 2017, American Physical Society.

to a mean relative error (MRE) of 3.8% and 2.3% for the absorption and emission energies, respectively. The Stokes shift is a much more sensitive quantity, resulting in a 40% mean relative error (MRE) with respect to experiment. The calculated transition energies match experiment within 0.3 eV for all the 15 materials. Moreover the slope of the fitting line for the absorption and emission energy is quite close to unity, which indicates a good predictive capability of the CDFT method for the transition energies of the luminescence center.

4.2.3 Trap state calculation

As we mentioned before, experimental measurements meet the difficulty for the atomic origin determination of trap state. Ab initio calculation indeed can also provide the valuable information on this subject. The present status of trap state calculation is based on the supercell method. The formation energy of promising trap site (with neutral or charged) in solids has been calculated using a well-established thermodynamic formalism [34].

$$\Delta E_f(q) = E_{def}(q) - E_{bulk} + q(\varepsilon_{VBM} + \mu_e) - \sum_i n_i \mu_i \tag{4.31}$$

where $\Delta E_f(q)$ is the formation energy of the trap site in a charge state q. $E_{def}(q)$ is the total energy of the system containing the oxygen vacancy in a charge state q, and E_{bulk} is the total energy of the ideal host with the same supercell size. The second term in Eq. (4.31) describes the dependence on the electron chemical potential, μ_e, which is measured with respect to the valence band maximum (VBM), ε_{VBM}. The formation

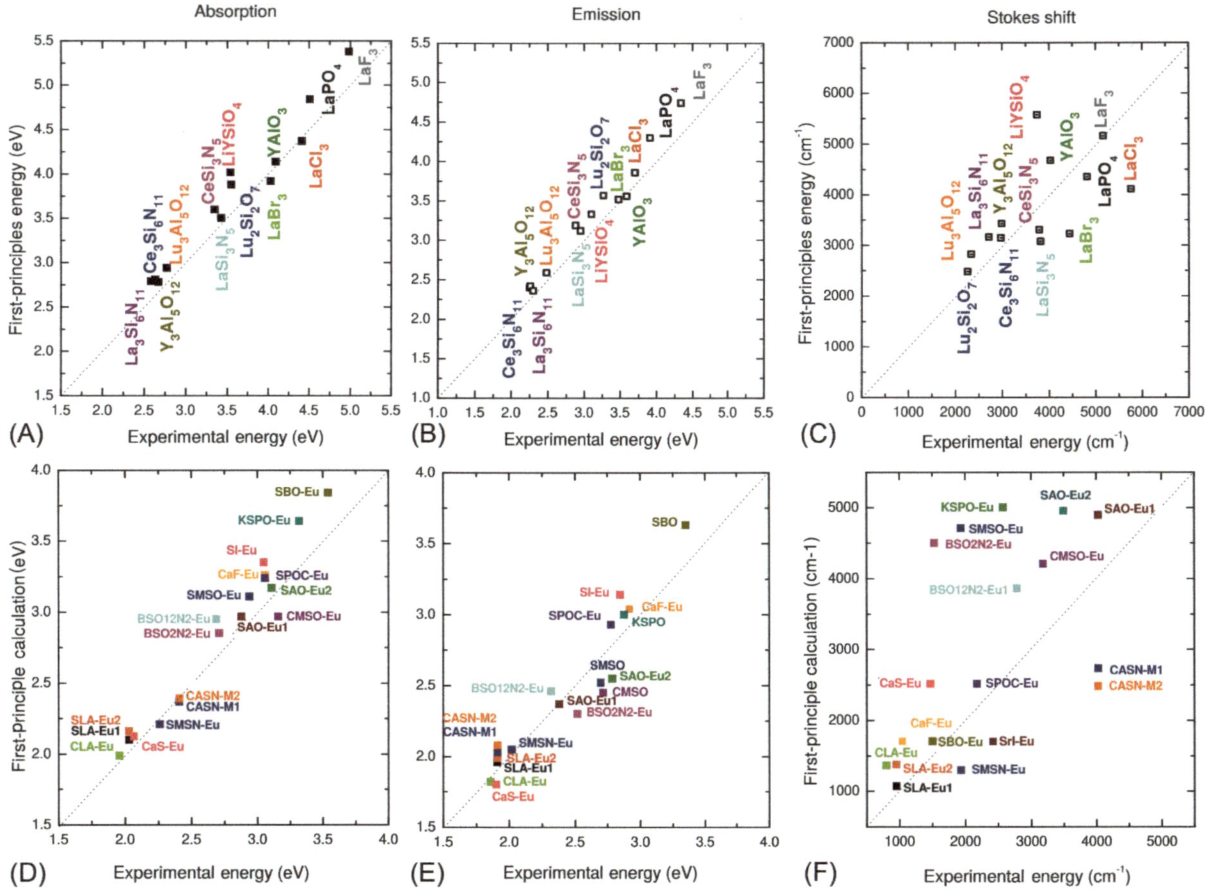

Fig. 4.18 Comparison between experimental results and first principles calculations of Ce^{3+}- and Eu^{2+}-doped phosphors: absorption energy, emission energy, and Stokes shift.

Reproduced with permission from Y. Jia, S. Ponce, A. Miglio, M. Mikami, X. Gonze, Assessment of first-principles and semiempirical methodologies for absorption and emission energies of Ce^{3+}-doped luminescent materials, Adv. Opt. Mater. 5 (2017), 1600997, Copyright 2017, Advanced Optical Materials, Wiley; Y. Jia, A. Miglio, S. Ponce, M. Mikami, X. Gonze, First-principles study of the luminescence of Eu^{2+}-doped phosphors, Phys. Rev. B 96 (2017), 125132, Copyright 2017, energy also depends on the chemical potential of the constituents as given by the last term, where the difference between the number of atoms of type i.

in the ideal cell with respect to the defect cell is denoted by Δn_i. In the calculations the chemical potential of trap site, μ_i, is determined from the exact synthesis condition. Accordingly the charge state of the trap can be determined from the relative thermodynamic transition energy level, compared with Fermi level:

$$\varepsilon\left(\frac{q}{q'}\right) = [\Delta E_f(q) - \Delta E_f(q')]/(q' - q) \tag{4.32}$$

In our previous work on scintillator materials, we have studied the stability of neutral and charged oxygen vacancies in rare earth oxyorthosilicate Lu_2SiO_5 with first principles calculations. The neutral oxygen vacancy in the $[SiO_4]$ tetrahedral site has the lowest formation energy, while vacancies in the interstitial oxygen site bonded to lutetium atoms only are more difficult to form [35]. Another interesting example is the ab initio study about the codoping effect of $LaBr_3$:Ce^{3+} [36]. The author present a mechanism that rationalizes such enhancement on the basis of first principles electronic structure calculations and point defect thermodynamics. Fig. 4.19 shows the main calculation results. It is shown that incorporation of Sr creates neutral V_{Br}-Sr_{La} complexes that can temporarily trap electrons. As a result, Auger quenching of free carriers is reduced, allowing for a more linear, albeit slower, scintillation light yield response. Experimental Stokes shifts can be related to different Ce_{La}-Sr_{La}-V_{Br} triple complex configurations.

4.2.4 Semiempirical Dorenbos' model

From experiment, efficient persistent phosphors can be obtained through codoping of rare earth ions. Such phenomena have been understood from the semiempirical model, proposed by Dorenbos, which are known as the redshift, charge transfer band, and chemical shift model, providing the information of transition energy, host reference binding energy (HRBE), and vacuum reference binding energy (VRBE) Schemes [37–40]. In the redshift model the energy of the first allowed 4f → 5d transition of the free RE ion is lowered by the crystalline environment, with a shift denoted D (A). This lowering is the sum of the spectroscopic redshift arising from the centroid shift of the RE-5d energy, $\varepsilon_C(A)$, and the crystal field splitting, $\varepsilon_{cfs}(A)$, of the RE-5d states. For the Ce^{3+} ion, the redshift D(A) can be written as

$$\mathrm{D}(\mathrm{A}) = \varepsilon_C(A) + \frac{\varepsilon_{cfs}(A)}{r(A)} - 1890cm^{-1} \tag{4.33}$$

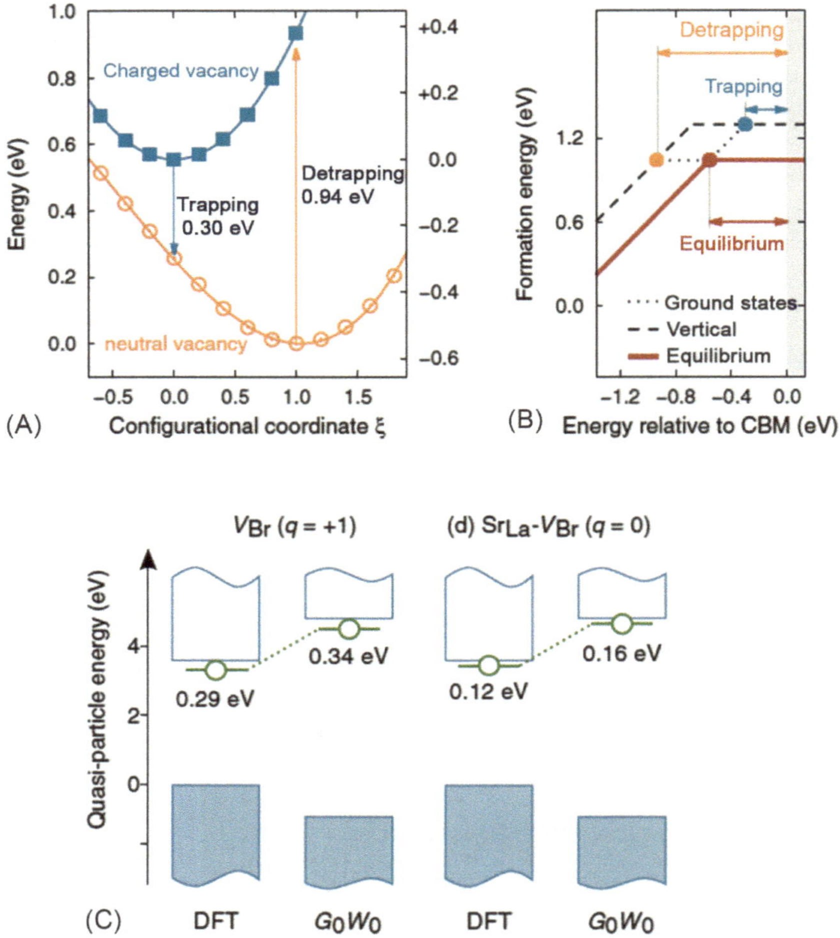

Fig. 4.19 (A) Configuration coordinate diagram illustrating the relation between neutral and charged Br vacancies. For convenience the electron chemical potential was assumed to coincide with the CBM ($\mu_e = \mu_{CBM}$) such that the energy differences indicated by the vertical arrows correspond directly to the transition levels indicated in the middle panel. (B) Formation energy of Br vacancy configurations for La-rich conditions as a function of the electron chemical potential in the vicinity of the conduction band edge. The *solid and dotted lines* are based on fully relaxed configurations, corresponding to minima in panel (A). *Dashed lines* indicate formation energies representing vertical transitions in panel (A). The *horizontal blue and orange arrows* thus correspond to the *vertical arrows* in panel (A). The red horizontal arrow is equivalent to the energy difference between the minima in panel (A). (C), (D) Quasiparticle spectra from DFT and G0W0 calculations for both (C) the isolated Br vacancy and (D) a Sr_{La}-V_{Br} complex. The position of the defect level relative to the CBM changes by less than 0.1 eV when comparing the two types of calculations. Note that in both cases the defect level is unoccupied in equilibrium.
From P. Erhart, B. Sadigh, A. Schleife, D. Aberg, First-principles study of codoping in lanthanum bromide, Phys. Rev. B 91 (2015), 165206.

in which the $\varepsilon_C(A)$ is the centroid shift of the Ce-5d energy relative to the free ion, defined as follows:

$$\varepsilon_C(A) = 1.44 \times 10^{17} \sum_{i=1}^{N} \alpha_{sp}^{i} R_i^6 \tag{4.34}$$

In the previous formula, α_{sp}^{i} is the spectroscopic polarization of anion i located at distance Ri from the Ce^{3+} ion in the relaxed structure. The summation is over all anions N in the coordination environment of RE ions. At present, quantitative relationships between α_{sp}^{i} and the electronegativity of the cations for oxides, nitrides, and fluorides have been proposed as

$$\alpha_{sp}^{O} = 0.33 + \frac{4.8}{\chi_{av}^2} \tag{4.35}$$

$$\alpha_{sp}^{N} = 0.87 + \frac{18.76}{\chi_{av}^2} \tag{4.36}$$

$$\alpha_{sp}^{F} = 0.15 + \frac{0.96}{\chi_{av}^2} \tag{4.37}$$

where the electronegativity is

$$\chi_{av} = \frac{1}{N} \sum_{i=1}^{M} \frac{Z_i \chi_i}{\gamma} \tag{4.38}$$

This formula is rationalized by considering that a cation of charge Z_i will bind on average with Z_i/γ anions of charge $-\gamma$. The summation is over all cations M in the compound, and N is the number of anions.

Another parameter affecting the spectroscopic redshift is the contribution from the crystal field shift, $\varepsilon_{cfs}(A)/r(A)$. The crystal field splitting $\varepsilon_{cfs}(A)$ is defined as the energy difference between the lowest and highest 5d levels. A fraction $1/r(A)$ contributes to the redshift, where $r(A)$ usually varies between 1.7 and 2.4. The $\varepsilon_{cfs}(A)$ is determined as

$$\varepsilon_{cfs}(A) = \frac{\beta}{R_{av}^2} \tag{4.39}$$

Here, β is a parameter related to the shape and size of the anion polyhedron coordinated to the Ce^{3+} ion, and R_{av} is the average distance between the Ce^{3+} ion and anions in the relaxed structure. Based on *D(A)* and the energy of the first 4f → 5d transition of Ce^{3+} as a free (gaseous) ion, 49,340 cm^{-1}, the transition energy of Ce^{3+} ion in compound A can be calculated as

$$E(A) = 49340\,\mathrm{cm}^{-1} - D(A) \tag{4.40}$$

Compared with the comprehensive study of Ce^{3+} ion in inorganic compounds, the related work for the Eu^{2+}-doped phosphors is limited because of the more complex electronic configuration of the Eu^{2+} ion. However, a similar quantitative expression for the Eu^{2+} ion can be expected since these two ions have similar 4f → 5d neutral excitations. A semiempirical relationship between the redshift for the 5d state of Eu^{2+} and Ce^{3+} ions, in the same host, has been determined:

$$D\left(Eu^{2+}, A\right) = 0.64 \times D\left(Ce^{3+}, A\right) - 0.233\,\mathrm{eV} \tag{4.41}$$

Compared with the redshift model, the building of HRBE energy scheme requires the information of lanthanide state with respect to either the valence band or the conduction band. The excitation of an electron from the valence band to the 4f shell of a trivalent lanthanide creates the divalent lanthanide plus a hole on the neighboring anion ligand. Then the information of charge transfer band (CTB) of trivalent lanthanides can provide the location of the 4f ground state level above the top of the valence band of divalent lanthanides. For the determination of HRBE energy scheme of trivalent ions, the needed information is the energy level barrier between the 5d state and conduction band, which are usually from the temperature-dependent photoluminescence measurement.

HRBE schemes were published for many compounds. They provide the coulomb correlation energy U(6, A) defined as

$$U(6, A) = E_{4f}(7, 2+, A) - E_{4f}(6, 3+, A) \tag{4.42}$$

From chemical shift model the 4f-VRBE energy in the divalent ions yields

$$E_{4f}(7, 2+, A) = -24.92 + \frac{18.05 - U(6, A)}{0.777 - 0.0353U(6, A)} \tag{4.43}$$

Fig. 4.20 shows the typical procedure and error bar for the usage of Dorenbos' model [41]. Here, it is also worthy to note that the building of HRBE energy scheme involves different relaxation geometries of lanthanide ions, which can lead to an error around 0.5 eV.

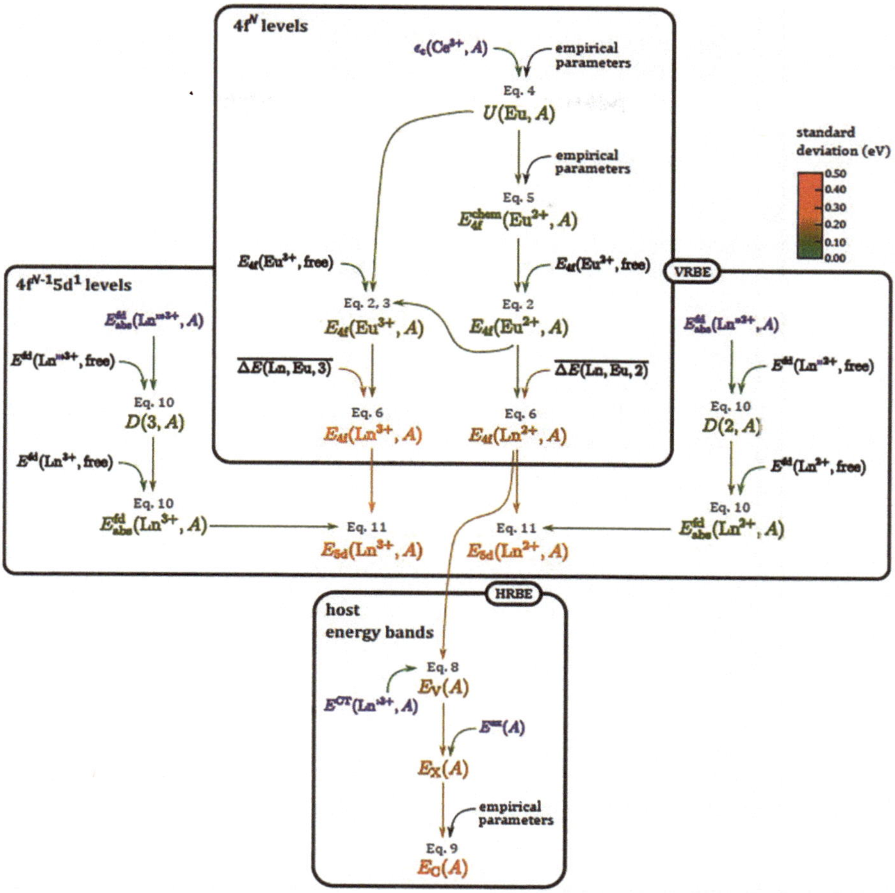

Fig. 4.20 Lanthanide energy level scheme roadmap. This flow diagram illustrates how VRBE and HRBE level schemes are obtained and which errors are accumulated. Empirical parameters are displayed in black, and experimental parameters in purple, and the color of calculated quantities maps their standard deviation. Ln', Ln", and Ln"' denote specific lanthanide ions from which input is needed, such as the experimental charge transfer and 4f-5d transition energies (for divalent and trivalent lanthanides).
Reproduced with permission from J.J. Joos, D. Poelman, P.F. Smet, Energy level modeling of lanthanide materials: review and uncertainty analysis, Phys. Chem. Chem. Phys 17 (2015), 19058–19078, Copyright 2015, Royal Society of Chemistry.

References

[1] F. Clabau, Phosphorescence: Mécanismes et nouveaux matériaux, PhD thesis, Institut des Matériaux Jean Rouxel, Nantes, 2005.
[2] S. Zhang, Spectroscopy of Rare Earth Ions: Spectral Property and Spectral Theory, Science Press (China), 2008.

[3] K. Van den Eeckhout, Persistent Luminescence: Kinetics and Compounds, PhD thesis, Ghent University, 2013.

[4] V. Bachmann, C. Ronda, A. Meijerink, Temperature quenching of yellow Ce^{3+} luminescence in YAG:Ce, Chem. Mater. 21 (2009) 2077–2084.

[5] J.M. Carvalho, C.C.S. Pedroso, I.P. Machado, J.H. Rodrigues, P. Głuchowski, M. Lastusaari, H.F. Brito, Persistent luminescence warm-light LEDs based on ti-doped RE_2O_2S materials prepared by rapid and energy-saving microwave-assisted synthesis, J. Mater. Chem. C 6 (2018) 8897–8905.

[6] L. Rodrigues, H. Brito, J. Holsa, R. Stefani, M. Felinto, M. Lastusaari, T. Laamanen, L. Nunes, Discovery of the persistent luminescence mechanism of $CdSiO_3$:Tb^{3+}, J. Phys. Chem. C 116 (2012) 11232–11240.

[7] E. Finley, M.W. Gaultois, J. Brgoch, Unlocking the key to persistent luminescence with X-ray absorption spectroscopy: a local structure investigation of Cr-substituted spinel-type phosphors, Phys. Chem. Chem. Phys. 21 (2019) 19349–19358.

[8] J.A. Weil, J.R. Bolton, Electron Paramagnetic Resonance: Elementary Theory and Practical Applications, John Wiley Sons, 2007.

[9] X. Qin, Y. Li, Y. Li, Y. Wu, R. Chen, K. Sharafudeen, Z. Ma, S. Liu, D. Wu, J. Qiu, Inducing NIR long persistent phosphorescence in Cr-doped perovskite titanate via redox, J. Alloys Compd. 666 (2016) 387–391.

[10] V.V. Laguta, M. Buryi, J. Rosa, D. Savchenko, J. Hybler, M. Nikl, S. Zazubovich, T. Karner, C.R. Stanek, K.J. McClellan, Electron and hole traps in yttrium orthosilicate single crystals: The critical role of Si-unbound oxygen, Phys. Rev. B 90 (2014) 064104.

[11] A.D. Baker, C.R. Brundle, in: C.R. Brundle, A.D. Baker (Eds.), Electron Spectroscopy: Theory, Techniques, and Applications, Vol. 1, Academic Press, London, 1977 (Chapter 1).

[12] A.F. Orchard, in: D. Briggs (Ed.), Handbook of X-Ray and Ultraviolet Photoelectron Spectroscopy, Heyden, London, 1977 (Chapter 1).

[13] N.V. Joshi, Photoconductivity: Art and Science Technology, Routledge, 2017.

[14] T. Nakanishi, Y. Katayama, J. Ueda, T. Honma, S. Tanabe, T. Komatsu, Fabrication of Eu: $SrAl_2O_4$-based glass ceramics using frozen sorbet method, J. Ceram. Soc. Jpn. 119 (2011) 609–615.

[15] Y. Katayama, J. Ueda, S. Tanabe, Photo-electronic properties and persistent luminescence in Pr^{3+} doped (Ca,Sr)TiO_3 ceramics, J. Lumin. 148 (2014) 290–295.

[16] T. Nakanishi, K. Watanabe, J. Ueda, K. Fushimi, S. Tanabe, Y. Hasegawa, Enhanced light storage of $SrAl_2O_4$ glass-ceramics controlled by selective europium reduction, J. Am. Ceram. Soc. 98 (2015) 423–429.

[17] Z.W. Pan, Y.Y. Lu, F. Liu, Sunlight-activated long-persistent luminescence in the near-infrared from Cr^{3+}-doped zinc gallogermanates, Nat. Mater. 11 (2012) 58–63.

[18] Y. Li, S.F. Zhou, G.P. Dong, M.Y. Peng, L. Wondraczek, J.R. Qiu, Anti-stokes fluorescent probe with incoherent excitation, Sci. Rep. 4 (2014) 04059.

[19] S.W. McKeever, Thermoluminescence of Solids, Vol. 3, Cambridge University Press, 1988.

[20] P.J. Wang, X.H. Xu, D.C. Zhou, X. Yu, J.B. Qiu, Sunlight activated long-lasting luminescence from $Ba_5Si_8O_{21}$: Eu^{2+},Dy^{3+} phosphor, Inorg. Chem. 54 (2015) 1690–1697.

[21] J.C. Zhang, X.L. Ma, Q.S. Qin, L.R. Shi, J.Y. Sun, M.J. Zhou, B.T. Liu, Y.H. Wang, The synthesis and afterglow luminescence properties of a novel red afterglow phosphor: SnO_2: Sm^{3+},Zr^{4+}, Mater. Chem. Phys. 136 (2012) 320–324.

[22] A.J.J. Bos, R.M. van Duijvenvoorde, E. van der Kolk, W. Drozdowski, P. Dorenbos, Thermoluminescence excitation spectroscopy: a versatile technique to study persistent luminescence phosphors, J. Lumin. 131 (2011) 1465–1471.

[23] T. Maldiney, A. Bessiere, J. Seguin, E. Teston, S.K. Sharma, B. Viana, A.J.J. Bos, P. Dorenbos, M. Bessodes, D. Gourier, D. Scherman, C. Richard, The in vivo activation of persistent nanophosphors for optical imaging of vascularization, tumours and grafted cells, Nat. Mater. 13 (2014) 418–426.
[24] Y. Li, S.F. Zhou, Y.Y. Li, K. Sharafudeen, Z.J. Ma, G.P. Dong, M.Y. Peng, J.R. Qiu, Long persistent and photo-stimulated luminescence in Cr^{3+}-doped Zn–Ga–Sn–O phosphors for deep and reproducible tissue imaging, J. Mater. Chem. C 2 (2014) 2657–2663.
[25] E.U. Condon, Nuclear motions associated with electron transitions in diatomic molecules, Phys. Rev. 32 (1928) 858.
[26] M. Born, K. Huang, Dynamical Theory of Crystal Lattices, Clarendon Press, 1954.
[27] A. Alkauskas, J.L. Lyons, D. Steiauf, C.G. Van de Walle, First-principles calculations of luminescence spectrum line shapes for defects in semiconductors: the example of GaN and ZnO, Phys. Rev. Lett. 109 (2012), 267401.
[28] B. Henderson, G.F. Imbusch, Optical Spectroscopy of Inorganic Solids, Vol. 44, Oxford University Press, 2006.
[29] R.M. Martin, Electronic Structure: Basic Theory and Practical Methods, Cambridge University Press, 2004.
[30] G. Onida, L. Reining, A. Rubio, Electronic excitations: density-functional versus many-body Green's-function approaches, Rev. Mod. Phys. 74 (2002) 601.
[31] Y. Jia, A. Miglio, S. Ponce, X. Gonze, M. Mikami, First-principles study of Ce^{3+}-doped lanthanum silicate nitride phosphors: neutral excitation, stokes shift, and luminescent center identification, Phys. Rev. B 93 (2016), 155111.
[32] Y. Jia, S. Ponce, A. Miglio, M. Mikami, X. Gonze, Assessment of first-principles and semiempirical methodologies for absorption and emission energies of Ce^{3+}-doped luminescent materials, Adv. Opt. Mater. 5 (2017) 1600997.
[33] Y. Jia, A. Miglio, S. Ponce, M. Mikami, X. Gonze, First-principles study of the luminescence of Eu^{2+}-doped phosphors, Phys. Rev. B 96 (2017) 125132.
[34] C. Freysoldt, B. Grabowski, T. Hickel, J. Neugebauer, G. Kresse, A. Janotti, C.G. Van de Walle, First-principles calculations for point defects in solids, Rev. Mod. Phys. 86 (2014) 253.
[35] Y. Jia, A. Miglio, X. Gonze, M. Mikami, Ab-initio study of oxygen vacancy stability in bulk and cerium-doped lutetium oxyorthosilicate, J. Lumin. 204 (2019) 499–505.
[36] P. Erhart, B. Sadigh, A. Schleife, D. Aberg, First-principles study of codoping in lanthanum bromide, Phys. Rev. B 91 (2015) 165206.
[37] P. Dorenbos, 5d−level energies of Ce^{3+} and the crystalline environment. II. Chloride, bromide, and iodide compounds, Phys. Rev. B 62 (2000), 15650.
[38] P. Dorenbos, 5d−level energies of Ce^{3+} and the crystalline environment. III. Oxides containing ionic complexes, Phys. Rev. B 64 (2001), 125117.
[39] P. Dorenbos, Relating the energy of the [Xe]$5d^1$ configuration of Ce^{3+} in inorganic compounds with anion polarizability and cation electronegativity, Phys. Rev. B 65 (2002), 235110.
[40] P. Dorenbos, Modeling the chemical shift of lanthanide 4f electron binding energies, Phys. Rev. B 85 (2012), 165107.
[41] J.J. Joos, D. Poelman, P.F. Smet, Energy level modeling of lanthanide materials: review and uncertainty analysis, Phys. Chem. Chem. Phys. 17 (2015) 19058–19078.

Persistent phosphors

5

5.1 Lanthanide-activated persistent phosphors

In the last century, zinc sulfide (ZnS) doped with copper (Cu) or cobalt (Co) was the most widely used persistent phosphor in various applications, such as watch dials, decoration, and toys [1, 2]. These materials give a broadband green emission centered at 540 nm, and the afterglow duration can last several hours. However, the persistent phosphorescence intensity is relatively weak, and the host materials are sensitive to air, which leads to the decrease of afterglow intensity and time. Radiative elements like tritium and promethium have been used to stimulate long glow lifetime up to 10 years, while the safety concerns have prohibited their applications on the large scale. To circumvent these problems, there has been a continuation on the research of new efficient, safe, and stable persistent phosphors with long duration time in the past decades. In 1997 Matsuzawa et al. discovered efficient $CaAl_2O_4:Eu^{2+}$, Nd^{3+} and $SrAl_2O_4:Eu^{2+},Dy^{3+}$ persistent phosphors, which give blue and green persistent phosphorescence, respectively [3]. In these phosphors, Eu^{2+} and Nd^{3+} (or Dy^{3+}) were used as emitter and trapping center, respectively. The benefit of using trivalent lanthanide ions as codopant is to control the trap depth and density through the creation of lattice defects (i.e., increase oxygen vacancies V_o) in the host crystal, which enhances the trapping and detrapping processes, leading to prolonged lifetime and more intense persistent phosphorescence. The green-emitting phosphor $SrAl_2O_4$: Eu^{2+},Dy^{3+} showed higher initial intensity, low sensitivity to moisture, and a much longer afterglow time compared with the commercially used ZnS:Cu,Co. Thus lanthanide-doped persistent phosphors have become the alternatives of the ZnS-based products and dominate the markets, ranging from display to decoration since then. Nowadays, many hosts doped with different activators such as rare earth ions and transition metal ions have been proposed to obtain efficient blue, green, red, and even near-infrared persistent phosphorescence for some advanced applications, such as night vision surveillance and in vivo medical imaging. In this section, recent advances in persistent phosphors doped with various rare earth ions (e.g., Eu^{2+}, Eu^{3+}, Dy^{3+}, Tb^{3+}, Sm^{3+}, Tm^{3+}, Pr^{3+}, Ce^{3+}, Yb^{2+}, Nd^{3+}, and Er^{3+}) will be summarized.

5.1.1 Eu^{2+}

The luminescence properties of divalent lanthanide ions with $4f^{n-1}5d$ electronic configurations are strongly influenced by lattice structures, as $5d$ orbitals are sensitive to the surrounding environment of host lattice. Therefore the $4f^n$–$4f^{n-1}5d$ electronic transitions are very different from 4f–4f transitions.

Eu^{2+} has the electron configuration of [Xe] $4f^6$ $6s^2$, and the electron transitions occur within the f-manifolds (f-f transitions). Eu^{2+} is the most widely used and efficient luminescent center, which emits broad visible (i.e., from blue to red emission and

Persistent Phosphors. https://doi.org/10.1016/B978-0-12-818637-4.00005-7

sometimes near-infrared emission) luminescence when it doped in different hosts. This broad emission peak shifts to longer or shorter wavelengths by changing the crystal field of the host (i.e., the exited states of $^{4}f_{6}^{5}d_{1}$ to lower level). It is well known that the process of persistent phosphorescence involves three key steps, namely, excitation, trapping, and detrapping (Fig. 5.1). Excitation refers to the process where charge carriers (electrons and holes) are generated upon photon irradiation. The charge carriers can be subsequently trapped at impurity levels, which can be considered as energy reservoirs. Note that the impurity levels could trap free charge carriers either

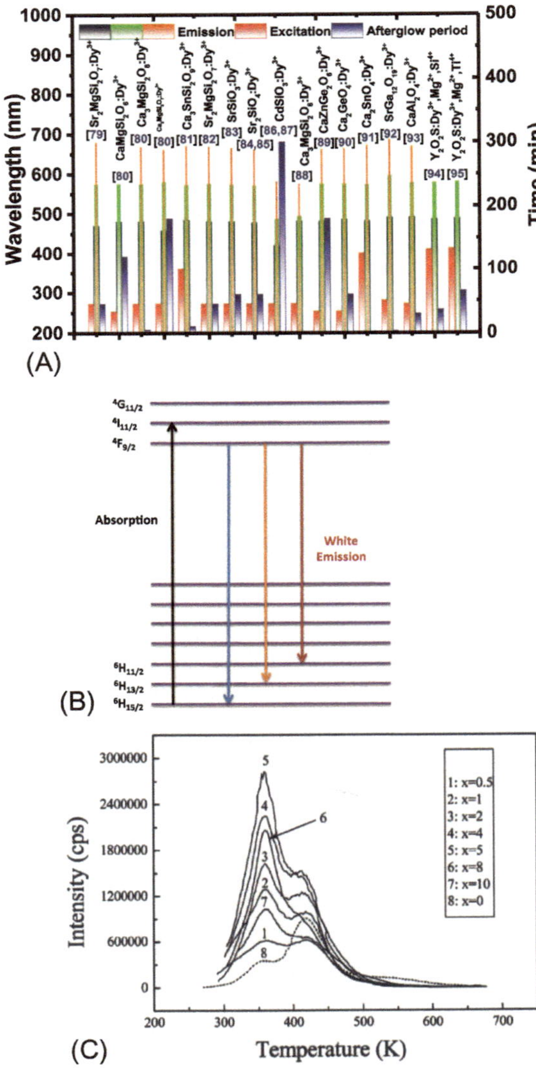

Fig. 5.1 Schematic for mechanisms of persistent phosphorescence in Eu^{2+}-doped persistent phosphors.

by quantum tunneling or through valence band–mediated/conduction band–mediated channels. These trapped charge carriers can be gradually released as photons responsible for the observed persistent phosphorescence. It is generally accepted that electrons are the main charge carriers. To effectively trap electrons, the most efficient way is to transport the electrons to the trapping center through the conduction band of the host. In principle the electrons can be transported to the host's conduction band through any of the following channels: (1) directly excited to the conduction band from their ground state upon irradiation, (2) pumped to the emitter's excited states above the conduction band, and (3) energy jumping from the gap states to the conduction band with the assistance of thermal fluctuation. The electron trapping probability is associated with the density of trapping centers. The trapping centers could be in various forms, including codopant, cation or anion vacancy, and antisite defect, which relies on the host.

After the development of $CaAl_2O_4:Eu^{2+},Nd^{3+}$ and $SrAl_2O_4:Eu^{2+},Dy^{3+}$ persistent phosphors in 1997, many researchers around the world attempted to synthesize different persistent phosphors (e.g., aluminates, silicates, and phosphates) codoped with Eu^{2+} and other Ln^{3+} ions (e.g., Ho^{3+}, Er^{3+}, and Tm^{3+}) rather than Nd^{3+} and Dy^{3+}, as shown in Fig. 5.2 [3–59]. All of these attempts were aimed to enhance the luminescence efficiency, increase the afterglow duration, and tune the emission band.

As shown in Fig. 5.2, Eu^{2+}-activated persistent phosphors are giving efficient blue persistent phosphorescence including $Sr_4Al_{14}O_{25}:Eu^{2+},Dy^{3+}$ [4] (Fig. 5.2A), $Sr_2MgSi_2O_7:Eu^{2+},Dy^{3+}$ [14] (Fig. 5.2B), $CaAl_2O_4:Eu^{2+},Dy^{3+}$ [3] (Fig. 5.2A), and $Sr_3MgSi_2O_8:Eu^{2+},Dy^{3+}$ [12] (Fig. 5.2B). Efficient green persistent phosphorescence are observed in $Ba_{0.92}Li_{0.08}Si_{1.92}P_{0.08}O_5:Eu^{2+},Pr^{3+}$ [53] (Fig. 5.2F), $BaZrSi_3O_9:Eu^{2+},Pr^{3+}$ [21] (Fig. 5.2B) $SrAl_2O_4:Eu^{2+},Dy^{3+}$ [3] (Fig. 5.2A), $Ba_2MgSi_2O_7:Eu^{2+},Tm^{3+}$ [20] (Fig. 5.2B), $Ca_6BaP_4O_{17}:Eu^{2+},Ho^{3+}$ [41] (Fig. 5.2C), and $Ca_2MgSi_2O_7:Eu^{2+}$ [23] (Fig. 5.2B). However, efficient yellow-orange to red persistent phosphorescence need special host with strong crystal field strength that reduces the lowest emitting level of the $^4f_6\ ^5d_1$ configuration. $BaSiO_3:Eu^{2+},Nd^{3+},Tm^{3+}$ [34] (Fig. 5.2B), $Sr_3SiO_5:Eu^{2+},Lu^{3+}$ [29, 30] (Fig. 5.2B), $Ca_2ZnSi_2O_7:Eu^{2+},Dy^{3+}$ [32] (Fig. 5.2B), and $Ca_2BO_3Cl:Eu^{2+},Dy^{3+}$ [50] (Fig. 5.2E) are good candidates, which gives strong afterglow intensity with long afterglow lifetime. In comparison, other reported persistent phosphors such as nitrides (Fig. 5.2D) exhibit relative weak emission intensity and short afterglow lifetime compared with the intensity and lifetime of aluminate- and silicate-based persistent phosphors. Based on the presented statistics, $Ba_{0.92}Li_{0.08}Si_{1.92}P_{0.08}O_5:Eu^{2+},Pr^{3+}$, $Ca_6BaP_4O_{17}:Eu^{2+},Ho^{3+}$, and $Ca_2BO_3Cl:Eu^{2+},Dy^{3+}$ may be the most efficient blue, green, and red persistent phosphors, respectively. These persistent phosphors possess afterglow duration more than 45h as result of the presence of high density of electron and/or hole traps.

5.1.2 Eu^{3+}

Trivalent lanthanide ions feature a [Xe] core electronic configuration with the addition of n 4*f* electrons, with *n* varying from 0 [for La^{3+}] to 14 [for Lu^{3+}]. In general, trivalent lanthanide-doped complexes show line emission features as the $4f^n$ subshell is

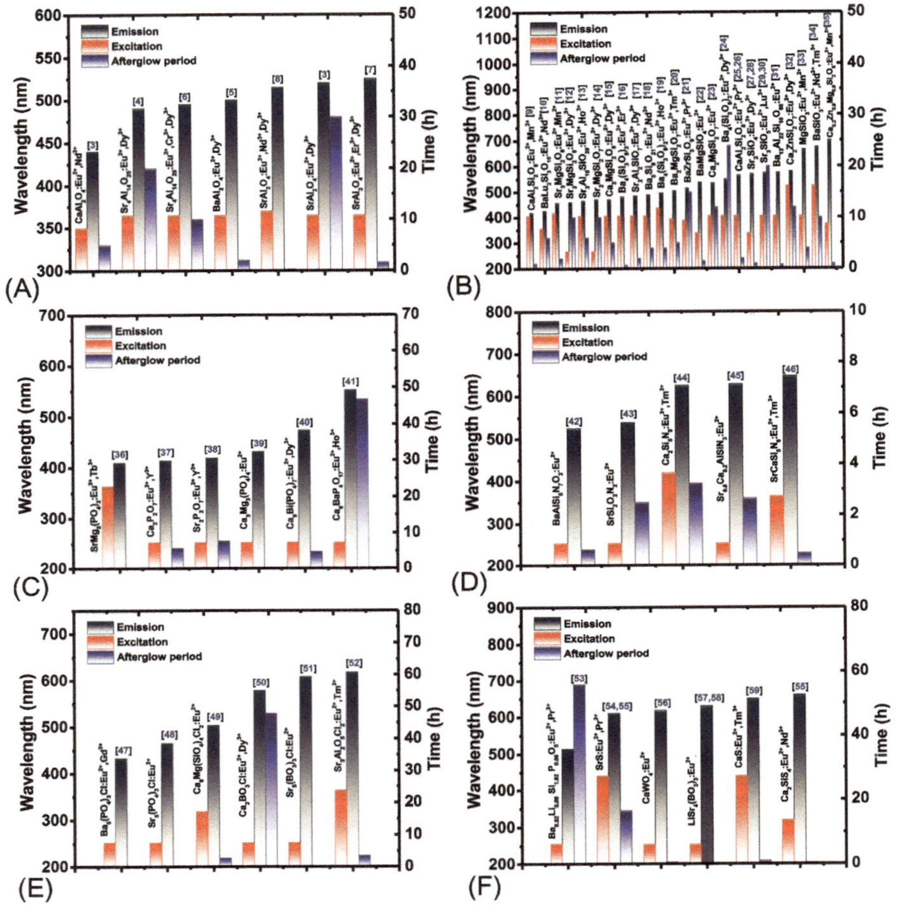

Fig. 5.2 Eu^{2+}-activated persistent phosphors of (A) aluminates, (B) silicates, (C) phosphates, (D) nitrides, (E) chlorides, and (F) others.

shielded by the $5s^2$ and $5p^6$ subshells. In many cases the influence of the crystal field on the optical properties of trivalent lanthanide ions can be regarded as a relatively small perturbation. In other words the weak environmental effect makes the energy levels of trivalent lanthanide ions relatively similar in different hosts. This enables to compare the energy levels of trivalent lanthanide ions doped in different hosts. Among trivalent lanthanide ions, Eu^{3+} has the electron configuration of [Xe] $4f^7$ $6s^2$. Eu^{3+} ions are considered as efficient red light-emitting ion due to the transitions from the excited $^5D_0$ level to the $^7F_J(J = 0, 1, 2, 3, 4)$ levels of $^4f_6$ configuration in the orange-red visible range (Fig. 5.3B) [60–62]. Unlike Eu^{2+}, the emission peak position of Eu^{3+} ions almost doesn't depend on the host crystal field, since Eu^{3+}-doped materials usually reveal similar spectrum. However, the peak intensity of Eu^{3+} ions is

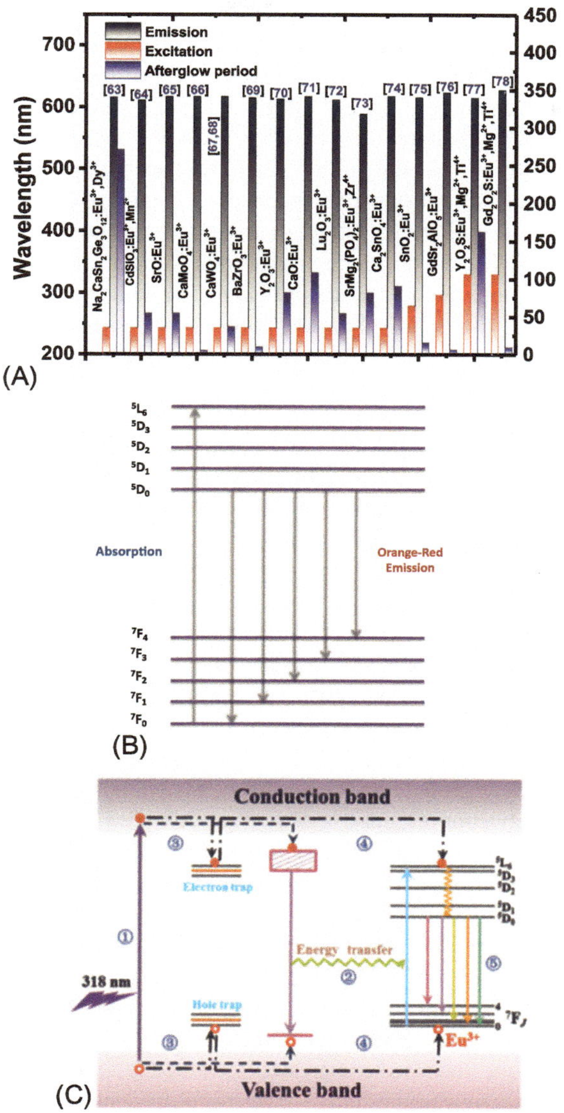

Fig. 5.3 (A) Eu^{3+}-activated persistent phosphors. (B) Energy level diagram of Eu^{3+}. (C) Schematic for mechanisms of persistent phosphorescence in $Na_2CaSn_2Ge_3O_{12}$:Eu^{3+}, Dy^{3+} persistent phosphors. Reproduced with permission from J. Xu, Z. Ju, X. Yang, X. Tang, W. Liu, Considerable improvement of $Na_2CaSn_2Ge_3O_{12}$:Eu^{3+} reddish orange long-persistent phosphorescence with co-doping of Dy^{3+}. Dyes Pigments, 2019, 162, 951–958, Copyright 2019, Elsevier.

dependent on the crystal field that reflects the site symmetry. So, Eu^{3+} is employed as spectroscopic probe to examine the local environment around rare earth ions in different crystals and glasses because it has exclusive emission bands, which are due to either magnetic or electric dipole transitions [61, 62]. Therefore the structural symmetry of the Eu^{3+} site can be assessed by calculating R-value, which is defined as the ratio between the integrated intensity of $^5D_0 \rightarrow {}^7F_2$ to $^5D_0 \rightarrow {}^7F_1$ emissions.

Few reports about Eu^{3+}-doped persistent phosphors have been published, as listed in Fig. 5.3A [63–78]. Most papers about persistent phosphors doped with Eu^{3+} have used oxide and oxysulfide as hosts, while Dy^{3+}, Zr^{4+}, Mg^{2+}, and Ti^{4}+ serve as trapping centers. In the case of oxysulfides such as Gd_2O_2S:Eu^{3+},Mg^{2+},Ti^{4+} [78] and Y_2O_2S: Eu^{3+},Mg^{2+},Ti^{4+} [77] phosphors, Eu^{3+} substitutes the site of Y^{3+}/Gd^{3+} as activators; however, Mg^{2+} and Ti^{4+} also substitute the site of Y^{3+}/Gd^{3+} to create traps that enhance the afterglow duration and persistent phosphorescence efficiency. Unfortunately the afterglow duration of Eu^{3+}-activated oxysulfide persistent phosphors is still relatively short (< 5 h), which can't fulfill the requirements of optical applications. In addition, the afterglow time of Eu^{3+}-doped oxide persistent phosphors is longer than oxysulfides, such as $Na_2CaSn_2Ge_3O_{12}$:Eu^{3+},Dy^{3+} phosphor with a 5-h persistent time after ceasing of ultraviolet light excitation [63]. Other oxide-doped Eu^{3+} are also reported, such as CaO:Eu^{3+} [71], Lu_2O_3:Eu^{3+} [72], $SrMg_2(PO_4)_2$:Eu^{3+},Zr^{4+} [73], Ca_2SnO_4:Eu^{3+} [74], and SnO_2:Eu^{3+} [75].

In $Na_2CaSn_2Ge_3O_{12}$:Eu^{3+},Dy^{3+} phosphor, Na^+ and Ca^{2+} occupy the same site in the $Na_2CaSn_2Ge_3O_{12}$ unit cell. Ca^{2+} and Na^+ have eight coordination with oxygen and similar ionic radius of Eu^{3+} and Dy^{3+}. It can infer that Eu^{3+} and Dy^{3+} tend to substitute Ca^{2+} or Na^+; this is because the ionic radii of Eu^{3+} and Dy^{3+} are close to that of Ca^{2+} and Na^+. In order to keep the $Na_2CaSn_2Ge_3O_{12}$:Eu^{3+},Dy^{3+} phosphor electrically neutral, the method which was generally used is as follows: (1) two Eu^{3+} (or Dy^{3+}) replace three Ca^{2+} which generates two positive charge defect Eu_{Ca}, Dy_{Ca} and one negative charge defects V_{Ca}. (2) One Eu^{3+} (or Dy^{3+}) replaces three Na^+ and hence it creates two negative charge defects V_{Na} and one positive charge defect Dy_{Na} (or Eu_{Na}). Therefore the persistent phosphorescence mechanism of $Na_2CaSn_2Ge_3O_{12}$: Eu^{3+},Dy^{3+} phosphor can be explained as follows: The internal defects (V_{Ca}, V_{Na}) function as the hole traps, and the foreign defects (Eu_{Ca}, Eu_{Na}, Dy_{Ca}, Dy_{Na}) serve as the electron traps. The persistent phosphorescence is caused by Eu^{3+} and Eu^{3+} that not only was used as emit center but also can provide trap. No light from Dy^{3+} was observed, indicating that Dy^{3+} is just a trap provider in this work. The mechanism of persistent phosphorescence in $Na_2CaSn_2Ge_3O_{12}$:Eu^{3+},Dy^{3+} phosphor was displayed in Fig. 5.3C. For illustration purposes the Eu^{3+} energy levels are put between the conduction band and valence band. At first (process 1), electrons can transit from the valence band to the conduction band under the excitation of 318-nm ultraviolet laser. Because of the emission spectrum of the $Na_2CaSn_2Ge_3O_{12}$ matrix and the excitation spectrum of Eu^{3+} is well overlapped, the absorbed energy of $Na_2CaSn_2Ge_3O_{12}$ matrix can be transferred to Eu^{3+} matrix through nonradiative transitions (process 2) and finally lead to the characteristic emissions of Eu^{3+} (process 5). The 318-nm ultraviolet light stimulates results in the generation of electrons and holes. Then the generated electrons and holes would be intercepted by energy matching traps from conduction band or valence band through nonradiative transition, respectively (process 3). After stopping the ultraviolet stimulation, afterglow luminescence of $Na_2CaSn_2Ge_3O_{12}$: Eu^{3+},Dy^{3+} phosphor is produced by the recombination of electrons and holes, which escaped from the traps upon thermal excitation at room temperature and pass the released energy to Eu^{3+} (process 4), make the 4*f* electron of Eu^{3+} transition from ground state to excited state, and then returned to the ground state (process 5). The afterglow luminescence of this material can sustain long time, which was caused by slow escape velocity of the charge carriers. This persistent phosphor possesses much higher density of traps in comparison with $Gd_2O_2S{:}Eu^{3+},Mg^{2+},Ti^{4+}$ and $Y_2O_2S{:}\ Eu^{3+}, Mg^{2+},Ti^{4+}$ persistent phosphors.

5.1.3 Dy^{3+}

As mentioned previously, Dy^{3+} ions are normally codoped into Eu^{2+}-activated persistent phosphors including aluminates and silicates to increase the afterglow lifetime of Eu^{2+} through increasing the number of defects. In this part, we discuss the persistent phosphorescence properties of Dy^{3+}-doped phosphors in which the Dy^{3+} ion acts as the activator. In general, white persistent phosphorescence can be obtained by mixing blue, green, and red persistent phosphors. But it is too difficult to keep the phosphor retain the persistent white light for long time, because of the change of both persistent phosphorescence intensity and decay time of different phosphors. Dy^{3+} has the electron configuration of [Xe] $4f^{10}\ 6s^2$. Alternatively, Dy^{3+}-activated persistent phosphors can give white light emission by the transitions of $^4F_{9/2} \rightarrow {}^6H_{15/2}$ (470–500 nm) in the blue region and the $^4F_{9/2} \rightarrow {}^6H_{13/2}, {}^6H_{11/2}$ transitions in the yellow-red region (570–600 nm) as shown in Fig. 5.4B. Thus we believe that Dy^{3+}-activated persistent phosphors are highly promising for the realization of white afterglow for the lighting applications [79–95]. As shown in Fig. 5.4A, previous studies have been focused on only silicate hosts singly doped with Dy^{3+} that show white afterglow in $Sr_2MgSi_2O_7{:}Dy^{3+}$ [79], $Sr_2SiO_4{:}Dy^{3+}$ [84, 85], $CdSiO_3{:}Dy^{3+}$ [86, 87], $CaMgSi_2O_6{:}Dy^{3+}$ [80], $Ca_3MgSi_2O_8{:}Dy^{3+}$ [80], $Ca_2MgSi_2O_7{:}Dy^{3+}$ [80], $Ca_3SnSi_2O_9{:}Dy^{3+}$ [81], $Sr_2MgSi_2O_7{:}Dy^{3+}$ [82], $SrSiO_3{:}Dy^{3+}$ [83], $Sr_2SiO_4{:}Dy^{3+}$ [84], and $Ca_3MgSi_2O_8{:}Dy^{3+}$ [92]. Other hosts such as aluminates ($CaAl_2O_4{:}Dy^{3+}$ [93]), gallates ($SrGa_{12}O_{19}{:}Dy^{3+}$ [92]), germinates ($Ca_2GeO_4{:}Dy^{3+}$ [90], $CaZnGe_2O_6{:}Dy^{3+}$ [89]), stannates ($Ca_2SnO_4{:}Dy^{3+}$ [91]), and oxysulfides ($Y_2O_2S{:}Dy^{3+},Mg^{2+},Si^{4+}$ [94], $Y_2O_2S{:}Dy^{3+},Mg^{2+},Ti^{4+}$ [95]) are also reported, which demonstrate persistent phosphorescence time less than 3 h. Among the oxide hosts under study, $CdSiO_3{:}Dy^{3+}$ crystal demonstrates better afterglow and persistent phosphorescence properties. $CdSiO_3{:}Dy^{3+}$ phosphor emits white afterglow with emission at 420, 486, and 580 nm, which can last for over 5 h after stopping the ultraviolet excitation at 254 nm. In another study, white afterglow emission was also observed in oxysulfide hosts, such as Y_2O_2S phosphor tridoped with Dy^{3+},Mg^{2+},Si^{4+} or Dy^{3+},Mg^{2+},Ti^{4+}. Unfortunately the afterglow duration doesn't exceed 1h at most.

In the case of $CdSiO_3{:}Dy^{3+}$ persistent phosphors, Dy^{3+} ions are expected to occupy only the Cd^{2+} sites in the $CdSiO_3$ host. To gain insight into the influence of different traps on the origin of the afterglow of the $CdSiO_3{:}Dy^{3+}$ persistent phosphors and to study the nature of the traps created under ultraviolet radiation, the thermoluminescence measurements were performed as shown in Fig. 5.4C. Fig. 5.4C represents the thermoluminescence spectra of stoichiometric and $0.5-10$ mol % Dy^{3+} ion–doped $CdSiO_3$ samples. As is shown in Fig. 5.4C, it is safe to say that there are at least two types of traps present in the $CdSiO_3{:}Dy^{3+}$ persistent phosphors for at least two peaks in the thermoluminescence spectra, located at 418 and 358 K, respectively.

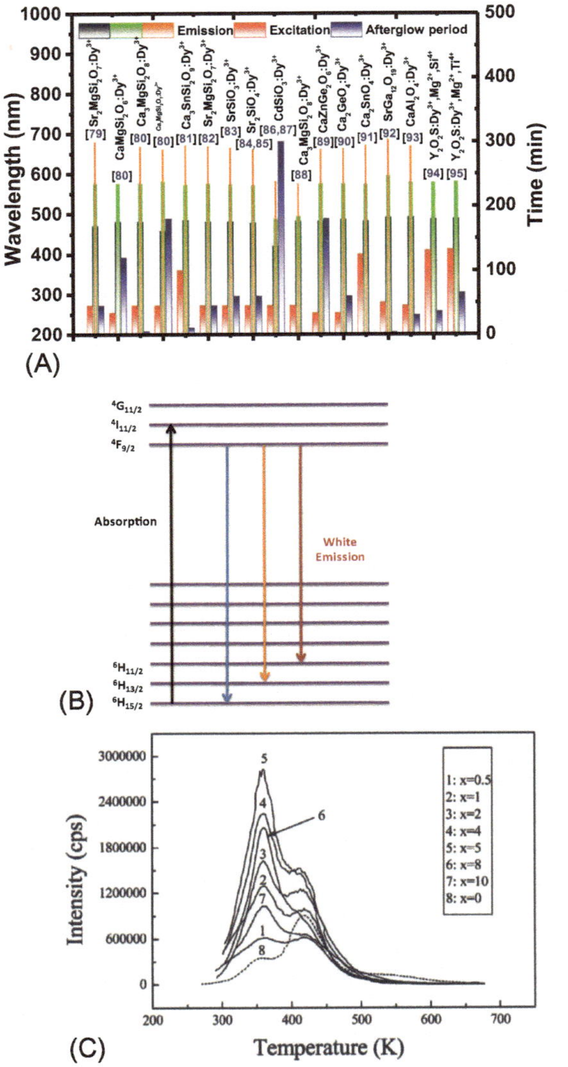

Fig. 5.4 (A) Dy^{3+}-activated persistent phosphors. (B) Energy level diagram of Dy^{3+}. (C) Thermoluminescence spectra of $CdSiO_3:Dy^{3+}$ phosphor with x mol% of Dy^{3+} ions.
Reproduced with permission from Y. Liu, B. Lei, C. Shi, Luminescent properties of a white afterglow phosphor $CdSiO_3:Dy^{3+}$. Chem. Mater., 2005, 17, 2108–2113, Copyright 2005, American Chemical Society.

The series of $Cd_{1-x}Dy_xSiO_3$ persistent phosphors have the same thermoluminescence position, indicating the presence of the same defects in the $CdSiO_3:Dy^{3+}$ and undoped samples. For the thermoluminescence peak at 358 K, the corresponding defect may be ascribed to the ($O_3 \equiv Si : O_3$) trap that usually appears in the high-purity SiO_2 material. To argue this assignation, the SiO_2 raw material was heated and measured under the same experimental conditions, and the result revealed that the SiO_2 raw materials have the same 358 K thermoluminescence peak. The amount of ($O_3 : Si : O_3$) trap is relatively smaller in the stoichiometric sample because the 358 K thermoluminescence peak in the glow curve 8 is weaker. Due to the nonequivalent substitution of Dy^{3+}

for Cd^{2+}, an excess of positive charge in the lattice must be compensated to maintain the electroneutrality of those phosphors. There are two possible patterns to fulfill the charge compensation of the Dy^{3+}-doped $CdSiO_3$ phosphor. One possible way is that two Dy^{3+} ions replace three Cd^{2+} ions to balance the charge of the phosphor, which create two Dy_{Cd} positive defects and one V_{cd} negative defect. Because the Dy_{Cd} positive site impurity trap is associated with the $(O_3 : Si : O_3)$ traps, the 358 K thermoluminescence peak intensity enhances greatly with the increase of Dy^{3+} concentration. The other possibility of the charge compensation is the vacancies of Cd^{2+} (V_{cd}) created during the synthesis process, which should be feasible because of the relatively high vapor pressure of the Cd^{2+} component.

Based on the aforementioned reason, another 418 K thermoluminescence peak is due to the vacancies of Cd^{2+} (V_{cd}) negative defect. On the basis of these glow curves in Fig. 5.4C, it is clear that the change of dopant concentration has a great effect on the thermoluminescence intensity, especially on the 358 K peak. The weaker 358 K thermoluminescence peak in the stoichiometric sample becomes predominant even when 0.5 mol % Dy^{3+} ions are introduced into the $CdSiO_3$ host lattice and the 358 K/418 K thermoluminescence intensity ratio increased consistently before the Dy^{3+} dopant up to 5%. This is because the introduction of Dy_{Cd} positive site impurity traps will inevitably increase the $(O_3 : Si : O_3)$ traps that are associated with Dy_{Cd} defect when Cd^{2+} ions are substituted by Dy^{3+}. As is seen from Fig. 5.4C, the thermoluminescence intensity reaches the maximal value when the Dy^{3+} dopant concentration is up to 5 mol %, indicating the electron traps have the optimal concentration at this dopant level.

5.1.4 Tb^{3+}

The electron configuration of Tb^{3+} ion is [Xe] $4f^9$ $6s^2$, and it always serves as an efficient activator giving green emission, owing to its $^5D_4$ to ^7F_J (J = 6, 5, 4, 3) transitions as shown in Fig. 5.5B [96–115]. As shown in Fig. 5.5A, the reported Tb^{3+}-doped green persistent phosphors include oxide phosphors of $CaYAl_3O_7$:Tb^{3+} [100], $SrLaGaO_4$:Tb^{3+} [105], Lu_2O_3:Tb^{3+} [96, 97], $CdSiO_3$:Tb^{3+} [114], $CaSnSiO_3$:Tb^{3+} [111], Ca_2SnO_4:Tb^{3+} [112], $CaZnGe_2O_6$: Tb^{3+} [110], and other hosts, which gives persistent phosphorescence shorter than 5 h. On the other hand, $SrGa_2O_4$:Tb^{3+} [103], Lu_2O_3:Tb^{3+},Ca^{2+} [108], and $NaAlSiO_4$:Tb^{3+}, Bi^{3+} [106] demonstrate efficient afterglow period of 20, 30, and 72 h, respectively.

In $SrGa_2O_4$:Tb^{3+} persistent phosphors, Tb^{3+} ions tend to prefer the Sr^{2+} sites due to the matching of their ionic radii. With respect to the intrinsic defects, because the high annealing temperatures may generate both O and Sr vacancies (V_O and V_{Sr}) in $SrGa_2O_4$, the roles of these defects are expected to be crucial to the persistent phosphorescence phenomena. In the literature, V_O is supposed to serve as an electron trap center, while V_{Sr} possibly acts as a hole trap center. The introduction of Tb^{3+} contributes to the increase number of the trapping centers without changing the depth of the traps that leads to an increase in the density of shallower traps and hence significantly improve the persistent phosphorescence properties of Tb^{3+}-doped $SrGa_2O_4$ samples. Meanwhile the increase in deeper one is critically important to prevent the thermal fading of carriers at room temperature.

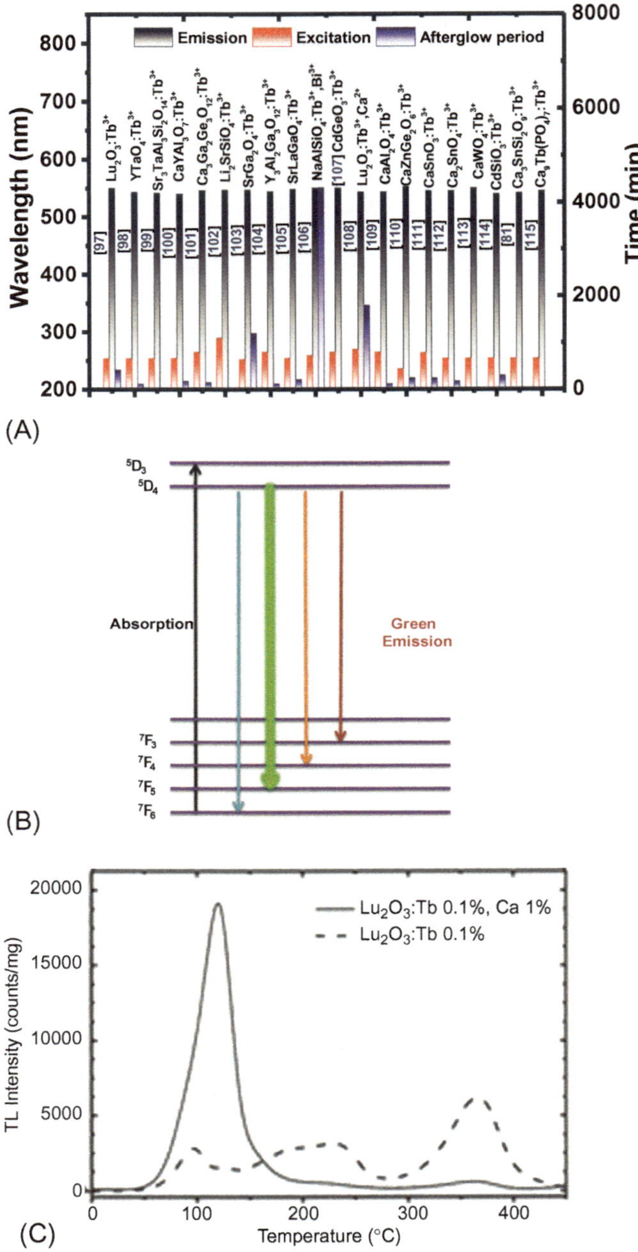

Fig. 5.5 (A) Tb^{3+}-activated persistent phosphors. (B) Energy level diagram of Tb^{3+}. (C) Thermoluminescence glow curves of the vacuum-sintered Lu_2O_3:Tb^{3+} and Lu_2O_3:Tb^{3+}:Ca^{2+} persistent phosphors.
Reproduced with permission from J. Trojan-Piegza, J. Niittykoski, J. Holsa, E. Zych, Thermoluminescence and kinetics of persistent luminescence of vacuum-sintered Tb^{3+}-doped and Tb^{3+}, Ca^{2+}-codoped Lu_2O_3 materials. Chem. Mater., 2008, 20, 2252–2261, Copyright 2008, American Chemical Society.

In $Lu_2O_3{:}Tb^{3+},Ca^{2+}$ persistent phosphors, the thermoluminescence glow curve (Fig. 5.5C) of the Lu_2O_3 materials doped with Tb^{3+} shows that the $Lu_2O_3{:}Tb^{3+}$ possess significantly more complex glow curves when compared with those containing Tb^{3+} and Ca^{2+}. The thermoluminescence glow curve exhibits three components located at ca. 100, 200, and 360 °C. The glow curve of $Lu_2O_3{:}Tb^{3+}$ reveals that there are several traps with rather different depths able to intercept electrons and/or holes freed upon the ultraviolet excitation. Some of the traps are relatively shallow (100 and 125 °C), while the others are much deeper (200 and 360 °C). Some indication as to the type of defects responsible for the carriers trapping in $Lu_2O_3{:}Tb^{3+}$ comes from the fact that the persistent phosphorescence appears only from materials heated at the reducing atmosphere of vacuum and only if the treatment is performed at temperatures around 1700 °C. Clearly, certain changes take place in the material when such high-temperature treatment is applied. Looking for these changes two effects, which are probable to happen, should be pointed out. First is the possibility to create vacancies, both metal and oxygen. Formation of such defects is strongly enhanced by the high temperature and vacuum. Both parameters facilitate the mobility of atoms, and those atoms that migrated to the surface may easily be expelled to vacuum, leaving increasing number of vacancies in the host. Another effect that is probable to occur upon high-temperature treatment is an aggregation of defects. This may take place since, as was already stated, the high-temperature enhances the migration of the various species present in the material.

Thermodynamically, aggregation and clustering of defects (vacancies, impurities, etc.) is usually advantageous. Evidently, appropriately high temperature not only increases the population of vacancies but also facilitates aggregation and clustering of the various defects present in the material. Summarizing, it can be expected that after the high-temperature vacuum treatment, cation and anion vacancies in $Lu_2O_3{:}Tb^{3+}$ ceramics will be rather abundant and their distribution may well not be random. Most of all, because of opposite net charges, one should expect that cation and anion vacancies may agglomerate and cluster. The presence of Tb^{3+} impurity may additionally enhance such aggregation and clustering. On the other hand the $Lu_2O_3{:}Tb^{3+},Ca^{2+}$ shows much simpler glow curve, essentially with only one band peaking around 125°C, suggesting the modifications in the trap structure due to Ca^{2+} codoping. The glow curve of $Lu_2O_3{:}Tb^{3+},Ca^{2+}$ consists essentially of only one, but slightly asymmetric, band peaking at 120 – 125 °C. The shape of these bands might suggest that they result from a superposition of two components, with the less intense one peaking slightly below 100 °C. Despite the very simple thermoluminescence band structure, the Ca^{2+} codoping enhanced the total thermoluminescence intensity of this material, implying that the density of shallower traps is increased, which is responsible for the lasting of green persistent phosphorescence for up to 30 h. In contrast, $NaAlSiO_4{:}Tb^{3+}$, Bi^{3+} persistent phosphors contain much dense shallow and deep traps owing to the replacement of Na^+ by Tb^{3+} and Bi^{3+}, resulting in longer persistent phosphorescence than $SrGa_2O_4{:}Tb^{3+}$ and $Lu_2O_3{:}Tb^{3+},Ca^{2+}$ persistent phosphors.

5.1.5 Sm^{3+}

The electron configuration of Sm^{3+} ion is [Xe] $4f^6$ $6s^2$, and it is prominent for producing an intense reddish-orange emission, which is ascribed to the typical transitions of $^4G_{5/2} \rightarrow {}^6H_J$ (J = 5/2, 7/2, 9/2) as shown in Fig. 5.6B. Indeed, important red-emitting phosphors doped with Sm^{3+} have been found in oxides and oxysulfides as shown in Fig. 5.6A [116–132]. Through systematic investigations of the afterglow properties of Sm^{3+}-doped phosphors, $BaGa_2O_4:Sm^{3+},Sr^{2+}$ [117], $Na_2CaSn_2Ge_3O_{12}:Sm^{3+}$ [122], $Ca_2SnO_4:Sm^{3+}$ [124], $Sr_2SnO_4:Sm^{3+},Dy^{3+}$ [125], $Zn_2P_2O_7:Sm^{3+},Li^+$ [129], $Y_2O_2S:Sm^{3+},Mg^{2+},Ti^{4+}$ [131], and $Sr_3Sn_2O_7:Sm^{3+}$ [127] phosphors have been developed, which gives intense reddish-orange afterglow from $^4G_{5/2}$ to ^6H_J (J = 5/2, 7/2, 9/2) transitions, and their afterglow could be seen with the naked eye in the dark clearly for more than 2 h and up to 7 h. On the other hand, orange afterglow was also observed in other hosts [116, 118–121]. However, the afterglow of these persistent phosphors only lasts for less than 1 h, which is shorter compared with the aforementioned Sm^{3+}-activated persistent phosphors.

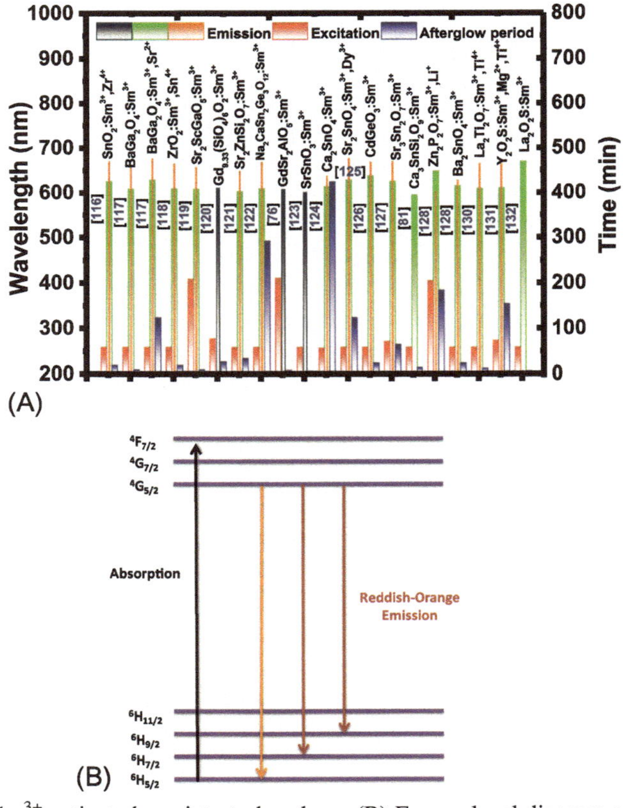

Fig. 5.6 (A) Sm^{3+}-activated persistent phosphors. (B) Energy level diagram of Sm^{3+}.

Among all of these persistent phosphors, the reddish-orange persistent phosphorescence time of Ca_2SnO_4:Sm^{3+} is more than 7 h, which is the longest red or reddish-orange long afterglow phosphor at the present time. The persistent phosphorescence mechanism of the long afterglow phosphor was studied by thermoluminescence and positron annihilation lifetime experiments. The experimental results suggest that the defects V_{Ca} and structure defects Sn^{3+} and Sn^{2+} may act as hole trapping centers, while the defects Sm_{Ca} may act as electron trapping centers. These trapping centers play an essential role for photoenergy storage in Ca_2SnO_4:Sm^{3+} persistent phosphors.

5.1.6 Tm^{3+}

Similar to Dy^{3+}, Tm^{3+} ions were used as trapping centers in several hosts. However, the research about Tm^{3+}-activated persistent phosphors is rare compared with that of Eu^{2+}-doped persistent phosphors. This is because of the difficulty of finding a suitable host for Tm^{3+} to yield the long persistent phosphorescence. Tm^{3+} has the electron configuration of [Xe] $4f^{13}$ $6s^2$. As shown in Fig. 5.7A, Tm^{3+} can be incorporated successfully into oxide and nitride hosts to generate blue and/or near-infrared persistent phosphorescence with short decay time $\leq$ 1 h.

Su et al. studied the blue persistent phosphorescence in Tm^{3+}-doped $Zn_2P_2O_7$ phosphor, which could last for about 1 h [133, 134]. The blue afterglow emission (365, 452, 478 nm) was ascribed to the transitions of $^1D_2 \rightarrow {}^3H_6$, $^1D_2 \rightarrow {}^3H_4$, and $^1G_4 \rightarrow {}^3H_6$ of Tm^{3+} (Fig. 5.7B). It should be noted that the aforementioned lanthanide-activated persistent phosphors usually emit light within the visible region. Despite high brightness and long afterglow lifetime, there are growing demands for persistent phosphors that gives afterglow emission in the near-infrared region. Liang et al. observed near-infrared (at 1195 nm) persistent phosphorescence for 20 min in Ca_2SnO_4:Tm^{3+} phosphor, which originates from the $^1G_4 \rightarrow {}^3H_4$ transition of Tm^{3+} [135]. Moreover, Sr_2SnO_4:Tm^{3+} showed near-infrared persistent phosphorescence (at 802 nm, $^3H_4 \rightarrow {}^3H_6$ transition) for 5 min [136]. As for Tm^{3+}-doped nitrides, Ba[$Mg_2Al_2N_4$]: Eu^{2+},Tm^{3+} showed near-infrared emission at 800 nm through the efficient energy transfer from Eu^{2+} to Tm^{3+} [137]. These persistent phosphors might be used as a persistent phosphorescence probe for in vivo imaging. However, we still need to improve the afterglow time and tune the excitation wavelength to meet the requirements of practical applications.

5.1.7 Pr^{3+}

Pr^{3+} is another ion for generating red light because it shows an emission by its electronic transitions of $^1D_2 \rightarrow {}^3H_4$, $^3P_0 \rightarrow {}^3H_6$, and $^3P_0 \rightarrow {}^3F_2$ (Fig. 5.8B) and it has the electron configuration of [Xe] $4f^3$ $6s^2$. The persistent phosphorescence properties of Pr^{3+}-doped persistent phosphors are summarized as shown in Fig. 5.8A [13, 76, 138–158].

As shown in Fig. 5.8A, the prolonged ($>$ 1 h) and efficient red afterglow of Pr^{3+} ions was observed in Pr^{3+}-doped $CaTiO_3$:Pr^{3+} [141], $Ca_2Ga_2GeO_7$:Pr^{3+} [143], and $La_2Ti_2O_7$:Pr^{3+} [145]. On the other hand, some reports observed that Pr^{3+} can emit

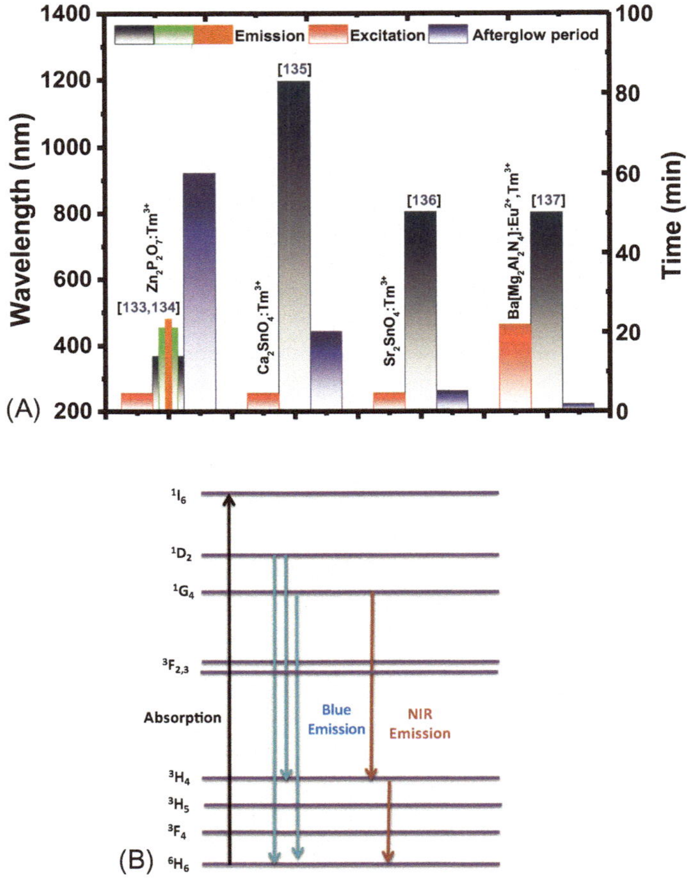

Fig. 5.7 (A) Tm^{3+}-activated persistent phosphors. (B) Energy level diagram of Tm^{3+}.

afterglow white light in $LiGa_5O_8$ [140], $Y_3Al_5O_{12}$ [148], $Ba_2Ga_2GeO_7$ [149], $CdGeO_3$ [150], Ca_2SnO_4 [151], and $CaSnO_3$ [152] phosphors. It is interesting to note that in the case of Ca_2SnO_4:Pr^{3+} [135] and $MgGeO_3$:Pr^{3+} [139] phosphors, Pr^{3+} ions show near-infrared (700–1200 nm) afterglow with long durations of 10 and 120 h, respectively. The near-infrared persistent phosphorescence properties in Ca_2SnO_4:Pr^{3+} and $MgGeO_3$:Pr^{3+} persistent phosphors are associated with the lattice defects (e.g., oxygen vacancies) as result of the substitution of Pr^{3+} to Ca^{2+} and Mg^{2+}, respectively. In addition, we can deduce that the trap distribution in the Ca_2SnO_4 and $MgGeO_3$ hosts is significantly affected by the introduction of Pr^{3+} ions, where the trap distribution in the phosphors becomes deeper and hence the near-infrared persistent phosphorescence can be clearly observed. Most of the reported persistent phosphors show visible and near-infrared persistent phosphorescence, and the number of persistent phosphors exhibiting ultraviolet persistent phosphorescence is not abundant. As known, two classes of centers, emitters and traps, are required in persistent phosphors. Emitters

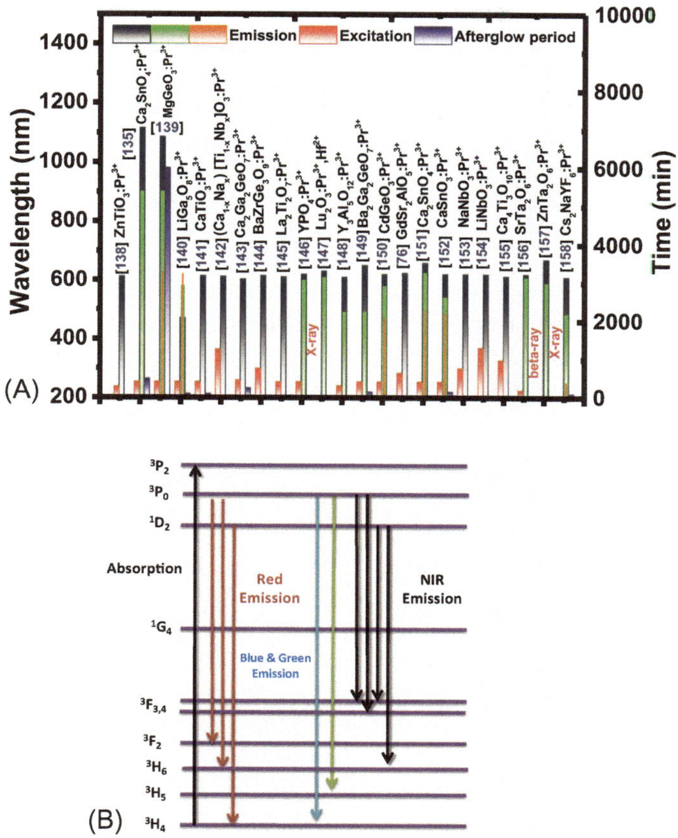

Fig. 5.8 (A) Pr^{3+}-activated persistent phosphors. (B) Energy level diagram of Pr^{3+}.

are optical centers that luminate after being excited, whereas traps, which are usually associated with defects in host lattices, store excitation energy and then release it slowly to the emitters by virtue of thermal stimulations. Since the emitter determines the emission wavelength, the first required condition for the design of ultraviolet persistent phosphorescence is the choice of a suitable emitter that can luminate in the ultraviolet region. It is noted that some rare earth ions (e.g., Pr^{3+}) possess strong, high-energy, interconfigurational transitions, which thus could be employed as candidate emitters in ultraviolet persistent phosphorescence. The second required condition is the selection of the candidate host materials with a large band gap and appropriate traps. In this respect, fluoride crystalline lattices can be considered because of their relatively large band gaps and the easy creation of anionic defects. The third condition is that the material system can be efficiently activated (i.e., charged) by suitable excitation sources. Given the large band gap required for the host, x-ray irradiation may be appropriate. Another requirement relates to the persistent time and intensity of the ultraviolet emission, which should be as long as possible and strong enough to satisfy some practical applications. Among Pr^{3+}-activated persistent phosphors, $Cs_2NaYF_6:Pr^{3+}$ phosphor gives strong ultraviolet persistent phosphorescence at 250 nm for 2 h after x-ray irradiation [158]. The development of Pr^{3+}-doped persistent phosphors in the ultraviolet and near-infrared regions not only expands the scope of afterglow phosphors but also offers a new opportunity for sensing and biomedicine [159].

5.1.8 Ce^{3+}

Ce^{3+} is the second important afterglow activator of the trivalent lanthanide series with the electronic configuration of [Xe] $4f^1$ $6s^2$, which produces a tunable broad emission band in the visible region, owing to the strong crystal field dependence of its 5d–4f transition energy [160–179]. As shown in Fig. 5.9A, Ce^{3+}-doped Lu_2SiO_5 [160], $Sr_2Al_2SiO_7$ [174], $CaAl_4O_7$ [162, 163], $Sr_3Al_2O_5Cl_2$ [165], and $BaMg_2Si_2O_7$ [175] show blue emission with long persistent times > 3 h after irradiation with ultraviolet light (250–360 nm). Ce^{3+} emits green persistent phosphorescence (> 2 h) when doped in $Y_3Al_{5-x}Ga_xO_{12}$ [166], $Gd_3Al_2Ga_3O_{12}$ [167, 168], $Ca_2Al_2SiO_7$ [169], $SrAl_2O_4$ [170, 171], $CaAl_2O_4$ [172], and $Y_3Sc_2Ga_3O_{12}$ [173] phosphors. One breakthrough in increasing the afterglow lifetime of Ce^{3+} ions was made by Tanabe et al. [164], who observed that the green afterglow duration of $Y_3Al_2Ga_3O_{12}$ codoped with Ce^{3+},Yb^{3+} can be extended up to 138.8 h because of the presence of Yb^{3+} in the host, which increases the trap density and depth. Most notably, this green persistent phosphorescence can be activated under the excitation by both blue light (i.e., 460 nm) and ultraviolet light. The Yb^{3+} ions act as electron traps by capturing an electron.

In this work, it was found that the Yb^{3+} trap makes a much deeper electron trap with a 1.01 eV depth than the Cr^{3+} electron trap with a 0.81 eV depth. This deeper Yb^{3+} trap provides a much slower detrapping rate of filled electron traps than the Cr^{3+}-codoped persistent phosphors (Fig. 5.9B and C). In the case of CaB_2O_4 doped with Ce^{3+} [161], researchers observed for the first time that Ce^{3+} ions emit strong ultraviolet persistent phosphorescence with afterglow duration of 15 h, where the crystal field effect is weak.

5.1.9 Yb^{2+}

Similar as Eu^{2+}, divalent Yb^{2+} ions have [Xe] $4f^{13}$ $6s^2$ electronic configuration that can produce a tunable broad emission band due to the strong crystal field dependence of its 5d–4f transitions. Eu^{2+} has been used as efficient emitter in several hosts, while the numbers of reports on Yb^{2+}-activated persistent phosphors are scarce, as shown in Fig. 5.10 [180–186]. In Yb^{2+}- and Dy^{3+}-codoped $BaAl_2O_4$ blue persistent phosphors, Yb^{2+} and Dy^{3+} enter into Ba^{2+} sites in which Yb^{2+} and Dy^{3+} serve as activators and trap centers, respectively. In addition, Yb^{2+} gives a broadband emission centered at 495 nm under the excitation of 365 nm. This afterglow emission can be assigned to the Yb^{2+} $4f^{13}5d^1 \rightarrow 4f^{14}$ transition, which lasts for 1.5 h [182]. Green persistent phosphorescence was observed in $SrS:Yb^{2+}$ [185], $Sr_4Al_{14}O_{25}:Yb^{2+},Dy^{3+}$ [180], and $SrAl_4O_7$: Yb^{2+},Dy^{3+} [180] phosphors, while red persistent phosphorescence

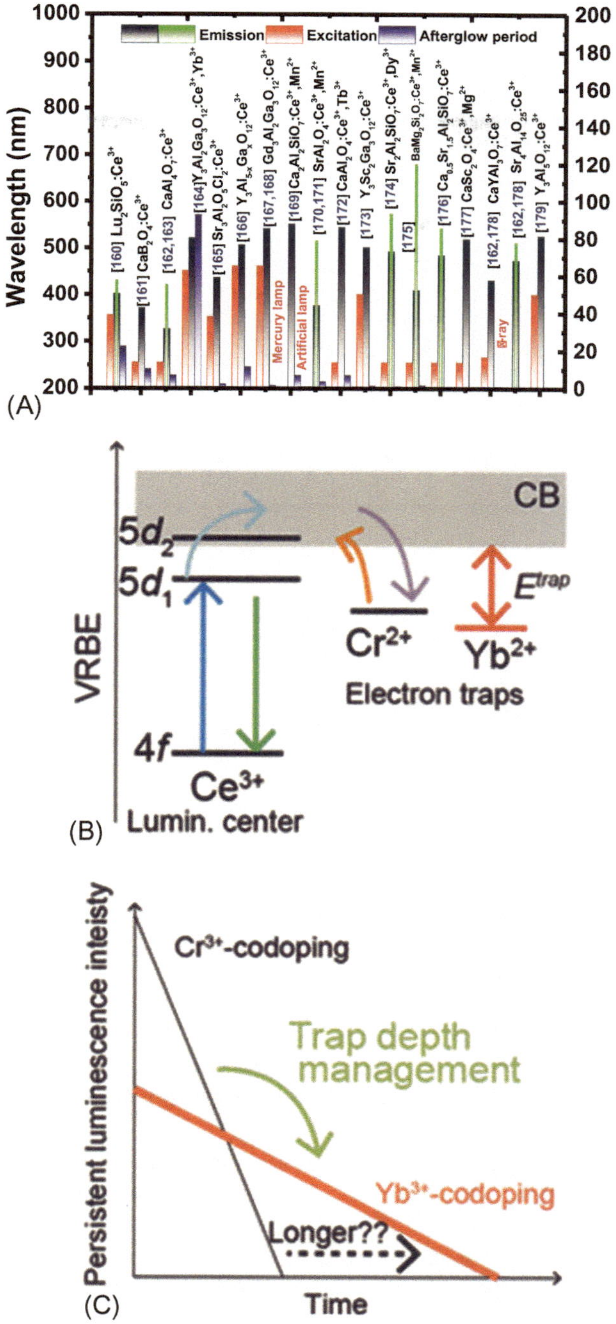

Fig. 5.9 (A) Ce^{3+}-activated persistent phosphors. (B) Schematic energy diagram of the relationship between Ce^{3+}:4f,5d levels and Cr^{3+} and Yb^{3+} electron traps and (C) schematic image of the persistent phosphorescence decay profile controlled by trap depth management. Reproduced with permission from J. Ueda, S. Miyano, S. Tanabe, Formation of deep electron traps by Yb^{3+} codoping leads to super-long persistent luminescence in Ce^{3+}-doped yttrium aluminum gallium garnet phosphors. ACS Appl. Mater. Interfaces, 2018, 10, 24, 20652–20660, Copyright 2018, American Chemical Society.

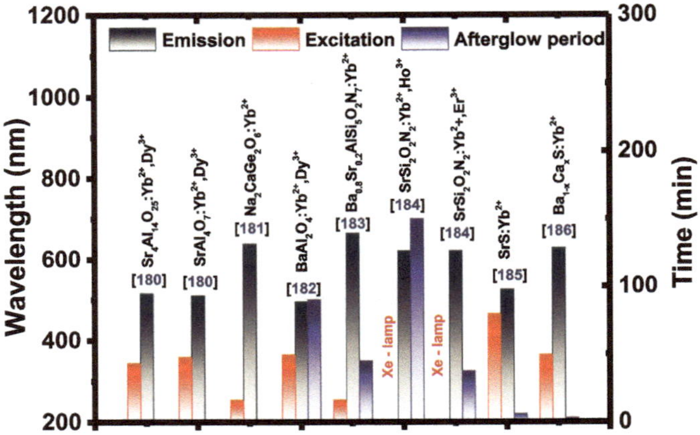

Fig. 5.10 Yb^{2+}-activated persistent phosphors.

was observed in $Na_2CaGe_2O_6:Yb^{2+}$ [181], $Ba_{0.8}Sr_{0.2}AlSi_5O_2N_7:Yb^{2+}$ [183], $Ba_{1-x}Ca_xS:Yb^{2+}$ [186], $SrSi_2O_2N_2$: Yb^{2+},Ho^{3+} [184], and $SrSi_2O_2N_2$: Yb^{2+},Er^{3+} [184]. The developed $SrSi_2O_2N_2:Yb^{2+},Ho^{3+}$ persistent phosphors with the large trap depth and strong crystal field effect showed great promise as optical information storage media, displaying efficient red persistent phosphorescence for 2.5 h.

5.1.10 Nd^{3+}

As for Nd^{3+}, Nd^{3+} has [Xe] $4f^4$ $6s^2$ electronic configuration. As shown in Fig. 5.11B, Nd^{3+} can emit blue and near-infrared emissions, which are assigned to the transitions of $^2P_{1/2}$, $^4G_{11/2} \rightarrow {}^4I_{9/2}$ and $^4F_{5/2,3/2} \rightarrow {}^4I_{13/2}$, $^4I_{11/2}$, $^4I_{9/2}$, respectively. However, most published reports (Fig. 5.11A) about Nd^{3+}-doped persistent phosphors focused only on the near-infrared persistent phosphorescence of Nd^{3+} in some hosts such as Sr_2SnO_4 and Ca_2SnO_4, which exhibits near-infrared persistent phosphorescence with lifetime of around 1 h [135, 187]. In addition, the near-infrared persistent phosphorescence can be also obtained by transferring the persistent phosphorescence energy from one activator (e.g., Eu^{2+} and Ce^{3+}) to another. Strong afterglow of $SrAl_2O_4:Eu^{2+},Dy^{3+},Nd^{3+}$ at 882 nm attributed to the Nd^{3+} 4f–4f transition ($^4F_{3/2} \rightarrow {}^4I_{9/2}$) was observed after irradiation by ultraviolet light at 365 nm owing to the persistent energy transfer from excited Eu^{2+} ions to Nd^{3+} ions [8]. In $Y_3Al_2Ga_3O_{12}:Ce^{3+},Nd^{3+}$, Nd^{3+} gives near-infrared persistent phosphorescence (808 nm $^4F_{5/2} \rightarrow {}^4I_{13/2}$, 1064 nm $^4F_{3/2} \rightarrow {}^4I_{11/2}$, 1335 nm $^4F_{3/2} \rightarrow {}^4I_{13/2}$) with a lifetime around 10 h [188]. Based on these results, Nd^{3+} and its hosts are not suitable to generate ultralong visible or near-infrared persistent phosphorescence to fulfill the practical application of in vivo bioimaging field. This is because of the existence of low density and less deeper traps in these hosts.

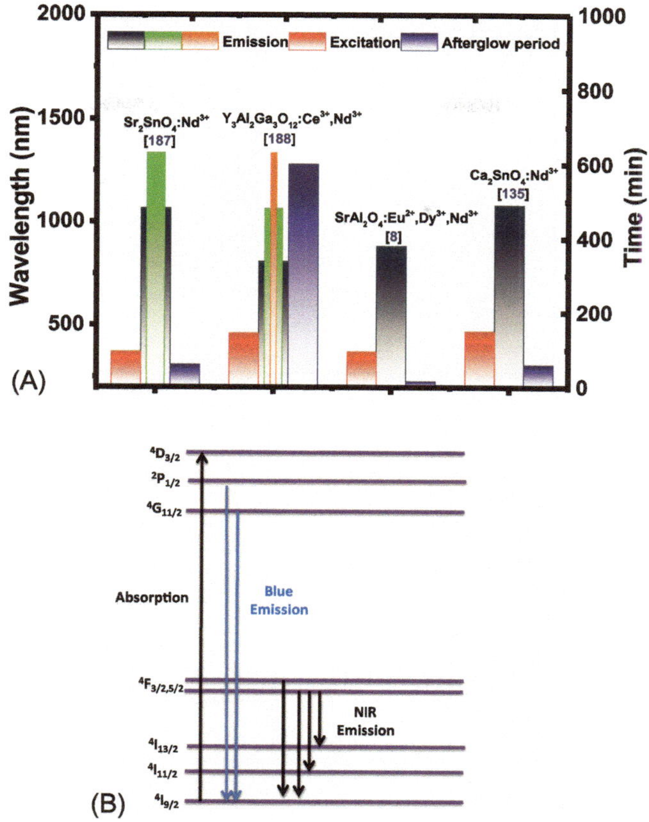

Fig. 5.11 (A) Nd^{3+}-activated persistent phosphors. (B) Energy level diagram of Nd^{3+}.

5.1.11 Er^{3+}

Er^{3+} has [Xe] $4f^{12}$ $6s^2$ electronic configuration, and it can exhibit green (550 nm), red (668), and near-infrared (1550 nm) luminescence as result of ${}^4S_{3/2}\rightarrow{}^4I_{9/2}$, ${}^4F_{9/2}\rightarrow{}^4I_{9/2}$, and ${}^4I_{11/2}\rightarrow{}^4I_{9/2}$ transitions, respectively (Fig. 5.12B) [60–62]. Persistent phosphors doped with Er^{3+} can be rarely found. Codoping of Er^{3+} with Yb^{3+} has been shown to be efficient in the generation of green and red upconversion under 980-nm excitation. As shown in Fig. 5.12A, red persistent phosphorescence (665 nm) for a few seconds can be observed in Er-doped ZrO_2 phosphor after irradiation by a 980-nm laser diode [189]. Er^{3+} also shows visible (red and green [550 and 668 nm]) and near-infrared (1533 nm) persistent phosphorescence in Ca_2SnO_4 phosphor, which lasts for 156 s and 10 h, respectively [135, 190]. The near-infrared persistent phosphorescence at 1530 nm can be activated via the persistent energy transfer process from Eu^{2+} to Er^{3+} in $SrAl_2O_4{:}Eu^{2+},Dy^{3+},Er^{3+}$ phosphor, which lasts for 100 min [7]. Under excitation of 455 nm, Cr^{3+} can also transfer its energy to Er^{3+}, producing near-infrared persistent phosphorescence at 1533 nm with duration of around 10 h in $Y_3Al_2Ga_3O_{12}$

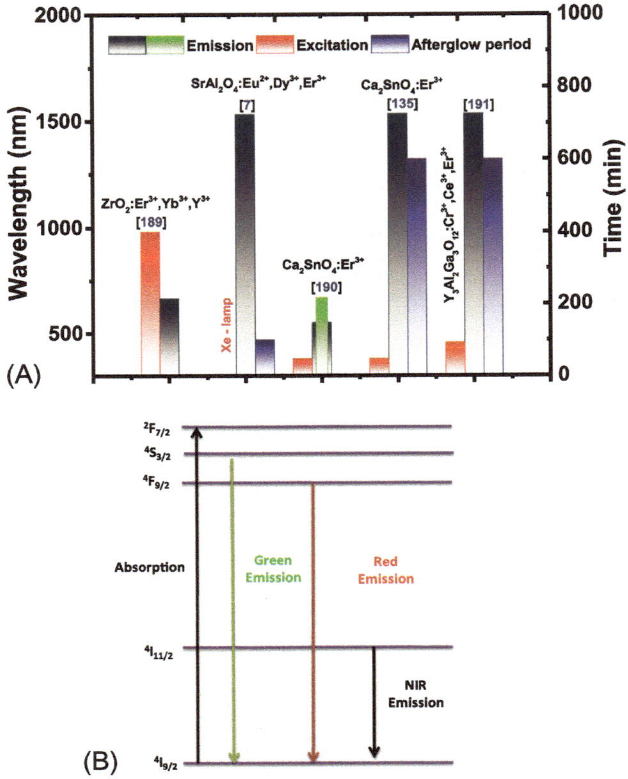

Fig. 5.12 (A) Er^{3+}-activated persistent phosphors. (B) Energy level diagram of Er^{3+}.

phosphor [191]. Finally, we can conclude that rare earth ion-doped persistent phosphors are promising materials for visible persistent phosphorescence with superlong afterglow period owing to the matching between the depth of traps and the luminescent excited state of rare earth ions as well as the possibility of creation of lattice defects with high density of traps. Due to the difficulty of extending of trap depth and make it more deeper, rare earth ion-activated persistent phosphors are not able to generate near-infrared persistent phosphorescence with long afterglow time. To achieve superlong near-infrared afterglow duration, efforts have been devoted in the development of the transition metal ion-doped materials, which will be discussed in the following parts.

5.2 Transition metal-activated persistent phosphors

Transition metal ions have the valence electrons residing in the incomplete d shell. Unlike 4f electrons, shielded by the outer shell, d electrons suffer large perturbation from the outer crystal field, much larger than that of spin–orbit interaction. The crystal field from coordination ligands dictates emission profile in transition metal ion–doped compounds as the ligands can remove the degeneracy of the orbitals and split them into different energy levels. Similar as lanthanide ions, transition metal ion-doped inorganic persistent phosphors have also been widely investigated by researchers. In these phosphors, transition metal ions like Mn^{2+}, Mn^{4+}, Cr^{3+}, and Cu^{+} serve as luminescent centers. The spectroscopic properties of Mn^{2+}, Mn^{4+}, and Cr^{3+} ions are the most widely studied among all transition metal ions. In this section, we focus on the persistent phosphors doped with transition metal ions (e.g., Mn^{2+}, Mn^{4+},Cr^{3+}, Cu^{+},Zr^{4+}, Ti^{4+}, and Ti^{3+}).

5.2.1 Mn^{2+}

Mn^{2+} has the electronic configuration of [Ar] $3d^5$ $4s^2$. More importantly, as shown in Fig. 5.13B, Mn^{2+} ion is able to emit both green and red light when it is tetrahedrally and octahedrally coordinated, respectively. The emission from the Mn^{2+} transition is ascribed to the parity-forbidden 3d–3d intraconfiguration transition from the lowest excited level $^4T_1(^4G)$ to the ground state $^6A_1(^6S)$. It is well known that Mn^{2+}-activated persistent phosphors are generally characterized by the high luminescence efficiency and brilliant luminescence color. Therefore Mn^{2+} ion has been considered as an efficient emitter in persistent phosphors. In the following part, we will discuss the luminescence properties of Mn^{2+}-activated persistent phosphors as shown in Fig. 5.13A [192–208]. As typical examples, $CaMgSi_2O_6$:Mn^{2+} [193], $MgGeO_3$:Mn^{2+} [205] and $CaZnGe_2O_6$:Mn^{2+} [206], $Ca_3(PO_4)_2$:Mn^{2+}/Dy^{3+} [200], and $Ca_{9.5}Mn(PO_4)_7$:Mn^{2+} [201] phosphors display good red persistent phosphorescence, while $Zn_3(PO_4)_2$:Mn^{2+}/Ga^{3+} [199] phosphor presents green and red persistent phosphorescence.

The reason is that two different coordination numbers (CN = 4 and 6) for Mn^{2+} are present in $Zn_3(PO_4)_2$:Mn^{2+}/Ga^{3+}, while only one coordination number (CN = 6) is available in red persistent phosphorescence. Other efficient Mn^{2+}-activated green persistent phosphors were also reported including $Ca_2Al_2SiO_7$:Ce^{3+},Mn^{2+} [195], Zn_2GeO_4:Mn^{2+} [203], Li_2ZnGeO_4:Mn^{2+} [202], Mg_2SnO_4: Mn^{2+} [207], and $Zn_3Ga_2Ge_2O_{10}$:Mn^{2+} [198]; however, $CdSiO_3$:Mn^{2+} [192], $Y_3Al_5O_{12}$:Mn^{2+} [196], and $Cd_2Ge_7O_{16}$:Mn^{2+} [204] present orange-red persistent phosphors. Apart from Mn^{2+}-doped oxides, Mn^{2+}-activated persistent phosphors are also observed in Mn^{2+}-doped glasses, glass ceramics, and nitrides (AlN:Mn^{2+}) [208]. In these Mn^{2+}-doped persistent phosphors, the lattice defects (e.g., types, depth, and density of traps) play a significant role to control the afterglow period. It can be clearly observed that all red persistent phosphors possess persistent phosphorescence time less than 3 h, suggesting that the depth and density of traps are not enough to prolong the persistent phosphorescence for longer time. However, some of green persistent phosphors (e.g., $Ca_2Al_2SiO_7$:Ce^{3+},Mn^{2+}, $Zn_3Ga_2Ge_2O_{10}$:Mn^{2+}, Li_2ZnGeO_4:Mn^{2+}, and Mg_2SnO_4:Mn^{2+}) demonstrate higher afterglow time (4–10 h) than that of red persistent phosphors.

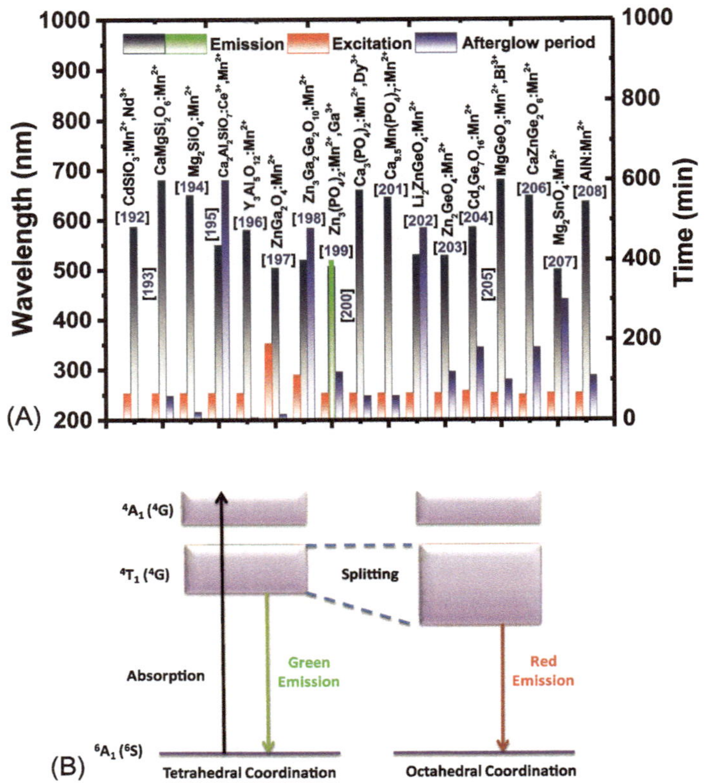

Fig. 5.13 (A) Mn^{2+}-activated persistent phosphors. (B) Energy level diagram of Mn^{2+} in tetrahedral and octahedral coordination.

5.2.2 *Cr^{3+}*

Cr^{3+} ion is one of the most investigated ions for persistent phosphors. The outer layer electron configuration of Cr^{3+} is [Ar] $3d^5\,4s^1$. According to the ligand field theory, the spectra terms of d^3 in a strong crystal field (i.e., octahedral sites) will split into $^4A_{2g}$, ^2E_g, $^2T_{1g}$, $^2T_{2g}$, and so on. Consequently, the ^2E and $^4T_2$ energy levels are defined as the lowest excited states that determined by the strength of the crystal field as shown in Fig. 5.14B. Commonly the emission of Cr^{3+} shows characteristic sharp R-line, due to the $^2E \rightarrow {}^4A_2$ transition. Meanwhile the transition of $^4T_2 \rightarrow {}^4A_2$ will result in broadband emission. Both $^2E \rightarrow {}^4A_2$ (R-line) and $^4T_2 \rightarrow {}^4A_2$ band are inhomogeneously broadened due to site-to-site disordering. The past two decades have witnessed a great success in developing persistent phosphors featuring visible light emission with high brightness and long afterglow lifetime. However, it was not until a few years ago that researchers discovered the persistent phosphors with near-infrared emission (700–2500 nm) that lasts from several minutes to several hundred hours. Recently, there has been a steady increase in the investigation of near-infrared emitting persistent phosphors,

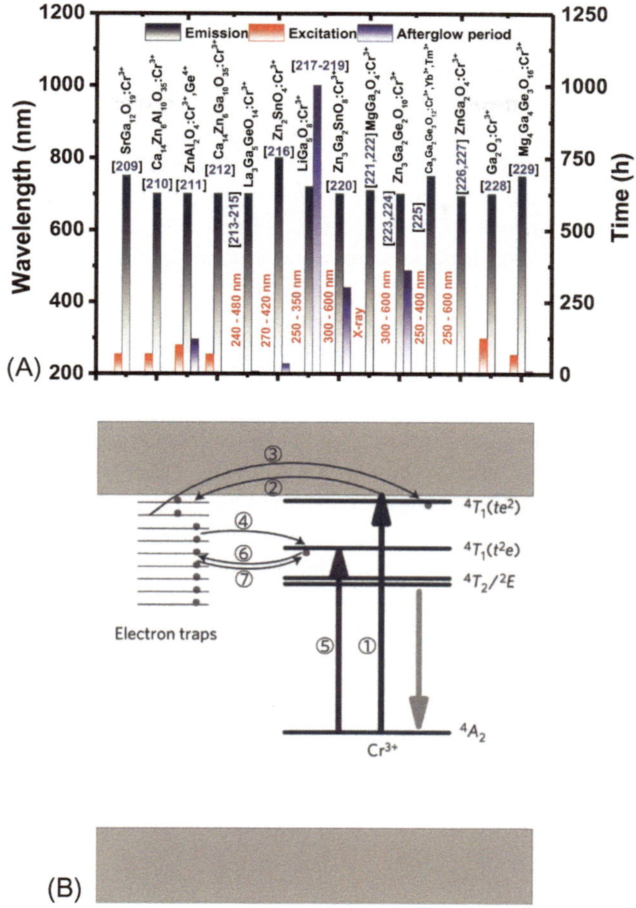

Fig. 5.14 (A) Cr^{3+}-activated persistent phosphors. (B) Schematic for mechanisms of persistent phosphorescence in $Zn_3Ga_2Ge_2O_{10}$:Cr^{3+} persistent phosphors.
Reproduced with permission from Z. W. Pan, Y. Y. Lu, F. Liu, Sunlight-activated long-persistent luminescence in the near-infrared from Cr^{3+}-doped zinc gallogermanates. Nat. Mater., 2012, 11, 58–63, Copyright 2012, Nature Publishing Group.

principally due to their potential applications in fields including labeling and security and defense technologies. In vivo imaging using near-infrared emitting persistent phosphors has been demonstrated recently, and it has stimulated tremendous interest in the search of materials and optimization of imaging techniques based on persistent phosphors. Optical imaging is a promising method to visualize biological tissues because it is a radioactive material-free technique. The near-infrared persistent phosphors are used because of the high transmittance of these signals through biological tissues. Cr^{3+} ion is usually used as the activator due to its capability to produce red and near-infrared emissions. Cr^{3+} ion is an ideal near-infrared emitter when doped in many oxide hosts [209–229]. As shown in Fig. 5.14A, Bessiere et al. first reported the persistent phosphorescence peaking at 695 nm in $ZnGa_2O_4$:Cr^{3+} crystals [226, 227]. In 2012, Pan et al. used $Zn_3Ga_2Ge_2O_{10}$ as the host and achieved a superlong near-infrared afterglow emission time of 360 h, which can be excited by solar spectrum [223, 224]. Qiu et al. have further improved the afterglow properties by partially substituting Zn and Sn for Ga to form a Zn–Ga–Sn–O solid solution. They demonstrated that $Zn_3Ga_2SnO_8$:0.5Cr^{3+} powder shows bright and persistent phosphorescence over 300 h in the near-infrared region, which may allow for an enabling long-term, reproducible, real-time, and reliable structural imaging of deep tissues [220]. Interestingly, all the known Cr^{3+}-activated persistent phosphors are gallate compounds. However, $LiGa_5O_8$:Cr^{3+} shows near-infrared persistent phosphorescence time of about 1000 h under ultraviolet excitation (250–350 nm), which is the longest near-infrared persistent phosphorescence so far [217–219]. In addition to $La_3Ga_5GeO_{14}$:Cr^{3+} [213–215] phosphor, the afterglow properties of many other Cr^{3+}-activated gallate persistent phosphors, such as $MgGa_2O_4$:Cr^{3+} [221, 222], $Ca_3Ga_2Ge_3O_{12}$:Cr^{3+} [225] and $SrGa_{12}O_{19}$:Cr^{3+} [209], $ZnGa_2O_4$:Cr^{3+} [226, 227], Ga_2O_3:Cr^{3+} [228], $Ca_{14}Zn_6Ga_{10}O_{35}$:Cr^{3+} [210], and $Mg_4Ga_4Ge_3O_{16}$:Cr^{3+} [229] have also been investigated because of the capability of Cr^{3+} ions entering into the Ga^{3+} sites in distorted octahedral coordination. The predominance of Cr^{3+}-activated gallates might suggest that only gallates can be used as hosts in Cr^{3+}-doped near-infrared persistent phosphors, but recently Qiu et al. developed a series of efficient near-infrared Cr^{3+}-doped nongallate persistent phosphors (Zn_2SnO_4:Cr^{3+}) with broad emission band (650–1200 nm, peaking at 800 nm) and long afterglow time >35 h [216]. Moreover, $Ca_{14}Zn_6Al_{10}O_{35}$:Cr^{3+} [212] and $ZnAl_2O_4$:Cr^{3+} [211] were also investigated for near-infrared persistent phosphors.

5.2.3 Mn^{4+}

Mn^{4+} is another highly desirable near-infrared activator because its 3d electronic configurations (i.e., [Ar] $3d^7$ $4s^2$) allows for strong near-infrared emission from 600 to 800 nm in different hosts, which also strongly depends on the crystal field environment of the host lattices. Due to the small ionic radius of Mn^{4+}, it is usually difficult to find suitable host for Mn^{4+} to create lattice defects and hence generate persistent phosphorescence. So the number of Mn^{4+}-doped persistent phosphors is limited. As shown in Fig. 5.15, Qiu et al. fabricated a series of Mn^{4+}-doped $MAlO_3$ (M = La, Gd) persistent phosphors with the emission maximum at around 730 nm and persistent phosphorescence duration of around 20 h [230]. In comparison, they found that Mn^{4+} gives near-infrared emission for less than 3 h in Mg_2GeO_4 and La_2MgGeO_6 phosphors [231, 232]. In Ge^{4+}- and Mn^{4+}-codoped $MAlO_3$ (M = La, Gd) persistent phosphors, Ge^{4+} and Mn^{4+} replace Al^{3+}. Ge^{4+} and Mn^{4+} serve as luminescent and trapping centers, respectively. So the lattice defects are created efficiently, and the near-infrared persistent phosphorescence was observed for 20 h. However, the replacing of Mg^{2+} by Yb^{3+} in Mn^{4+}-doped Mg_2GeO_4 doesn't show significant enhancement in the near-infrared persistent phosphorescence, and the afterglow time is less than 3 h, suggesting

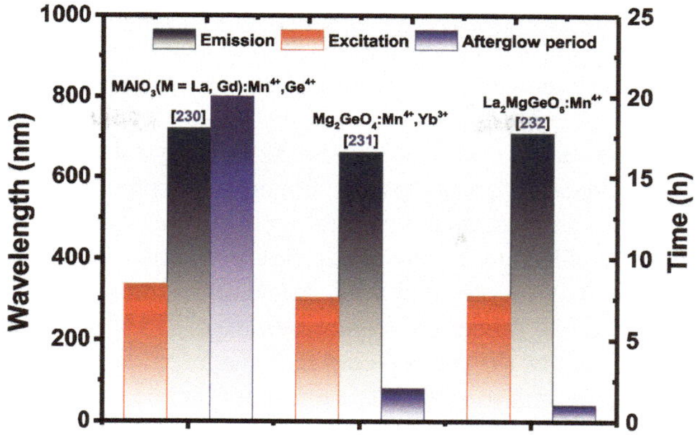

Fig. 5.15 Mn^{4+}-activated persistent phosphors.

that it is better to replace the central ions of tetrahedral sites in the host lattice by trapping center ions to make the trap depth more deeper.

5.2.4 Ti^{4+}

The electronic configuration of Ti^{4+} is [Ar] $3d^2$ $4s^2$. Ti^{4+} ion has been widely used as a codopant because of its high oxidation number that stabilizes oxygen vacancies when it replaces lower oxidation number ions in the host. However, we find that Ti^{4+} (Fig. 5.16) can be also used as phosphorescent center in Mg_2SnO_4 [233] and ZrO_2 [234] hosts by emitting blue broad emission, which lasts for 2.5 h in Mg_2SnO_4: Ti^{4+}. In Mg_2SnO_4:Ti^{4+}, Mg_2SnO_4 exhibits green persistent phosphorescence, which originates from oxygen vacancies. When Ti^{4+} ions were doped, the green persistent phosphorescence turns to bluish-white persistent phosphorescence under ultraviolet irradiation. The centers responsible for the green emission are considered to be the

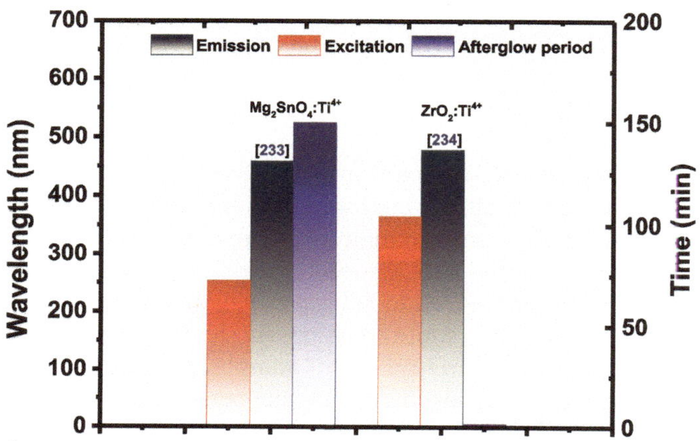

Fig. 5.16 Ti^{4+}-activated persistent phosphors.

F centers (oxygen vacancies), and the blue centers are the TiO_6 complex. As result of the substitution of Mg by Ti^{4+}, defects such as Sn_{Mg}, O_i, and V_O have been created in the band gap with different depths that is responsible for the observed persistent phosphorescence.

The thermoluminescence curve of Mg_2SnO_4:Ti^{4+} crystal exhibits three peaks locating at 85, 146, and 213 °C, respectively. The most intensive peak has a maximum at 146 °C. The thermoluminescence glow curve also reveals that there may be three trap levels with rather different depths in Mg_2SnO_4:Ti^{4+} that intercept the electrons and/or holes freed upon ultraviolet excitation. Some of the traps are relatively shallow (85 and 146 °C), while the others are somewhat deeper (213 °C). These trap levels (i.e., Sn_{Mg}, O_i and V_O) allow the trapped carriers to be freed slowly by thermal energy at room temperature, leading to long persistent luminescence.

5.2.5 Cu^+

The electronic configuration of Cu^+ is [Ar] $3d^{10}$ $4s^1$. The emission of Cu^+ that from excited s state to ground d state ($d^9s \rightarrow d^{10}$) transition is partially allowed in solids. The s excited state of Cu^+ usually splits into singlet ($^1T_{2g}$, ^1E_g) and triplet ($^3T_{2g}$, ^3E_g) parts, and then the triplets may be split further by spin-orbit interactions. Hence, Cu^+ exhibits broad excitation in ultraviolet and visible regions. Cu^+ is the first luminescent center that has been used to achieve persistent phosphorescence in ZnS host [1]. In ZnS host, Cu^+ substitutes Zn, demonstrating long green- persistent phosphorescence centered at 526 nm for 3 h after removing the ultraviolet excitation (360 nm). However, in 2002, Qiu et al. observed persistent phosphorescence at 510 nm for 1 h in a Sn^{2+},Cu^{2+}-codoped Na_2O–CaO–SiO_2 glass at room temperature under ultraviolet illumination of 254 nm [235]. They observed that after compressing the glass under 3, 6, and 9 GPa, the persistent phosphorescence shifted to 465 nm and its decay rate became shorter. The optical absorption spectra of the samples changed after compression, showing that Cu^{2+} ions were reduced to Cu^+ ions. The high-pressure treatment also resulted in a lower energy shift in the absorption edge. They suggested that Sn^{2+} ions act as hole trapping centers, while oxygen vacancies surrounded by Ca^{2+} ions and active sites in the glass matrix serve as electron trapping centers.

5.2.6 Ti^{3+}

Ti^{3+} has been widely used as defect center in several hosts. In fact the intrinsic thermal defects and extrinsic defects often work together to improve the trap distribution in most persistent phosphors. The electronic configuration of Ti^{3+} is [Ar] $3d^3$ $4s^2$. Ti^{3+} has many unique properties such as broad absorption and emission bands as well as color tuning from blue and yellow to red and even near-infrared depending on the chemical environment, enabling white emission. As shown in Fig. 5.17, Holsa et al. studied the optical properties of ZrO_2:Ti^{3+},Lu^{3+} phosphor [236]. In this phosphor, Ti^{3+} serves as luminescent center producing broad blue persistent phosphorescence with duration of around 800 s. In addition, Ti^{3+} emits broad yellow afterglow emission with duration of 5 h in Gd_2O_2S and Y_2O_2S [237–239]. The introduction of Mg^{2+} obviously

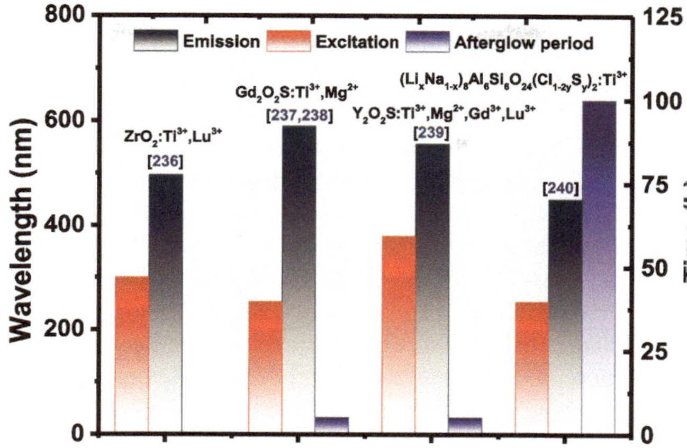

Fig. 5.17 Ti^{3+} -activated persistent phosphors.

extends the afterglow time. Mg^{2+} ion replaces Gd^{3+} and Y^{3+} in Gd_2O_2S and Y_2O_2S, and an oxygen vacancy was created as a result of charge compensation. These defects directly influence the trap distribution of the Gd_2O_2S and Y_2O_2S hosts, acting as traps for electrons (V_O) or holes ($M_{Gd,Y}$) in the lattice. Surprisingly, Ti^{3+} demonstrates broad blue white afterglow emission in $(Li_xNa_{1-x})_8Al_6Si_6O_{24}(Cl_{1-2y}S_y)_2$, which lasts for 100 h [240].

5.3 Main group element-activated persistent phosphors

Similar as lanthanide ions and transition metal ions, main group elements such as Bi^{3+} and Bi^{2+} have been also used as activators in persistent phosphors. In this part, we discuss the optical properties of the main group element–activated (e.g., Bi^{3+}, Bi^{2+}, Pb^{2+}, and Sb^{3+}) persistent phosphors.

5.3.1 Bi^{3+}

Bi^{3+} has the electronic configuration of [Xe] $6s^2\ 4f^{14}\ 5d^{10}$. One of the most important features of this ion is its luminescence in the visible spectral region when excited by ultraviolet and visible optical sources, arising from their multiplex electronic structures. In general the absorption and emission bands for Bi^{3+} are much broader than those of f–f transitions of rare earth ions, owing to the unsheltered outer electrons. The ground state of the free ion is $^1S_0$, whereas the 6s6p excited states give rise to the triplet levels [$^3P_0$, $^3P_1$, $^3P_2$] and the $^1P_1$ singlet state. The luminescence spectrum of Bi^{3+} usually shows a rather broad blue or green band ranging from 400 to 600 nm as result of the transition of $^3P_0 \rightarrow {}^1S_0$. Several Bi^{3+}-activated crystals have been reported for their potential applications in white light-emitting diodes [241–251]. Though their emission bands cover broad spectral regions from blue to red depending on the coordination environment of the active centers, the reported Bi^{3+}-activated persistent phosphors (Fig. 5.18), including $CaWO_4:Bi^{3+}$ [244], $ZnGa_2O_4:Bi^{3+}$ [242], $Zn_2GeO_4:Bi^{3+}$ [243], $CdSiO_3:Bi^{3+}$ [241], $CaZnGe_2O_6:Bi^{3+}$ [245], $Lu_2CaGeO_6:Bi^{3+}$ [246], $BaGa_2O_4:Bi^{3+}$ [247], $KGaGeO_4:Bi^{3+}$ [248], $Sr_3Ga_4O_9:Bi^{3+}$ [249], $LiYGeO_4:Bi^{3+}$ [250], and $CaGa_2O_4:Bi^{3+}$ [251], all offered an ultraviolet-blue-green-red afterglow emission with lifetimes from several seconds to over 600 h. It is worthy to note that significant development was recently achieved in Bi^{3+}-activated persistent phosphors as reported in Ref. [247, 250], where Bi^{3+} demonstrated orange and ultraviolet persistent phosphorescence with afterglow duration of 600 and 300 h in $BaGa_2O_4$ and $LiYGeO_4$ hosts, respectively. For $BaGa_2O_4$: Bi^{3+}, the effective positive charge of Bi_{Ba} may be generated when substituting Bi^{3+} for Ba^{2+}. Simultaneously, for compensating the extra positive charge, an equivalent concentration of negative effective charge may be generated. Moreover, in the sintering process, some intrinsic negative effective charge defects, such as O_I, V_{Ba}, and V_{Ga}, and positive effective charge defects, such as V_o, Ba_I, and Ga_I, may exist in the phosphor. Under ultraviolet light, these defects can act as traps, which could store charge carriers. The trapped charge carriers can be released from the traps gradually and combined around the Bi^{3+} emission center with energy release.

5.3.2 Bi^{2+}

In addition to Bi^{3+}, Bi^{2+} has [Xe] $6s^2$ $4f^{14}$ $5d^{10}$ $6p^1$ electronic configuration. Its electronic configuration yields a $^2P_{1/2}$ ground state and a $^2P_{3/2}$ excited state. The luminescent behaviors of Bi^{2+} are also strongly influenced by the nearby chemical environment, and it generates emissions from red to near-infrared spectral regions. Recently, we synthesized new persistent phosphor, $SrSnO_3:Bi^{2+}$. This phosphor shows deep-red-near-infrared persistent phosphorescence with a broad emission band covering a spectral region from 700 to 900 nm and a peak at 808 nm, and it exhibits a persistence time of longer than 30 min [252]. This phosphor can be excited by both ultraviolet and visible wavelengths. It is well known that trapping centers play an essential and crucial role for photoenergy storage in persistent phosphorescence

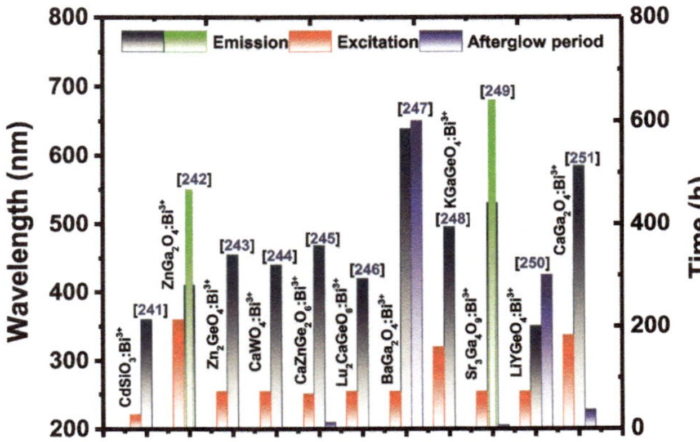

Fig. 5.18 Bi^{3+}-activated persistent phosphors.

process. The thermoluminescence spectrum is very useful for evaluating the trap depth, which is an important factor affecting the time of afterglow. The thermoluminescence measurement of $SrSnO_3$:Bi^{2+} shows two broad bands with maxima at 335 and 583 K, corresponding to the shallow and deep traps, respectively. At room temperature, shallow traps are easily emptied, whereas deep traps are difficult to be emptied, and parts of captured electrons remain stored there. If a trap is too deep, the captured electrons cannot escape, and no persistent phosphorescence is detected. Engineering a suitable trap depth is essential for achieving room temperature persistent phosphorescence. This phosphor has broken the domination of conventional Cr^{3+}-, Mn^{2+}-, and Mn^{4+}-activated persistent phosphors and provides a new perspective on the near-infrared persistent phosphorescence of Bi-doped persistent phosphors. Meanwhile the characteristic operational wave band offers the possibility of potentially being applied in vivo bioimaging of this near-infrared phosphorescent material.

5.3.3 Pb^{2+}

Pb^{2+} has the electronic configuration of [Xe] $6s^2\ 4f^{14}\ 5d^{10}$ in which $^3P_1$ and $^1S_0$ are the excited and ground states, respectively. For ultraviolet persistent phosphorescence, Pb^{2+} is a promising ultraviolet luminescent center, which typically shows a broad ultraviolet emission band (220–400 nm, generally dominated by the $^3P_1 \rightarrow {}^1S_0$ transition), and the emission peak is tunable depending on the host lattices. As shown in Fig. 5.19, to generate persistent phosphorescence in the ultraviolet spectral region, Liang et al. developed a new ultraviolet persistent phosphor, $Sr_2MgGe_2O_7$:Pb^{2+} [253]. The $Sr_2MgGe_2O_7$:Pb^{2+} phosphor exhibits strong persistent phosphorescence peaking at 370 nm and a long afterglow time of >12 h after ultraviolet excitation. The $Sr_2MgGe_2O_7$:Pb^{2+} persistent phosphor also exhibits a photostimulated persistent phosphorescence capability, in which the persistent phosphorescence of a high-energy light (250–310 nm) preirradiated sample can be rejuvenated after a short time, with low energy visible or near-infrared light stimulation. The successful development of the $Sr_2MgGe_2O_7$:Pb^{2+} persistent phosphor opens a new avenue for persistent phosphors research and applications. In comparison, CaO:Pb^{2+} and SrO:Pb^{2+} show ultraviolet persistent phosphorescence at 360 and 390 nm with lifetime of 1 h, respectively [65].

5.4 Defect-activated persistent phosphors

It has been generally accepted that during the solid-state reaction, intrinsic defects are formed in the lattice. The intrinsic defects can be cation vacancies, anion vacancies, and interstitial ions, which are easily generated during high-temperature treatment. These lattice defects can serve as electron or hole traps in the materials. However, these traps can be tuned by changing the material composition, impurity type, and dopant concentration. The traps that are generated by codopants and/or intrinsic lattice defects can serve as luminescent centers in many hosts, such as $CdSiO_3$, $CaSnSiO_5$, $Ca_3SnSi_2O_9$, $Ca_2ZrSi_4O_{12}$, Ca_2SnO_4, Mg_2SnO_4, and $BaZrO_3$ [254–260]. As shown in Fig. 5.20, the defect-activated phosphors like $CdSiO_3$ [254], $CaSnSiO_5$ [255],

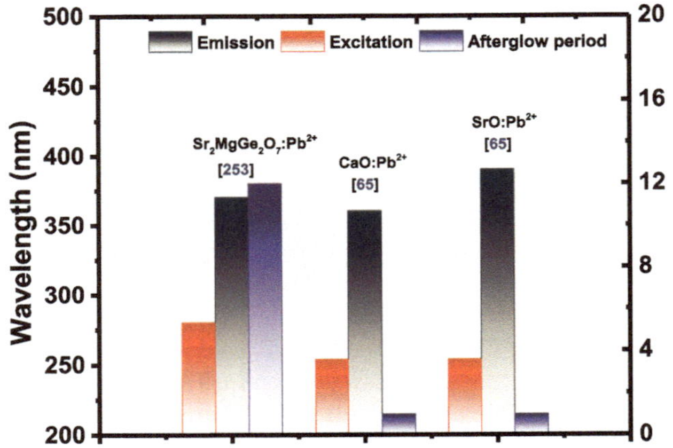

Fig. 5.19 Pb^{2+}-activated persistent phosphors.

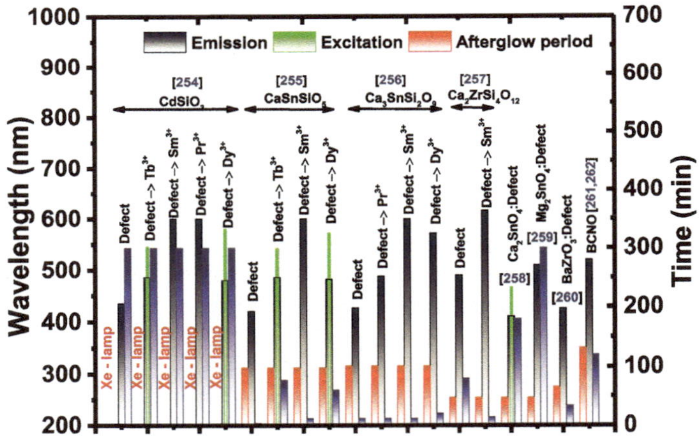

Fig. 5.20 Defected-activated persistent phosphors. BCNO refers to boron-carbon oxynitride.

$Ca_3SnSi_2O_9$ [261], $Ca_2ZrSi_4O_{12}$ [257], Ca_2SnO_4 [258,] and $BaZrO_3$ [260] show blue persistent phosphorescence, which lasts for 5 h, 200 s, 700 s, 4800 s, 3 h, and 2000 s, respectively. These defect-activated persistent phosphors can efficiently transfer their energy to trivalent lanthanide ions such as Tb^{3+}, Sm^{3+}, Pr^{3+}, and Dy^{3+}. Furthermore, efficient green defect-activated persistent phosphors such as Mg_2SnO_4 [259] and boron-carbon oxynitride [262, 263] were also reported, which demonstrates broad green emission at 515 and 520 nm for 5 and 2 h, respectively.

5.5 Influence of different hosts

The process of persistent phosphorescence occurs through a sequence of electronic transitions involving energy bands of the host lattice and energy states of dopants, as well as the trapping processes by the intrinsic defects. The emission wavelength of activator strongly depends on the energy levels of the emitter and the surrounding crystal field environment of the host crystal. The afterglow intensity and decay time both depend on the traps, including trap type, depth, and density, which are generated by codopants and/or intrinsic lattice defects. The intrinsic defects refer to the thermal defects, including cation vacancies, anion vacancies, and interstitial ions, which are easily generated during high-temperature treatment. These lattice defects can serve as electron or hole traps in the material. These traps can be tuned by changing the material composition, impurity type, and dopant concentration. In this part, we will discuss the effect of hosts (e.g., oxides, sulfides, nitrides, halides, and glasses) and luminescent centers on the luminescence intensity, wavelength, and phosphorescence duration.

5.5.1 Oxides

5.5.1.1 Binary oxides

Alkaline earth oxides and lanthanide oxides have been widely used as hosts to produce persistent phosphorescence in different spectral regions. Because of the large ionic radius of lanthanide ions and alkaline earth ions, rare earth ions can be easily doped into the cation sites. In 2003, Lin et al. prepared Eu^{3+}-activated Y_2O_3 phosphor codoped with Ti^{4+} and Mg^{2+} ions [70]. Red long afterglow was also observed in the Y_2O_3:Eu^{3+}, Ti^{4+}, Mg^{2+} after excitation by 254 nm light, and the afterglow duration can reach 1.5 h. The afterglow may be ascribed to some suitable complex trap centers produced by the introduction of Ti^{4+} and Mg^{2+} ions in the Y_2O_3 host lattice. In addition, orange persistent phosphorescence from Eu^{3+}-activated CaO and SrO phosphors was observed after the materials had been irradiated with ultraviolet light [65]. The orange persistent phosphorescence resulted from $^5D_0$ to $^7F_1$ transitions of the Eu^{3+} ions. The persistent phosphorescence lasted for over 1 h. Trapped-electron centers and hole centers introduced into the CaO and SrO hosts by the ultraviolet light are considered to be responsible for generating the persistent phosphorescence. In another report, it was shown that the Tb^{3+},M^{2+} codoping (M = Ca, Sr, Ba) of the Lu_2O_3 ceramics, prepared by heat treatment at temperatures around 1700 °C in a reducing atmosphere, showed surprisingly long-lasting and efficient persistent phosphorescence [97, 108]. As shown in Fig. 5.21, the photoluminescence could be observed even for 10 – 20 h after ceasing the irradiation of the materials with ultraviolet photons ($\lambda \leq 330$ nm). The green persistent phosphorescence is exclusively due to emission from the Tb^{3+} ions. It is postulated that the individual trapping sites cluster during the high-temperature sintering that may serve as electron traps and lead to the prolong of phosphorescence.

5.5.1.2 Silicates

Silicates are the most widely used hosts for Eu^{2+}-activated phosphors. Among these hosts, Guo et al. developed a novel cyan persistent phosphor, $BaZrSi_3O_9$:Eu^{2+},Pr^{3+} [21]. The Pr^{3+} ion is the trapping center. The key finding of this work is the cyan persistent phosphorescence of $BaZrSi_3O_9$:Eu^{2+},Pr^{3+}, which could be observed by the

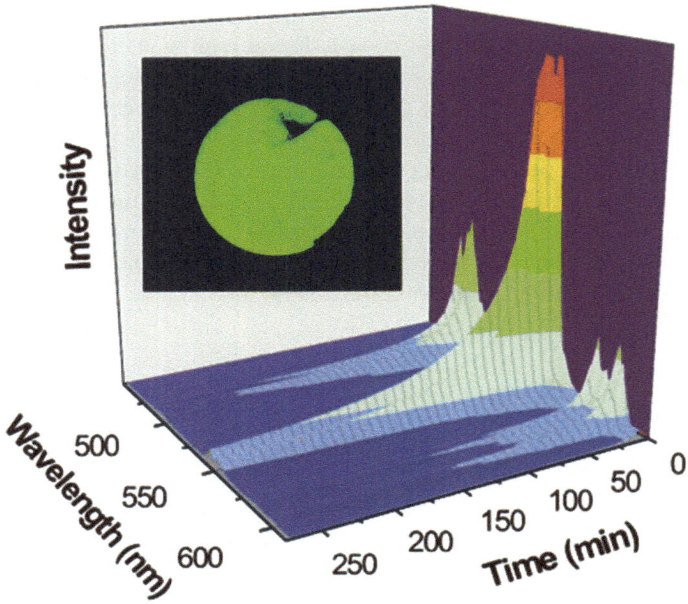

Fig. 5.21 3-D photoluminescence decay curve of the persistent phosphorescence of the investigated ceramics of Lu_2O_3:Tb^{3+},M^{2+} (M = Ca, Sr, Ba) after the irradiation of the materials with the ultraviolet radiation at 270 nm. The insets show the digital photo under ultraviolet light irradiation.
Reproduced with permission from J. Trojan-Piegza, E. Zych, J. Holsa, J. Niittykoski, Spectroscopic properties of persistent luminescence phosphors: Lu_2O_3:Tb^{3+},M^{2+} (M= Ca, Sr, Ba). J. Phys. Chem. C, 2009, 113, 20493–20498, Copyright 2009, American Chemical Society.

naked eye for 15 h in the dark after switching off the irradiation source. Moreover, they analyzed the reason why the $BaZrSi_3O_9$:Eu^{2+},Pr^{3+} phosphor showed such excellent persistent phosphorescence by exploring the distribution of traps in $BaZrSi_3O_9$: Eu^{2+},Pr^{3+} with the help of the thermoluminescence spectra. In 2019 a series of cyan light-emitting persistent phosphors, such as $Ba_{1-x}Li_xSi_{2-x}P_xO_5$:0.008$Eu^{2+}$,0.01$Pr^{3+}$, were designed by Feng et al. [53]. As shown in Fig. 5.22A, under 345-nm light excitation, the as-prepared $Ba_{1-x}Li_xSi_{2-x}P_xO_5$:0.008$Eu^{2+}$,0.01$Pr^{3+}$ presents a strong cyan emission located at 514 nm, and the afterglow time of samples can be increased from 38 h to about 56 h ($Ba_{0.92}Li_{0.08}$ $Si_{1.92}$ $P_{0.08}O_5$:0.008Eu^{2+}, 0.01Pr^{3+}) after switching off the excitation source (Fig. 5.22B). The persistent time of this novel phosphor is two times longer than that of the green persistent phosphor $SrAl_2O_4$:Eu^{3+}:Dy^{3+}.

In the case of silicate host doped with Tb^{3+}, Zhou et al. synthesized $Na_{1-x}AlSiO_4$: Tb^{3+},$x$$Bi^{3+}$ phosphor with green persistent phosphorescence by using Bi^{3+} as trapping centers [106]. In this work the influence of metal ion Bi^{3+} on the optical storage properties of $NaAlSiO_4$: Tb^{3+} phosphor was investigated in details. As shown in

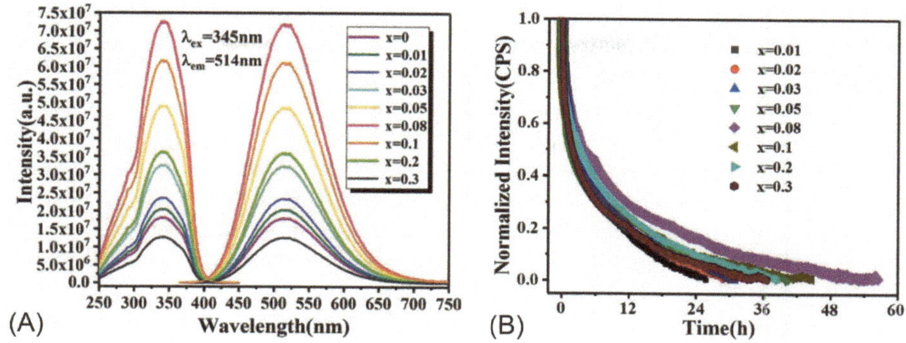

Fig. 5.22 (A) Excitation and emission spectra. (B) Persistent phosphorescence decay curves of $Ba_{1-x}Li_xSi_{2-x}P_xO_5$:0.008$Eu^{2+}$,0.01$Pr^{3+}$($0 \leq x \leq 0.3$).
Reproduced with permission from P. Feng, G. Li, H. Guo, D. Liu, Q. Ye, Y. Wang, Identifying a cyan ultralong persistent phosphorescence (Ba, Li) (Si, Ge, P)$_2O_5$:Eu^{2+}, Pr^{3+} via solid solution strategy. J. Phys. Chem. C, 2019, 123, 5, 3102–3109, Copyright 2019, American Chemical Society.

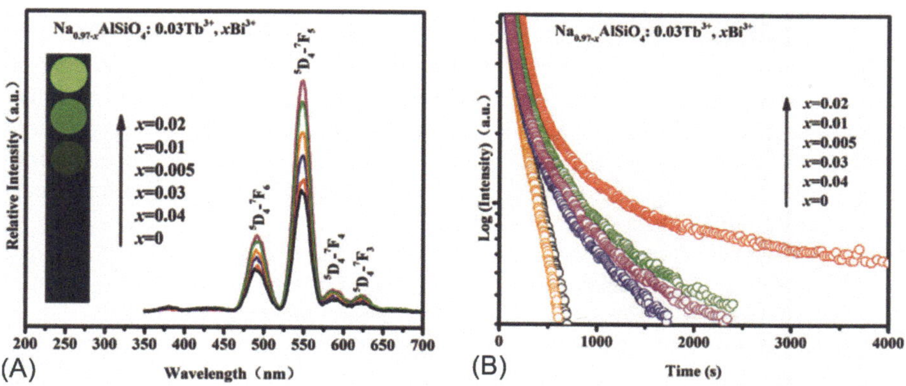

Fig. 5.23 (A) Persistent phosphorescence spectra of $Na_{0.97-x}AlSiO_4$: 0.03Tb^{3+}, $x$$Bi^{3+}$. The inset shows the persistent phosphorescence images of $Na_{0.97-x}AlSiO_4$: 0.03Tb^{3+}, $x$$Bi^{3+}$ activated with irradiation by a ultraviolet lamp. (B) Persistent phosphorescence decay curves of $Na_{0.97-x}AlSiO_4$: 0.03Tb^{3+}, $x$$Bi^{3+}$.
Reproduced with permission from J. Zhou, X. Yu, T. Wang, D. Zhou, J. Qiu, Improved optical storage properties of $NaAlSiO_4$:Tb^{3+} induced by Bi^{3+}. Opt. Mater., 2016, 57, 140–145, Copyright 2016, Elsevier.

Fig. 5.23A, with increasing concentration of Bi^{3+}, no characteristic emission of Bi^{3+} ions is observed, indicating that Bi^{3+} ions don't serve as emitting centers in the persistent phosphors but may act as trapping centers. Fig. 5.23B displays the persistent phosphorescence decay curves of $Na_{0.97-x}AlSiO_4$: 0.03Tb^{3+}, $x$$Bi^{3+}$ ($x = 0$, 0.005, 0.01, 0.02, 0.03, and 0.04) measured at room temperature. All the samples show persistent phosphorescence with different persistent times, and the decay curves consist of a fast decay and a consequent slow decay with a long decay tail, implying the existence of various trap depths. Compared with Tb^{3+} singly doped sample, longer persistent time is observed in the phosphors codoped with Bi^{3+}. The thermoluminescence curves suggest that the emitter Tb^{3+} introduced two types of traps located at 350 (T_A) and 440K (T_B) (Fig. 5.24B). More importantly, Bi^{3+} as a codopant ion introduced a new trap peaking at 390K (T_C) (Fig. 5.24A and C), which contributed to the improved persistent phosphorescence properties. Besides, owing to the existence of deep and stable trap T_B, green emission can still be observed after 72 h after excitation was stopped (Fig. 5.24D). It is known that the lattice defects acting as traps play an essential role in persistent phosphorescence process. Hence, it could be safe to say that more defects are created with the incorporation of Bi^{3+}, which act as trapping centers and have a significant influence on the performance of persistent phosphors.

5.5.1.3 Aluminates

Aluminates are another category of important hosts that have been widely used for developing persistent phosphors activated with rare earth ions and/or transition metal ions. In the last two decades, $Sr_4Al_{14}O_{25}$: Eu^{2+},Dy^{3+}, $SrAl_2O_4$: Eu^{2+},Dy^{3+}, and $CaAl_2O_4$: Eu^{2+},Nd^{3+} were the most widely investigated hosts after their discovery in 1996 because of their excellent afterglow properties [3, 4]. In comparison the persistent phosphors based on other aluminate hosts have also been developed that give red to near-infrared persistent phosphorescence. For example, Zhang et al. prepared Cr^{3+}-doped aluminate persistent phosphors (i.e., $Zn_{1+x}Al_{2-2x}Ge_xO_4$:Cr^{3+}) by germanium substitution in the original $ZnAl_2O_4$:Cr^{3+} host, which showed superlong near-infrared persistent phosphorescence [211].

Fig. 5.25 shows the emission and excitation spectra of the $ZnAl_2O_4$:Cr^{3+} and $Zn_{1.2}Al_{1.6}Ge_{0.2}O_4$:$Cr^{3+}$ under excitation at 387 nm. Both samples reveal similar emission spectrum resulting from the $^2E \rightarrow {}^4A_2$ spin-forbidden transition of Cr^{3+} in an octahedral crystal field. The R-lines located at 686 nm are related to the Cr^{3+} in unperturbed octahedral sites, while the N2 lines at 694 nm are assigned to Cr^{3+} ions with a neighboring antisite defect. Importantly the emission band from $Zn_{1.2}Al_{1.6}Ge_{0.2}O_4$:$Cr^{3+}$ is broader than the emission band from $ZnAl_2O_4$:Cr^{3+} as result of the increase of disorder around Cr^{3+} ions, which are induced by the incorporation of Ge. The excitation spectra of $ZnAl_2O_4$:Cr^{3+} and $Zn_{1.2}Al_{1.6}Ge_{0.2}O_4$:$Cr^{3+}$ at 686-nm emission show a very broad spectrum, which is attributed to the $^4A_2 \rightarrow {}^4T_2$(530 nm) and $^4A_2 \rightarrow {}^4T_1$ (^4F) (385 nm) transitions of Cr^{3+} ions. Unlike the negligible near-infrared persistent phosphorescence of $ZnAl_2O_4$:Cr^{3+} upon ultraviolet or visible light excitation, the Ge-substituted phosphors feature strong and superlong persistent phosphorescence at approximately 650–750 nm for more than 120 h as shown in Fig. 5.26. The relation between the Ge substitution and the defect trapping states was investigated systematically in this work. In this report the experimental results combined with the first principles calculations reveal that Ge_{Zn} and Ge_{Al} in the spinel structure would introduce shallow and deep defect states in the band gap to serve as new efficient traps, which are mainly responsible for the strong and superlong

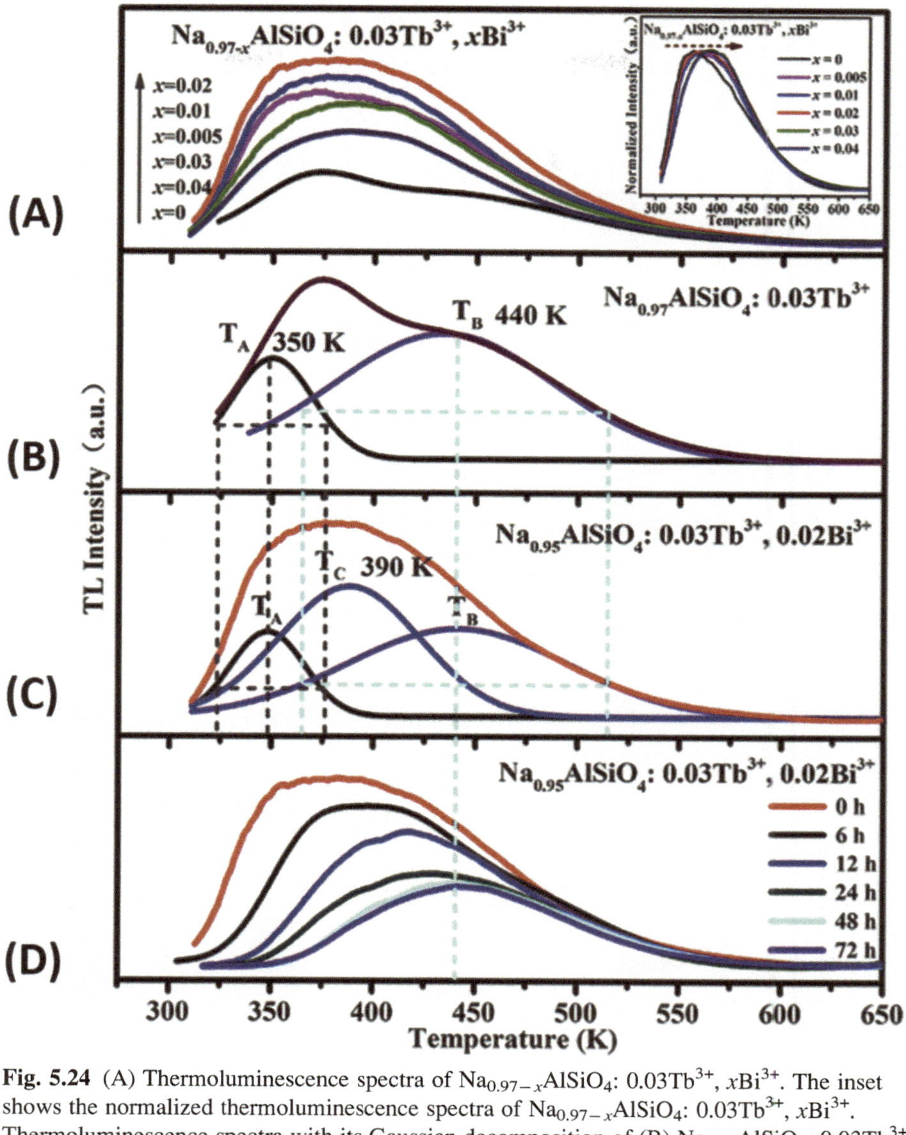

Fig. 5.24 (A) Thermoluminescence spectra of $Na_{0.97-x}AlSiO_4$: $0.03Tb^{3+}$, xBi^{3+}. The inset shows the normalized thermoluminescence spectra of $Na_{0.97-x}AlSiO_4$: $0.03Tb^{3+}$, xBi^{3+}. Thermoluminescence spectra with its Gaussian decomposition of (B) $Na_{0.97}AlSiO_4$: $0.03Tb^{3+}$ and (C) $Na_{0.95}AlSiO_4$: $0.03Tb^{3+}$, $0.02Bi^{3+}$. (D) Thermoluminescence spectra of $Na_{0.95}AlSiO_4$: $0.03Tb^{3+}$, $0.02Bi^{3+}$ placed in dark place for different durations.
Reproduced with permission from J. Zhou, X. Yu, T. Wang, D. Zhou, J. Qiu, Improved optical storage properties of $NaAlSiO_4$:Tb^{3+} induced by Bi^{3+}. Opt. Mater., 2016, 57, 140–145, Copyright 2016, Elsevier.

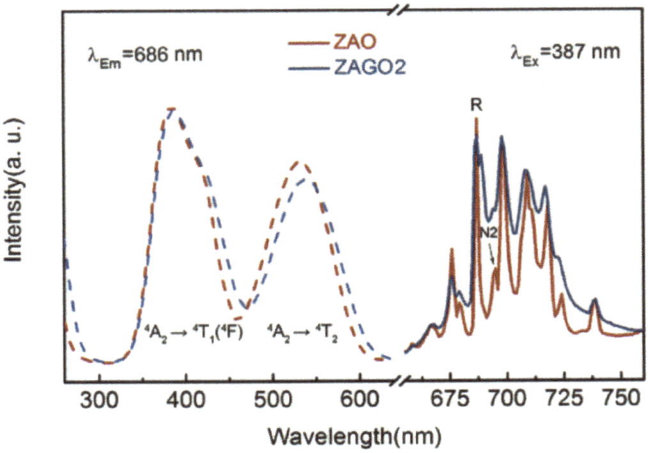

Fig. 5.25 Photoluminescence spectra of $ZnAl_2O_4:Cr^{3+}$ (ZAO) and $Zn_{1.2}Al_{1.6}Ge_{0.2}O_4:Cr^{3+}$ (ZAGO2) samples (ZAGO2) acquired under excitation at 387 nm and excitation spectra of ZAO and ZAGO2 obtained with monitoring wavelength of 686 nm.
Reproduced with permission from Y. Zhang, R. Huang, H. Li, D. Hou, Z. Lin, J. Song, Y. Guob, H. Lin, C. Song, Z. Lin, J. Robertson, Germanium substitution endowing Cr^{3+}-doped zinc aluminate phosphors with bright and super-long near-infrared persistent luminescence. Acta Mater., 2018, 155, 214–221, Copyright 2018, Elsevier.

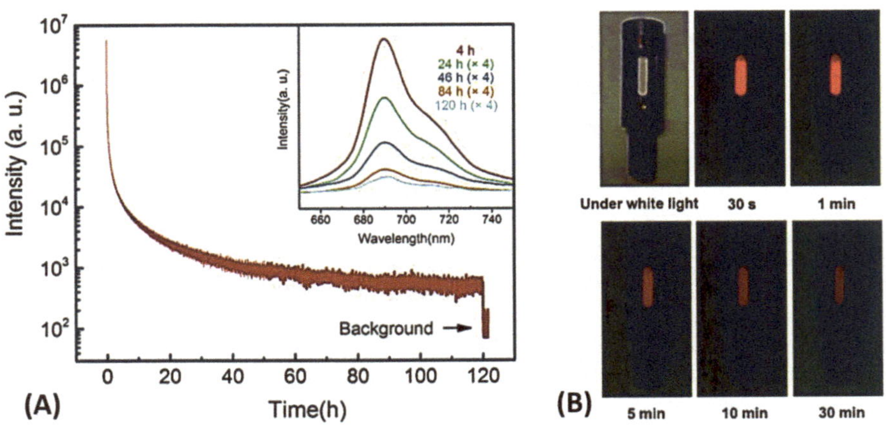

Fig. 5.26 (A) Persistent phosphorescence decay curve of the $Zn_{1.2}Al_{1.6}Ge_{0.2}O_4:Cr^{3+}$ sample at 690-nm emission with a decay time of more than 120 h after irradiation by a 280-nm Xe lamp for 10 min. The inset shows the persistent phosphorescence spectra acquired at different decay times. (B) Near-infrared afterglow images (photograph parameters: manual/ISO 1600/exposure time of 10 s) at different afterglow times (30 s–30 min).
Reproduced with permission from Y. Zhang, R. Huang, H. Li, D. Hou, Z. Lin, J. Song, Y. Guob, H. Lin, C. Song, Z. Lin, J. Robertson, Germanium substitution endowing Cr^{3+}-doped zinc aluminate phosphors with bright and super-long near-infrared persistent luminescence. Acta Mater., 2018, 155, 214–221, Copyright 2018, Elsevier.

persistent phosphorescence upon ultraviolet or visible light excitation. This phosphor may find potential applications in biomedical imaging and night vision surveillance.

Qiu et al. fabricated a series of Mn^{4+}-doped $MAlO_3$ (M = La, Gd) persistent phosphors with the emission maximum at around 730 nm and afterglow duration of 20 h [230]. They proposed a holistic design concept for near-infrared persistent phosphorescence and successfully fabricate a series of Mn^{4+}-doped $MAlO_3$ (M = La, Gd) persistent phosphors. By optimizing the defect types and trap depths, an improvement of persistent time to 20 h is realized by codoping Ge^{4+}/Mn^{4+}. Through many experiments the optimal composition for the lanthanum aluminate phosphor was found to be $LaAlO_3$:0.1%Mn^{4+},0.9%Ge^{4+}. After irradiation with a Xe lamp for 10 min, $LaAlO_3$:0.1%Mn^{4+},0.9%Ge^{4+} demonstrates long afterglow duration for more than 20 h (Fig. 5.27A). Fig. 5.27B shows the afterglow intensity I_{10s} of sample $LaAlO_3$:0.1%Mn^{4+},0.9%Ge^{4+} monitored at 731 nm as a function of the excitation wavelength. It is clear that the near-infrared persistent phosphorescence can be effectively achieved under UV illumination (300–400 nm). The most effective excitation wavelength for the persistent phosphorescence is around 330 nm, corresponding to the absorption maximum (336 nm).

Fig. 5.28A shows the emission and excitation spectra of sample $GdAlO_3$:0.1% Mn^{4+}, 0.9%Ge^{4+}. Under the excitation at 326 nm, the sample exhibits six sharp emissions peaking at 680, 685, 693, 698, 704, and 719 nm. When monitoring at all emissions, the excitation spectra all consist of two bands at 326 and 480 nm. After being excited at 325 nm for 10 min, persistent phosphorescence with two main emission bands at 704 and 719 nm is detected in the afterglow spectra. When monitored at

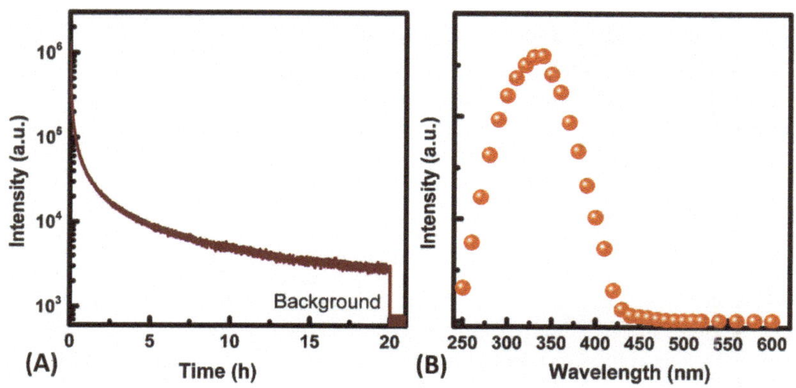

Fig. 5.27 Persistent phosphorescence properties of $LaAlO_3$:0.1%Mn^{4+},0.9%Ge^{4+}. (A) Afterglow intensity monitored at 731 nm as a function of time. The phosphor was preirradiated by a Xe lamp for 10 min. (B) Afterglow intensity I_{10s} monitored at 731 nm as a function of the excitation wavelength over the 250–600 nm spectral range.
Reproduced with permission from Y. Li, Y. Y. Li, K. Sharafudeen, G. P. Dong, S. F. Zhou, Z. J. Ma, M. Y. Peng, J. R. Qiu, A strategy for developing near infrared long-persistent phosphors: taking $MAlO_3$: Mn^{4+}, Ge^{4+} (M = La, Gd) as an example. J. Mater. Chem. C, 2014, 2, 2019–2027, Copyright 2014, The Royal Society of Chemistry.

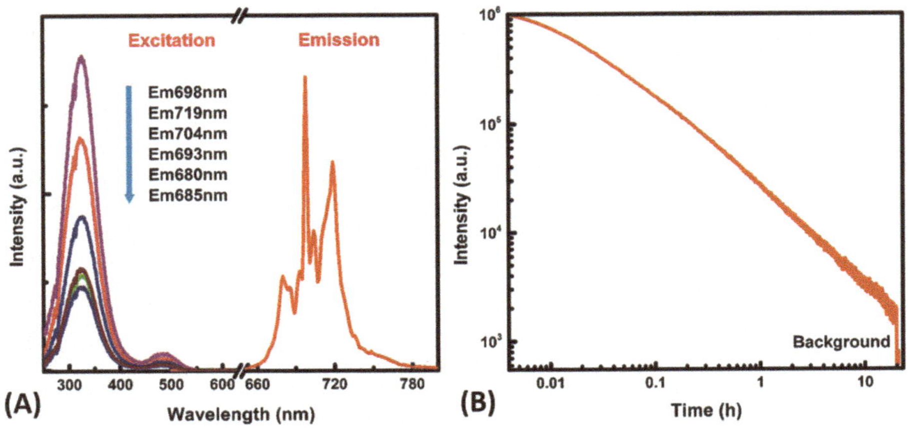

Fig. 5.28 Photoluminescence and persistent phosphorescence properties of $GdAlO_3$:0.1% Mn^{4+},0.9%Ge^{4+}. (A) Emission spectra excited at 326 nm and the excitation spectra monitored at 680, 685, 693, 698, 704, and 719 nm emissions. (B) Afterglow intensity monitored at 719 nm as a function of time. The phosphor was preexcited at 325 nm for 10 min. Reproduced with permission from Y. Li, Y. Y. Li, K. Sharafudeen, G. P. Dong, S. F. Zhou, Z. J. Ma, M. Y. Peng, J. R. Qiu, A strategy for developing near infrared long-persistent phosphors: taking $MAlO_3$: Mn^{4+}, Ge^{4+} (M = La, Gd) as an example. J. Mater. Chem. C, 2014, 2, 2019–2027, Copyright 2014, The Royal Society of Chemistry.

719 nm as a function of time after irradiation at 325 nm for 10 min, a persistent phosphorescence duration longer than 20 h is also observed (Fig. 5.28B), but the afterglow intensity of sample $GdAlO_3$:0.1%Mn^{4+},0.9%Ge^{4+} at 24 h after the stoppage of irradiation is weaker than that of sample $LaAlO_3$:0.1%Mn^{4+},0.9%Ge^{4+}(Fig. 5.29). The obtained imaging of deep tissues reveals that the new luminescent indicators will open the possibility of advanced optical imaging, with high resolution and weak light disturbance, for factual assessment of the structural and functional processes in cells, tissue, and other complex systems.

5.5.1.4 Gallates

Gallates are the best host for Cr^{3+} ions to produce visible and near-infrared persistent phosphorescence. In 2012, Pan et al. discovered the Cr^{3+}-doped zinc gallogermanate phosphor, which showed incredibly long near-infrared afterglow (650–1000 nm) over 360 h [223]. This study indicates that $Zn_3Ga_2Ge_2O_{10}$:0:5% Cr^{3+} phosphor exhibits strong near-infrared emission (650–1000 nm) and long afterglow lifetime. It is noteworthy that these persistent phosphors can be rapidly, effectively, and repeatedly charged through exposure to sunlight under any of the outdoor conditions. As shown in Fig. 5.30, under excitation at 400 nm, the doped sample shows a broadened emission band covering spectral region from 650 to 1000 nm, peaking at 696 nm, which can be ascribed to ${}^4T_2 \rightarrow {}^4A_2$ transition of Cr^{3+}. By comparing the excitation spectrum of the phosphor and the spectrum of the solar light, a large degree of overlap between

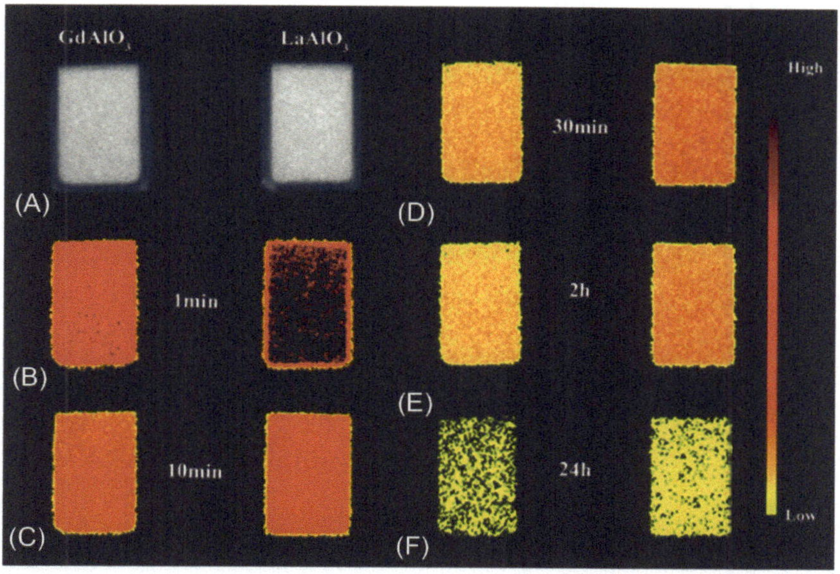

Fig. 5.29 Persistent phosphorescence imaging of samples $LaAlO_3$:0.1%Mn^{4+},0.9%Ge^{4+}(right) and $GdAlO_3$:0.1%Mn^{4+},0.9%Ge^{4+} (left). (A) Preirradiation images and (B) 1, (C) 10, (D) 30, (E) 120, and (F) 1440 min postirradiation fluorescence images. The samples were preexcited at 325 nm for 10 min.
Reproduced with permission from Y. Li, Y. Y. Li, K. Sharafudeen, G. P. Dong, S. F. Zhou, Z. J. Ma, M. Y. Peng, J. R. Qiu, A strategy for developing near infrared long-persistent phosphors: taking $MAlO_3$: Mn^{4+}, Ge^{4+} (M = La, Gd) as an example. J. Mater. Chem. C, 2014, 2, 2019–2027, Copyright 2014, The Royal Society of Chemistry.

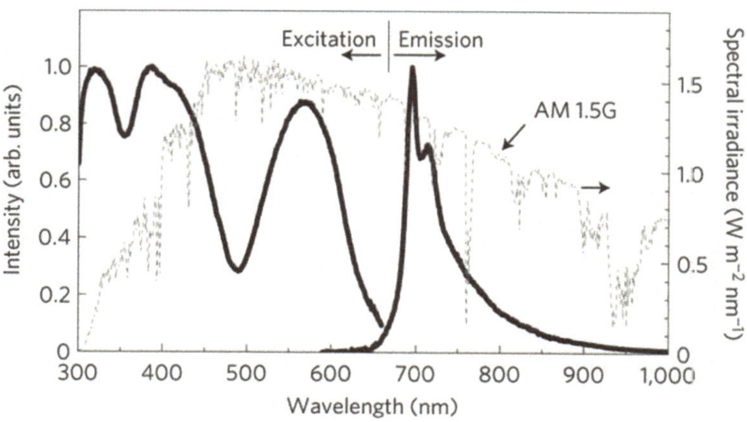

Fig. 5.30 Normalized excitation (monitored at 713 nm) and emission (at 400 nm) spectra of $Zn_3Ga_2Ge_2O_{10}$:0:5%Cr^{3+} phosphor. The *gray dashed line* is the solar spectrum.
Reproduced with permission from Z. W. Pan, Y. Y. Lu, F. Liu, Sunlight-activated long-persistent luminescence in the near-infrared from Cr^{3+}-doped zinc gallogermanates. Nat. Mater., 2012, 11, 58–63, Copyright 2012, Nature Publishing Group.

these two spectra from 300 to 650 nm indicates that the $Zn_3Ga_2Ge_2O_{10}$:0:5% Cr^{3+} phosphor could be effectively activated through solar radiation. Fig. 5.31A shows the persistent phosphorescence decay curve of a $Zn_3Ga_2Ge_2O_{10}$:0:5% Cr^{3+} ceramic disk monitored at 713 nm after irradiation by 365 nm light for 5 min. Unexpectedly the afterglow intensity is still higher than the light limit that is perceivable by naked eyes even after 100 h. The decay curve plotted in the bottom inset of Fig. 5.31A indicates a good linear relationship between reciprocal afterglow intensity I^{-1} and time t at the 20–100 h time period, suggesting that the long near-infrared persistent phosphorescence in $Zn_3Ga_2Ge_2O_{10}$:0:5% Cr^{3+} phosphor probably arises from tunneling-related processes. Although the emission intensity decreases with increasing decay

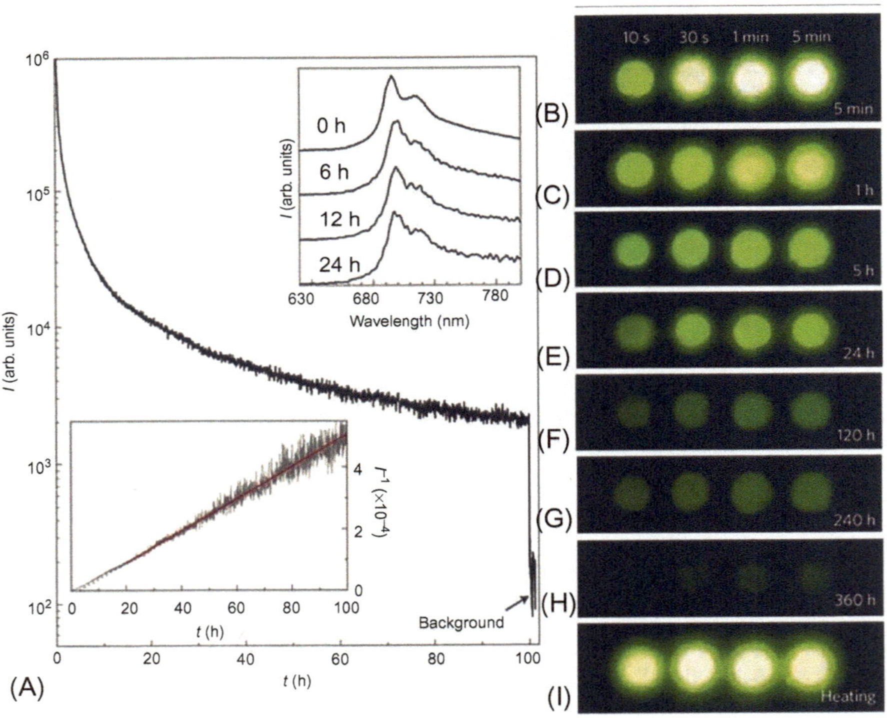

Fig. 5.31 Near-infrared afterglow decay of $Zn_3Ga_2Ge_2O_{10}$:0:5% Cr^{3+} disks excited by a 365-nm ultraviolet lamp. (A) Seven hundred thirteen nanometer afterglow decay curve of the phosphor after 5 min irradiation. The inset located at the bottom plots the same data, which describes the correlation of reciprocal of persistent phosphorescence intensity (I^{-1}) with time (t). The upper inset displays four afterglow spectra recorded at 0, 6, 12, and 24 h after removal of excitation source. (B–H) Near-infrared images of four phosphor disks taken at different times (5 min to 360 h) after irradiation with 365 nm light for 10 s to 5 min. (I) Near-infrared image of the four 360-h decayed disks upon heating to 400°C on the hot plate.
Reproduced with permission from Z. W. Pan, Y. Y. Lu, F. Liu, Sunlight-activated long-persistent luminescence in the near-infrared from Cr^{3+}-doped zinc gallogermanates. Nat. Mater., 2012, 11, 58–63, Copyright 2012, Nature Publishing Group.

time (Fig. 5.31B–H), the afterglow remains visible to naked eye even after 360 h. More interestingly the decayed phosphors can regain their luminescence properties upon heat stimulation (Fig. 5.31I). On the basis of the superior optical properties, these near-infrared persistent phosphors can be potentially applied for light harvesting in solar energy and in vivo bioimaging.

In 2013 Liu et al. observed an intense persistent phosphorescence signal peaking at 716 nm, and this result can be repeatedly obtained in a period of more than 1000 h in ultraviolet light (250–360 nm) preirradiated $LiGa_5O_8:Cr^{3+}$ phosphor [219]. Fig. 5.32A shows the normalized photoluminescence emission and excitation spectra of a $LiGa_5O_8:Cr^{3+}$ phosphor disk at room temperature. Under excitation at 400 nm the material exhibits a narrowband emission peaking at 716nm. The photoluminescence excitation spectrum monitored at 716-nm emission covers a very broad spectral region (from ~250 to ~660nm) and consists of three main excitation bands originating from the inner transitions of Cr^{3+}, including the 300 nm ($^4A_2 \rightarrow {}^4T_1(^4P)$ transition), the 415 nm ($^4A_2 \rightarrow {}^4T_1(^4F)$ transition), and the 605 nm ($^4A_2 \rightarrow {}^4T_2(^4F)$ transition). The excitation by ultraviolet light can also induce very long lasting near-infrared persistent phosphorescence in $LiGa_5O_8:Cr^{3+}$ phosphor with a persistence time of longer than 1000 h (Fig. 5.32B). To evaluate the influence of different excitation wavelengths on persistent phosphorescence, they measured the persistent phosphorescence decay curves after the excitation of monochromatic light with different wavelengths between 250 and 600 nm. It is clear that the near-infrared persistent phosphorescence can be effectively achieved by ultraviolet irradiation between 250 and 360nm and the intensity increases as the excitation moves to shorter wavelengths. The super long lasting near-infrared persistent phosphorescence of the $LiGa_5O_8:Cr^{3+}$ phosphor disks was also visually evaluated using a night vision monocular in a dark room that shows a decay time up to 1008 h after exposure to a 4-W 254-nm ultraviolet lamp for 10 s, 30 s, 1 min, and 5 min (Fig. 5.32C–J). The $LiGa_5O_8:Cr^{3+}$ phosphor has promising applications in optical information storage, night vision surveillance, and in vivo bioimaging.

In 2014, Li et al. have also observed red persistent phosphorescence in a new compound, $Zn_3Ga_2SnO_8:Cr^{3+}$ phosphor [220]. Fig. 5.33 shows the normalized excitation and emission spectra of $Zn_3Ga_2SnO_8:0.5Cr^{3+}$ and $Zn_3Ga_2GeO_8:0.5Cr^{3+}$. Under the excitation wavelength of 365 nm, $Zn_3Ga_2SnO_8:0.5Cr^{3+}$ and $Zn_3Ga_2GeO_8:0.5Cr^{3+}$ phosphors exhibit two sharp emission peaks at 696 nm due to the transition of Cr^{3+} ($^2E \rightarrow {}^4A_2$) and 705 nm, attributed to the vibrational sidebands, and a broad emission band from 600 to 800 nm peaking at 713 nm by the transition of Cr^{3+} ($^4T_2 \rightarrow {}^4A_2$). Fig. 5.34A shows the afterglow intensity monitored at 696 nm as a function of time (purple line). For comparison the luminescence decay of $Zn_3Ga_2GeO_8:0.5Cr^{3+}$ powders is measured (orange line) and compared with the luminescence decay of $Zn_3Ga_2GeO_8:0.5Cr^{3+}$ powders reported previously (blue dot). It can be seen that $Zn_3Ga_2SnO_8:0.5Cr^{3+}$ phosphor shows about three times increase in afterglow intensity compared with $Zn_3Ga_2GeO_8:0.5Cr^{3+}$ powders at 5 h after the stoppage of irradiation, affirming longer afterglow duration of $Zn_3Ga_2SnO_8:0.5Cr^{3+}$ phosphor. To further study the persistent time of two phosphors, digital photos and near-infrared images of $Zn_3Ga_2GeO_8:0.5Cr^{3+}$ (left) and $Zn_3Ga_2SnO_8:0.5Cr^{3+}$ (right) powders after

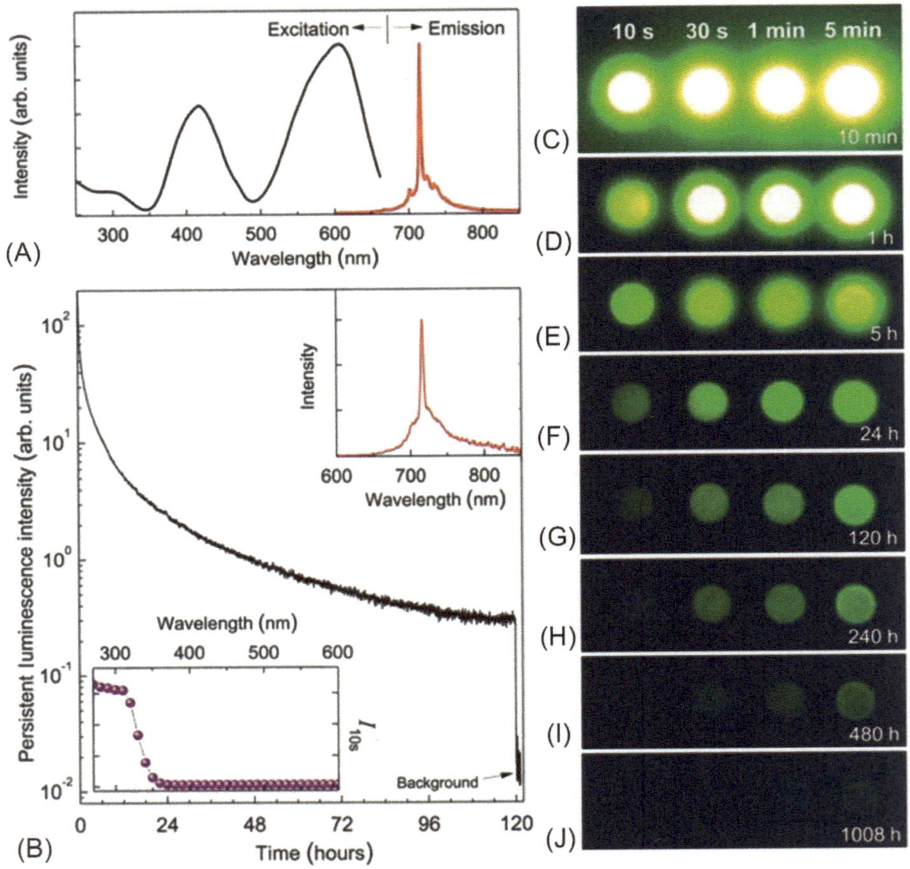

Fig. 5.32 Photoluminescence and persistent phosphorescence of $LiGa_5O_8:Cr^{3+}$ phosphor disks at room temperature. (A) Normalized excitation and emission spectra for photoluminescence under 400 nm light excitation and the excitation spectrum is obtained by monitoring 716 nm emission. (B) Near-infrared persistent phosphorescence decay curve monitored at 716 nm after irradiation by 300 nm light for 20 min. The upper inset shows the persistent phosphorescence emission spectrum recorded at 1 h after the stoppage of the irradiation. The bottom inset is the persistent phosphorescence excitation spectrum obtained by plotting the persistent phosphorescence intensity (I_{10s}) monitored at 716 nm as a function of the excitation wavelengths over the 250–600 nm spectral range. (C–J) Near-infrared images of four phosphor disks taken at different persistent afterglow times (10 min to 1080 h) after irradiated by a 254-nm lamp for 10 s to 5 min.
Reproduced with permission from F. Liu, W. Z. Yan, Y. J. Chuang, Z. P. Zhen, J. Xie, Z. W. Pan, Photostimulated near-infrared persistent luminescence as a new optical read-out from Cr^{3+}-doped $LiGa_5O_8$. Sci. Rep., 2013, 3, 01554, Copyright 2013, Nature Publishing Group.

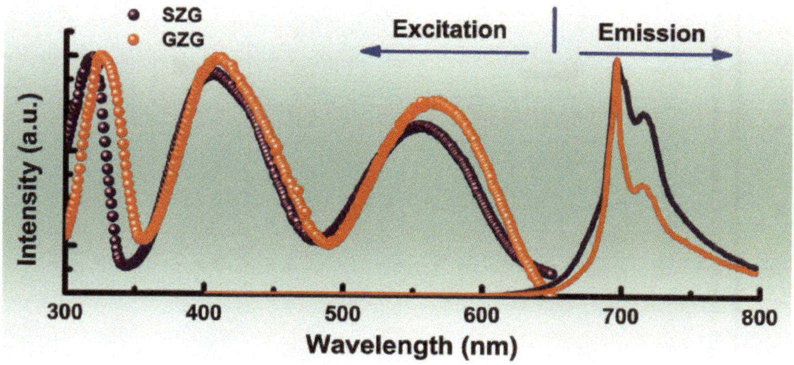

Fig. 5.33 Photoluminescence spectra of $Zn_3Ga_2SnO_8{:}0.5Cr^{3+}$ (SZG) and $Zn_3Ga_2GeO_8{:}0.5Cr^{3+}$ (GZG) powders. The emission spectra were acquired by excitation at 365 nm, and the excitation spectra were obtained by monitoring at 696 nm emission.
Reproduced with permission from Y. Li, S. F. Zhou, Y. Y. Li, K. Sharafudeen, Z. J. Ma, G. P. Dong, M. Y. Peng, J. R. Qiu, Long persistent and photo-stimulated luminescence in Cr^{3+}-doped Zn–Ga–Sn–O phosphors for deep and reproducible tissue imaging. J. Mater. Chem. C, 2014, 2, 2657–2663, Copyright 2014, The Royal Society of Chemistry.

being pressed are acquired at different time intervals (from 5 min to 300 h) after being irradiated by using a Xe lamp for 30 min (Fig. 5.34B). In contrast to the signals from $Zn_3Ga_2GeO_8{:}0.5Cr^{3+}$ powders, those from $Zn_3Ga_2SnO_8{:}0.5Cr^{3+}$ powders could be clearly detected even 300 h after the irradiation. To synthetically describe the afterglow properties of the $Zn_3Ga_2SnO_8{:}0.5Cr^{3+}$ phosphors, Fig. 5.34C shows the afterglow intensity I_{10s} of $Zn_3Ga_2SnO_8{:}0.5Cr^{3+}$ powders monitored at 696 nm as a function of the excitation wavelengths, indicating that phosphor can only be less effectively excited in the wavelength region from 430 to 600 nm but can be efficiently excited by the light in the spectral range from 250 to 430 nm.

To explain the nature of defects in Cr^{3+}-doped $Zn_3Ga_2SnO_8$ and $Zn_3Ga_2GeO_8$, electron spin resonance spectra of $Zn_3Ga_2SnO_8{:}0.5Cr^{3+}$ powders at different time intervals (30 min, 6 h, 24 h and 96 h) were measured after irradiation by a Xe lamp for 10 min (Fig. 5.35). Signals with $g = 2.004$, $g = 2.08$, and $g = 2.45$ were detected before and after the irradiation. Signals with $g = 2.08$ and $g = 2.45$ remained unchanged at different time intervals; only the signal intensity with $g = 2.004$ was found to decrease. The signal with $g = 2.08$ was attributed to the Cr^{3+}–Cr^{3+} exchange-coupled pairs or small clusters of several exchange-coupled pairs. The line in the electron spin resonance spectra at $g = 2.0024$ was attributed to the Zn vacancy in zinc oxide. The electron spin resonance signals due to singly negatively charged Zn vacancy (V_{Zn}) and zinc vacancy interstitial zinc complex ($V_{Zn}{:}Zn_i$) exhibit a g value very close to 2 in zinc oxide phosphor. The signal with $g = 2.0046$ can be assigned to a single ionized interstitial Zn (Zn_i, monovalent Zn or Zn^+) in Zn_2GeO_4 phosphor. To sum up, considering the point that the value of the signal with $g = 2.004$ was close to the g-factor ($g = 2$) of a single electron ion,

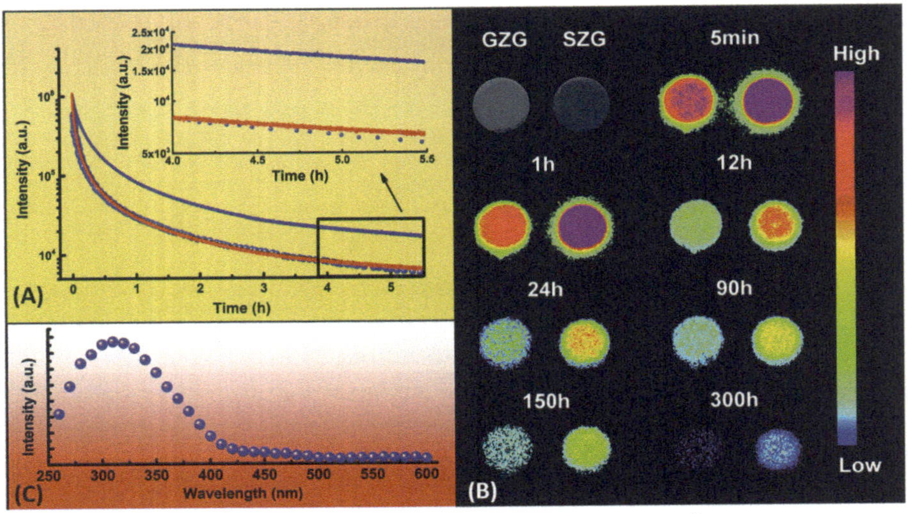

Fig. 5.34 Persistent phosphorescence properties of $Zn_3Ga_2SnO_8{:}0.5Cr^{3+}$(SZG) and $Zn_3Ga_2GeO_8{:}0.5Cr^{3+}$(GZG) powders. (A) Afterglow intensity monitored at 696 nm as a function of time of SZG powders (*purple line*) and GZG powders (*orange line*). The inset shows the magnified view of afterglow decays from 4 to 5.5 h. Before the measurements, two phosphors were preirradiated using a Xe lamp for 5 min. (B) Digital photos and near-infrared images of GZG (left) and SZG (right) powders after being pressed acquired at different time intervals (from 5 min to 300 h) after the irradiation using a Xe lamp for 30 min. During the process of imaging, the camera parameters stay the same. (C) Afterglow intensity I_{10s} of SZG powders monitored at 696 nm as a function of the excitation wavelengths over the 250–600 nm spectral range; the sample was preirradiated using a Xe lamp for 5 min.
Reproduced with permission from Y. Li, S. F. Zhou, Y. Y. Li, K. Sharafudeen, Z. J. Ma, G. P. Dong, M. Y. Peng, J. R. Qiu, Long persistent and photo-stimulated luminescence in Cr^{3+}-doped Zn–Ga–Sn–O phosphors for deep and reproducible tissue imaging. J. Mater. Chem. C, 2014, 2, 2657–2663, Copyright 2014, The Royal Society of Chemistry.

it should be attributed to the defect agglomerate composed of single ionized interstitial Zn and the Zn vacancy capturing an electron. For comparison the electron spin resonance signals of $Zn_3Ga_2GeO_8{:}0.5Cr^{3+}$ powders are also measured, and the similar signals with $g = 1.996$, $g = 2.08$, and $g = 2.45$ reveal the parallel trap types in $Zn_3Ga_2SnO_8{:}0.5Cr^{3+}$ and $Zn_3Ga_2GeO_8{:}0.5Cr^{3+}$ powders.

5.5.1.5 Phosphates

Phosphates doped with lanthanide ions are promising hosts for solid-state lighting. Guo et al. developed a new yellow-emitting persistent phosphor doped with Eu^{2+} (i.e., $Ca_6BaP_4O_{17}{:}Eu^{2+},Ho^{3+}$) [41]. The incorporation of Ho^{3+} ions, which act as trap centers, largely extends the trap depth and obviously improves the persistent phosphorescence behavior of the phosphor. As shown in Fig. 5.36A, the photoluminescence and persistent phosphorescence spectra of $Ca_6BaP_4O_{17}{:}0.02Eu^{2+},0.015Ho^{3+}$ reveal

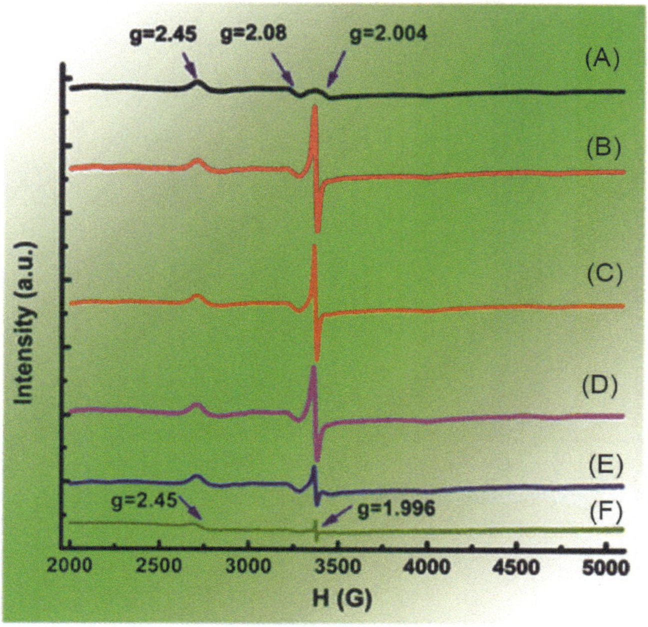

Fig. 5.35 Electron spin resonance spectra of (A–E) $Zn_3Ga_2SnO_8$:0.5Cr^{3+} and (F) $Zn_3Ga_2GeO_8$:0.5Cr^{3+} powders measured at 100 K. (A) Before irradiation, (B–E) 0.5, 6, 24, and 96 h after the stoppage of irradiation, and (F) 96 h after the stoppage of irradiation. Two phosphors were preirradiated by a Xe lamp for 5 min.
Reproduced with permission from Y. Li, S. F. Zhou, Y. Y. Li, K. Sharafudeen, Z. J. Ma, G. P. Dong, M. Y. Peng, J. R. Qiu, Long persistent and photo-stimulated luminescence in Cr^{3+}-doped Zn–Ga–Sn–O phosphors for deep and reproducible tissue imaging. J. Mater. Chem. C, 2014, 2, 2657–2663, Copyright 2014, The Royal Society of Chemistry.

only one asymmetric broad emission band located at 553 nm, attributed to the 5d–4f transitions of Eu^{2+} ions in two different Ca^{2+} sites. As shown in Fig. 5.36B, after 15 min of 370 excitation, the initial persistent phosphorescence intensity of $Ca_6BaP_4O_{17}$:0.02Eu^{2+},0.015Ho^{3+} can reach about 0.13 mcd m^{-2}, and its persistent phosphorescence can last for more than 47 h above the recognizable intensity level (0.32 mcd m^{-2}). In addition, we can clearly observe that Ho^{3+} concentration of 0.015 is the optimal value to achieve longest afterglow time. These results imply that the phosphor has the potential to become a novel commercial phosphor used in the field of emergency lighting and display.

However, Pang et al. synthesized a series of new reddish-orange phosphor by codoping Li^+ ions into Sm^{3+}-activated α-$Zn_2P_2O_7$ and characterized their luminescence properties together with the effect of Li^+ codoping on both photoluminescence and persistent phosphorescence [129]. The inset of Fig. 5.37 showed that both the photoluminescence and persistent phosphorescence originated from characteristic reddish-orange emissions of Sm^{3+} from its 4f–4f transitions of $^4G_{5/2}\rightarrow{}^6H_{5/2}$, $^4G_{5/2}\rightarrow{}^6H_{7/2}$, $^4G_{5/2}\rightarrow{}^6H_{9/2}$, and $^4G_{5/2}\rightarrow{}^6H_{11/2}$. Besides the remarkable enhancement

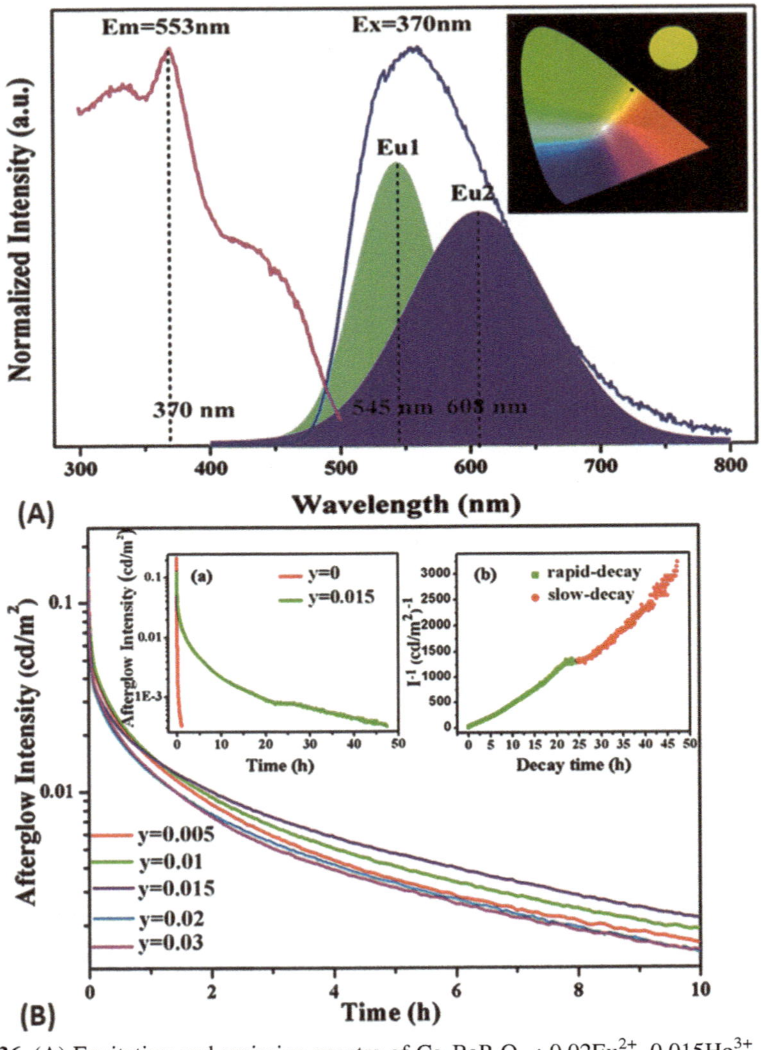

Fig. 5.36 (A) Excitation and emission spectra of $Ca_6BaP_4O_{17}$: $0.02Eu^{2+}$, $0.015Ho^{3+}$. Inset: international commission on illumination diagram for the $Ca_6BaP_4O_{17}$: $0.02Eu^{2+}$, $0.015Ho^{3+}$ phosphor excited at 370 nm. (B) Afterglow decay curves of $Ca_6BaP_4O_{17}$: $0.02Eu^{2+}$, yHo^{3+} ($0.05 \leq y \leq 0.03$). Inset: (A) afterglow decay curves of $Ca_6BaP_4O_{17}$: $0.02Eu^{2+}$ and $Ca_6BaP_4O_{17}$: $0.02Eu^{2+}$, $0.015Ho^{3+}$ and (B) plot of reciprocal afterglow intensity (I^{-1}) versus time (t) of $Ca_6BaP_4O_{17}$: $0.02Eu^{2+}$, $0.015Ho^{3+}$ after excitation for 15 min.

Reproduced with permission from H. Guo, W. Chen, W. Zeng, G. Li, Y. Wang, Y. Li, Y. Li, X. Ding, Structure and luminescence properties of a novel yellow super long-lasting phosphate phosphor $Ca_6BaP_4O_{17}$: Eu^{2+}, Ho^{3+}. J. Mater. Chem. C, 2015, 3, 5844–5850, Copyright 2015, The Royal Society of Chemistry.

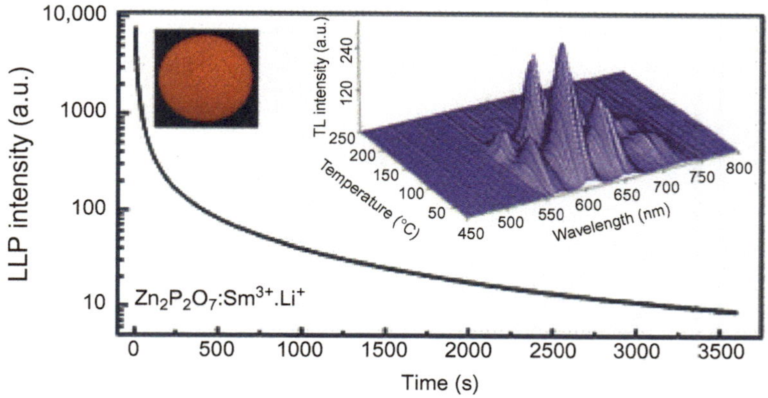

Fig. 5.37 Persistent phosphorescence decay curve of 604 nm emission of $Zn_2P_2O_7:Sm^{3+},Li^+$. Inset: the optical images under ultraviolet excitation and the 3-D thermoluminescence emission spectrum of $Zn_2P_2O_7:Sm^{3+},Li^+$.
Reproduced with permission from R. Pang, W. Z. Sun, J. P. Fu, H. F. Li, Y. L. Jia, D. Li, L. H. Jiang, S. Zhang, C. Y. Li, Luminescence properties of a novel reddish orange long-lasting phosphorescence phosphor $Zn_2P_2O_7$: Sm^{3+}, Li^+. RSC Adv., 2015, 5, 82704–82710, Copyright 2015, The Royal Society of Chemistry.

in the photoluminescence intensity of Sm^{3+}, codoping of Li^+ into the crystal lattice also promoted the persistent phosphorescence performance by modifying the defect levels in the phosphors. They observed that the optimal long afterglow performance was observed at Li^+ concentration of 2 mol%. Fig. 5.37 shows bright reddish-orange persistent phosphorescence that could last for more than 3 h in the dark. Four peaks appeared in its thermoluminescence curve (inset of Fig. 5.37), and the one at around 350 K was proved to be responsible for the occurrence of persistent phosphorescence. The release of the captured electrons in the defect levels corresponding to this thermoluminescence peak at room temperature to emission centers of Sm^{3+} underwent a tunneling process.

5.5.1.6 Stannates

To extend the persistent phosphorescence wavelength to shortwave infrared (900–1700 nm), Liang et al. reported that some trivalent lanthanides can act as efficient emitters for shortwave infrared persistent phosphorescence [135]. Using Ca_2SnO_4 as a host, they achieved strong shortwave infrared persistent phosphorescence from a group of trivalent lanthanide (e.g., Nd^{3+}, Pr^{3+}, Yb^{3+}, Ho^{3+}, Tm^{3+}, and Er^{3+}) with the emission wavelengths tunable from ~900 to ~1550 nm, as shown in Fig. 5.38A and B. Under the excitation of ultraviolet or visible light, the Ca_2SnO_4 phosphors doped with trivalent lanthanide exhibit strong photoluminescence in the 750–1550 nm wavelength range, which mostly falls into the shortwave infrared spectral region. Under 380 nm light excitation, the $Ca_2SnO_4:Er^{3+}$ phosphor shows two

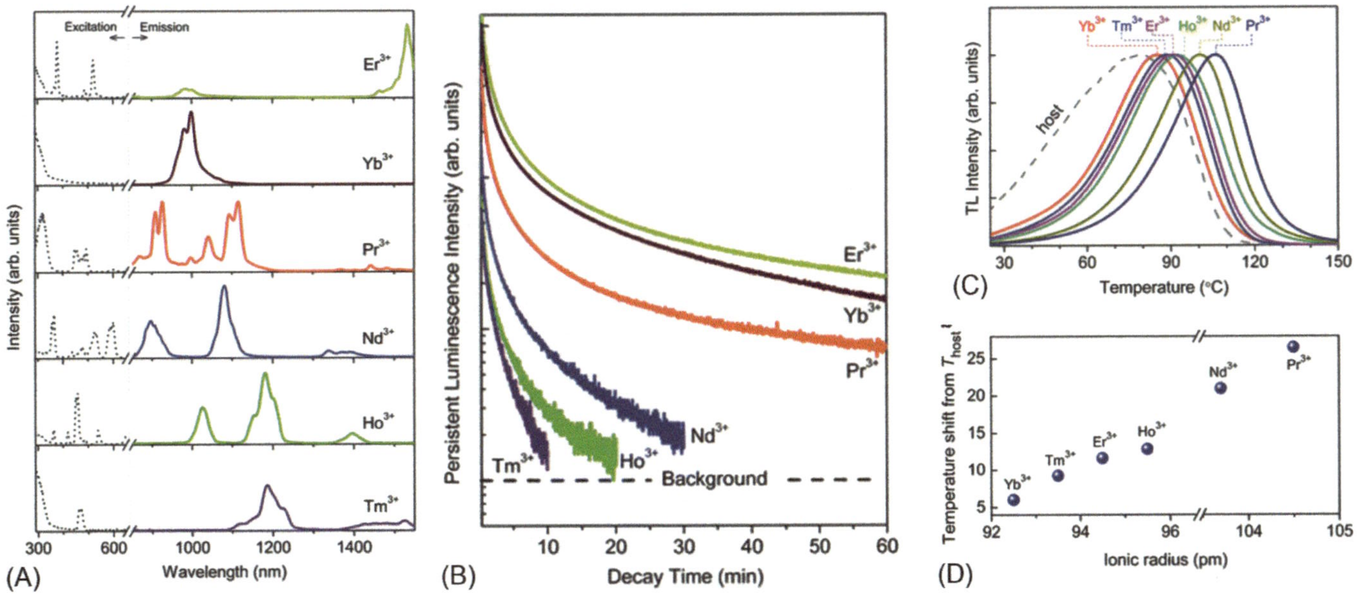

Fig. 5.38 (A) Photoluminescence emission and excitation spectra. (B) Persistent phosphorescence decay curves. (C) Thermoluminescence spectra of Ca_2SnO_4:Ln^{3+} (Ln = Er, Yb, Pr, Nd, Ho, Tm) persistent phosphors. (D) Shift of the thermoluminescence peak position (i.e., temperature) of Ca_2SnO_4: Ln^{3+} relative to that of the Ca_2SnO_4 host.

Reproduced with permission from J. Y. Kuang, Y. L. Liu, J. X. Zhang, Effects of RE^{3+} as a co-dopant in blue-emitting long-lasting phosphors, $Sr_3Al_{10}SiO_20$: Eu^{2+}. J. Mater. Sci., 2006, 41, 5500–5503, Copyright 2017, The Royal Society of Chemistry.

photoluminescence emission bands in the shortwave infrared region—a weak band peaking at 985 nm and a strong band peaking at ~1533 nm—which are attributed to the $^4I_{11/2} \rightarrow {}^4I_{15/2}$ and $^4I_{13/2} \rightarrow {}^4I_{15/2}$ transitions of the Er^{3+} ion, respectively. Under 300-nm light excitation, $Ca_2SnO_4:Yb^{3+}$ phosphor exhibits two emission bands peaking at 980 and 1000 nm, owing to the transitions between the $^2F_{7/2}$ and $^2F_{5/2}$ manifolds of the Yb^{3+} ion, respectively. Under 450-nm light excitation the $Ca_2SnO_4:Pr^{3+}$ phosphor shows five main emission lines in the shortwave infrared region, which correspond to the Pr^{3+} ions $^1D_2 \rightarrow {}^3H_6,{}^3F_2$ transition at 925 nm, $^1G_4 \rightarrow {}^3H_4$ transition at 997 nm, $^1D_2 \rightarrow {}^3F_3$ transition at 1040 nm, $^1D_2 \rightarrow {}^3F_4$ transition at 1115 nm, and $^1D_2 \rightarrow {}^1G_4$ transition at 1440 nm. Under 365-nm light excitation, the $Ca_2SnO_4:Nd^{3+}$ phosphor gives characteristic emissions of Nd^{3+} ion, showing three emission bands peaking at 900, 1080, and 1350 nm due to the $^4F_{3/2} \rightarrow {}^4I_J$ ($J = 9/2$, 11/2, and 13/2) transitions, respectively. Under 360-nm light excitation, $Ca_2SnO_4:Ho^{3+}$ phosphor presents three emission peaks at 1026, 1180, and 1395 nm, which correspond to the $^5S_2,{}^5F_4 \rightarrow {}^5I_6$, $^5I_6 \rightarrow {}^5I_8$, and $^5F_5 \rightarrow {}^5I_6$ optical transitions of Ho^{3+}, respectively. Under 470-nm light excitation the emission lines of the $Ca_2SnO_4:Tm^{3+}$ phosphor are observed at 1195 nm for the $^1G_4 \rightarrow {}^3H_4$ transition and 1460 nm for the $^3H_4 \rightarrow {}^3F_4$ transition of Tm^{3+} ion. Fig. 5.38B shows the persistent phosphorescence decay curves of the Ca_2SnO_4 phosphors doped with trivalent lanthanide after irradiation by a 254-nm ultraviolet lamp for 10 min. The result shows decay times of 10 h for $Ca_2SnO_4:Er^{3+}$, $Ca_2SnO_4:Yb^{3+}$, and $Ca_2SnO_4:Pr^{3+}$ and 1 h, 30 min, and 20 min for $Ca_2SnO_4:Nd^{3+}$, $Ca_2SnO_4:Ho^{3+}$, and $Ca_2SnO_4:Tm^{3+}$, respectively. From Fig. 5.38C and D, they suggested that the trap distribution in the Ca_2SnO_4 host is significantly affected by the introduction of trivalent lanthanide ions; as the ionic radius of trivalent lanthanide ions increases, the traps in the phosphors become deeper (i.e., moving to a higher-temperature region). This change in the thermoluminescence band may result from lattice distortions caused by the difference in the trivalent lanthanide ion sizes, as trivalent lanthanide ions are closely coupled to the traps. This study greatly expands the field of persistent phosphorescence in the shortwave infrared spectral region, and the trivalent lanthanide-activated Ca_2SnO_4 phosphors are expected to find promising applications in a variety of areas, such as biomedical imaging and night vision surveillance.

Another stannate host was discussed by Qiu et al. in 2015 [216]. They developed a series of efficient near-infrared Cr^{3+}-doped nongallate persistent phosphors ($Zn_2SnO_4:Cr^{3+}$ and $Zn_{2-x}Al_{2x}Sn_{1-x}O_4:Cr^{3+}$), which showed broad emission band (650–1200 nm, peaking at 800 nm) and long afterglow time of 35 h (Fig. 5.39C–E). Zn_2SnO_4 with the inverse spinel crystal structure was employed as the host, in which the Zn^{2+} and Sn^{4+} sites with the distorted octahedral coordination are available for doping of Cr^{3+} ions (Fig. 5.39A). In Zn_2SnO_4, all of the Sn octahedral (SnVI) cations and half of the Zn octahedral (ZnVI) cations occupy octahedral sites, whereas the other half of the Zn tetrahedral (ZnIV) cations occupied tetrahedral sites. Based on this arrangement, in addition to the intrinsic Zn vacancies and Zn interstitials, some antisite defects (such as [SnVI] and [ZnVI] cations, which can freely exchange places in octahedral sites) and nonequivalent substitutional defects (SnCr or ZnCr) are expected to be generated. The Zn and Sn octahedra surrounding the Cr^{3+} centers present weaker crystal

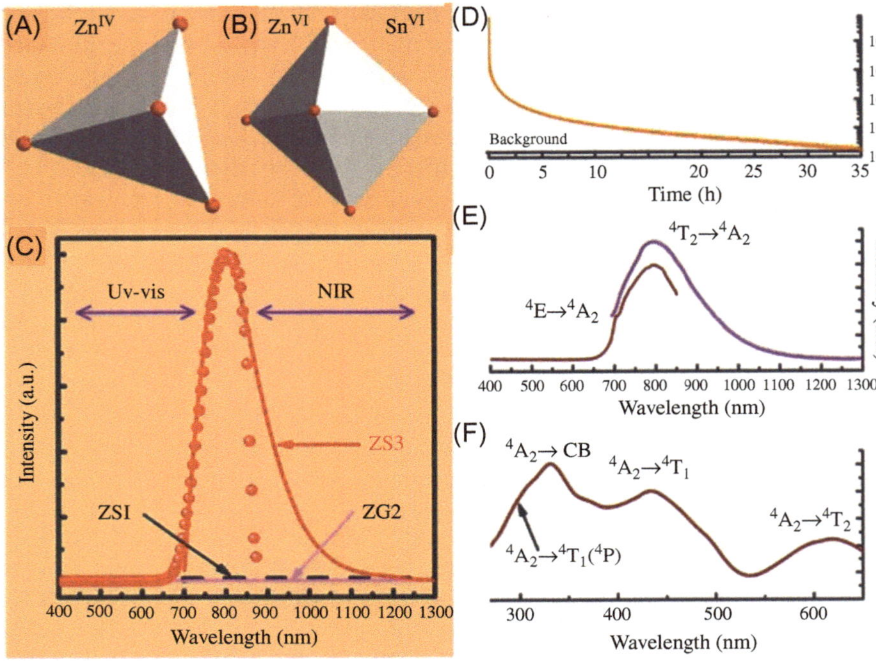

Fig. 5.39 Persistent phosphorescence of Zn_2SnO_4: Cr^{3+} phosphor. (A, B) Octahedral and tetrahedral coordination in the crystal structure of Zn_2SnO_4. (C) Normalized visible near-infrared persistent phosphorescence spectra of Zn_2SiO_4 (*black dotted line*), Zn_2GeO_4 (*pink line*), and Zn_2SnO_4 (*red dot and curve*) phosphors. The *red dot line* and *curve* are measured using Si and InGaAs detectors, respectively. The measurements were taken 1 min after irradiation ceased. (D) Afterglow decay curve monitored at 800-nm emission. (E, F) Normalized photoluminescence spectra in the visible and near-infrared regions excited at 330 nm and photoluminescence excitation spectra monitored at 800 nm for the Zn_2SnO_4: Cr^{3+} phosphor. Before the measurements the phosphors were preannealed at 600 K.
Reproduced with permission from Y. Li, Y. Y. Li, R. C. Chen, K. Sharafudeen, S. F. Zhou, M. Gecevicius, H. H. Wang, G. P. Dong, Y. L. Wu, X. X. Qin, J. R. Qiu, Tailoring of the trap distribution and crystal field in Cr^{3+}-doped non-gallate phosphors with near-infrared long-persistence phosphorescence. NPG Asia Mater., 2015, 7, e180, Copyright 2015, Nature Publishing Group.

field strength. As we know the electronic configuration of Cr^{3+} was affected strongly by the arrangement of surrounding ligands. The dominant emission peak of a Cr^{3+} center experienced a shift due to the change in crystal field strength as a result of Al doping and the formation of $Zn_{2-x}Al_{2x}Sn_{1-x}O_4$:$Cr^{3+}$ solid solution. Al^{3+} ions are incorporated into the crystal phase by replacing octahedral Zn^{2+} and Sn^{4+} in the disordered ZnO_6 and SnO_6 octahedron. Therefore the crystal field strength surrounding Cr^{3+} increases as the Cr^{3+}–O^{2-} distances are reduced due to the substitution of Zn^{2+} (74 pm) and Sn^{4+} (69 pm) with the smaller Al^{3+} (53 pm) ions. The increase of crystal field strength thus allows for the tuning in the photoluminescence peak (from 722 to 800 nm) and afterglow time up to 24 h, as shown in Fig. 5.40.

5.5.1.7 Germanates

Germanates are other promising hosts to generate near-infrared persistent phosphorescence for bioimaging applications. Liang et al. developed red/near-infrared/shortwave infrared multiband persistent phosphorescence by doping Pr^{3+} into germinate host (i.e., $MgGeO_3$). Under 254-nm light excitation, Fig. 5.41A reveals three intense emission bands at around 625, 900, and 1085 nm, which are assigned to $^1D_2 \rightarrow ^3H_4$, $^1D_2 \rightarrow ^3H_6$, $^3F_2$, and $^1D_2 \rightarrow ^3F_{3,4}$ transitions of Pr^{3+} ions, respectively. They also observed very long persistent phosphorescence time (>120 h) for the emission bans after stopping the ultraviolet light excitation, as shown in Fig. 5.41B and the inset of Fig. 5.41A [139].

However, in 2019, Shi et al. developed a superlong ultraviolet A region persistent luminescent material, $LiYGeO_4:Bi^{3+}$ [250]. After irradiation at 254 nm for 10 min, the ultraviolet persistent phosphorescence of $LiYGeO_4:Bi^{3+}$ peaking at 350 nm can persist for more than 300 h (Fig. 5.42). In addition, the $LiYGeO_4:Bi^{3+}$ also possesses a strong ultraviolet emission with a power density of 11.83 mW m^{-2} at 10 s after the removal of the excitation source. As shown in Fig. 5.42B, even after 300 h, the ultraviolet persistent emission can still be clearly detected, which is the same as the photoluminescence emission spectrum observed for Bi^{3+} centers. Furthermore, $LiYGeO_4$: Bi^{3+} also exhibits photostimulated persistent phosphorescence, which can be easily activated with a red light-emitting diode or near-infrared laser irradiation after the long-term decay. The observed superlong ultraviolet A region persistent phosphorescence behavior in $LiYGeO_4:Bi^{3+}$ phosphor greatly expands the application of persistent luminescent materials in the biomedicine, energy, environmental, and catalysis fields. To extend the exploration of ultraviolet persistent luminescent materials, Liang et al. developed a new ultraviolet persistent phosphor, $Sr_2MgGe_2O_7:Pb^{2+}$, which exhibits single-band persistent phosphorescence peaking at 370 nm and a long persistence time of more than 12 h as shown in Fig. 5.43 [253]. The successful development of the $Sr_2MgGe_2O_7:Pb^{2+}$ persistent phosphor opens a new avenue for rare earth–free low-cost persistent phosphors research and its applications.

5.5.1.8 Wolframites

In contrast with silicates, aluminates, gallates, and stannates, wolframite-based phosphors are rarely investigated in more details by researchers to examine whether it can generate persistent phosphorescence. A series of Eu^{3+}-/Sm^{3+}-codoped $CaWO_4$ phosphors were synthesized by Kang et al. [67]. The effect of the doping concentration of Sm^{3+} on the luminescence properties of the obtained products was investigated in this work, and the optimal Sm^{3+} concentration was experimentally determined to be 0.5%. The photoluminescence properties indicate that there is an efficient energy transfer from Sm^{3+} to Eu^{3+}. Red long afterglow originated from the $^5D_0 \rightarrow ^7F_J$ (J=0, 1, 2, 3, 4)

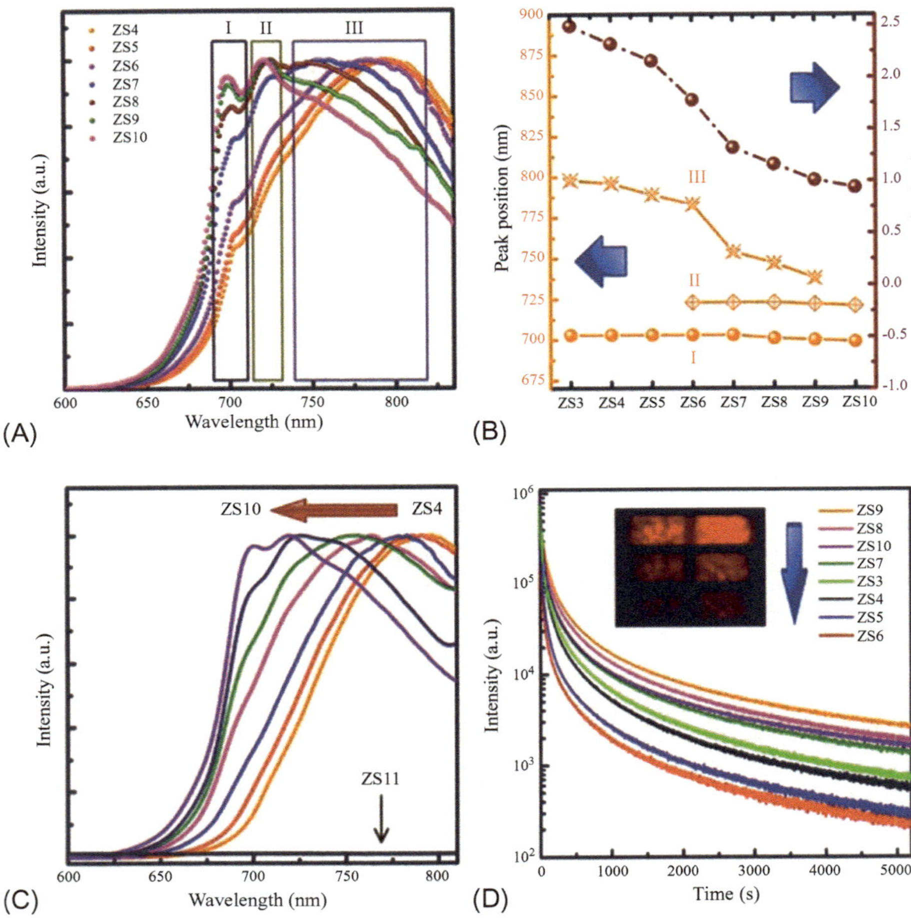

Fig. 5.40 (A) Normalized photoluminescence spectra as a function of Al concentration excited at 320 nm (Sample formulas are $Zn_{2\text{-}x}Al_{2x}Sn_{1\text{-}x}O_4$, $x=0.001$ (ZS4), 0.005(ZS5), 0.01(ZS6), 0.05 (ZS7), 0.1(ZS8), 0.2(ZS9), or 0.4(ZS10)). (B) Dependence of emission peak positions ([I] [II] and [III] labeled in Fig. 3.21A (orange dots) and intensity ratio of the emission peaks [III]/[I] (brown dots) on Al concentration. (C) Normalized persistent phosphorescence spectra for samples with different Al concentration. The measurements were taken at 1 min after irradiation ceased. (D) Persistence luminescence decay curves monitored at, respectively, optimized emission wavelengths of the samples Zn_2SnO_4 and ZS4–ZS10 (at wavelength 800, 798, 792, 783, 753, 741, 722, and 722 nm). The inset shows near-infrared images of samples Zn_2SnO_4 (left) and ZS9 (right) acquired at different time intervals (15 min, 12 and 24 h); camera parameters were unchanged during the imaging process.

Reproduced with permission from Y. Li, Y. Y. Li, R. C. Chen, K. Sharafudeen, S. F. Zhou, M. Gecevicius, H. H. Wang, G. P. Dong, Y. L. Wu, X. X. Qin, J. R. Qiu, Tailoring of the trap distribution and crystal field in Cr^{3+}-doped non-gallate phosphors with near-infrared long-persistence phosphorescence. NPG Asia Mater., 2015, 7, e180, Copyright 2015, Nature Publishing Group.

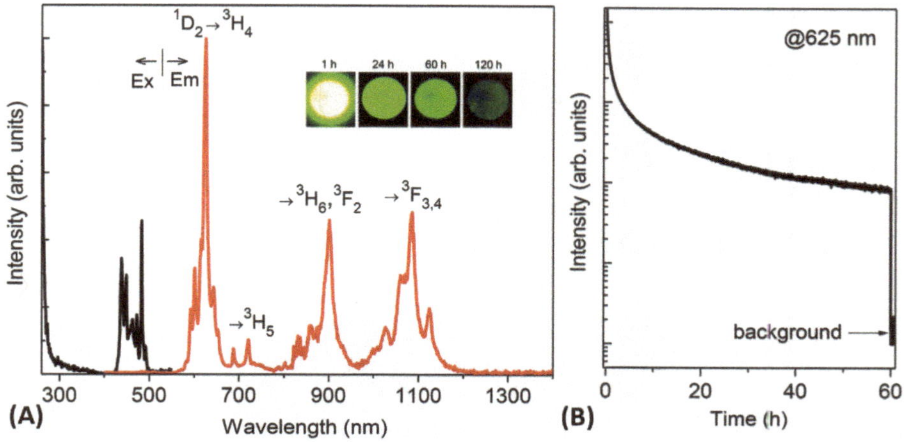

Fig. 5.41 (A) Photoluminescence emission and excitation spectra of the $MgGeO_3:Pr^{3+}$ phosphor. The emission spectrum (right panel) was obtained under 254-nm light excitation. The excitation spectrum (left panel) was acquired by monitoring at 625-nm emission. The inset is persistent phosphorescence images taken using a night vision monocular at different decay times from 1 to 120 h. (B) Persistent phosphorescence decay curve of the $MgGeO_3:Pr^{3+}$ phosphor obtained by monitoring at 625-nm emission.
Reproduced with permission from Y. Liang, F. Liu, Y. Chen, X. Wang, K. Sun, Z. Pan, Red/near-infrared/short-wave infrared multi-band persistent luminescence in Pr^{3+}-doped persistent phosphors. Dalton Trans., 2017, 46, 11149–11153, Copyright 2017, American Chemical Society.

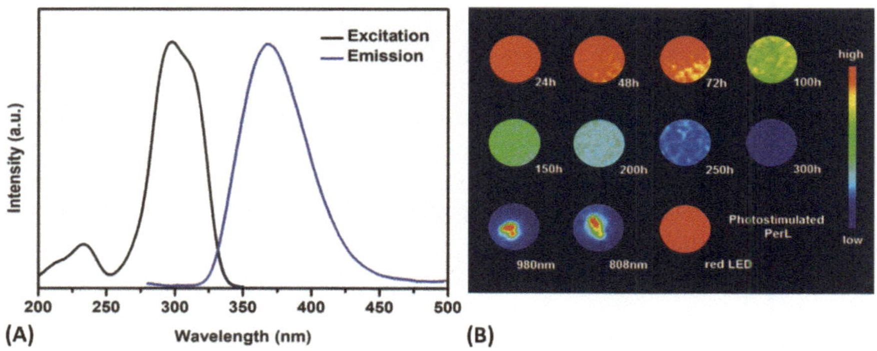

Fig. 5.42 (A) Photoluminescence excitation and emission spectra of $LiYGeO_4:Bi^{3+}$.The emission spectrum was recorded under 254-nm light excitation, and the excitation spectrum was acquired by monitoring the emission at 365 nm. (B) Ultraviolet A persistent phosphorescence images of $LiYGeO_4:Bi^{3+}$ acquired at different persistent phosphorescence decay times (24–300 h) after 254-nm ultraviolet lamp excitation for 10 min; bottom row: photostimulated persistent phosphorescence images of $LiYGeO_4:Bi^{3+}$ acquired after a 300 h decay time.
Reproduced with permission from J. Shi, X. Sun, S. Zheng, X. Fu, Y. Yang, J. Wang, H. Zhang, Super-long persistent luminescence in the ultraviolet A region from a Bi^{3+}-doped $LiYGeO_4$ phosphor. Adv. Opt. Mater., 2019, 7, 1900526, Copyright 2019, Wiley.

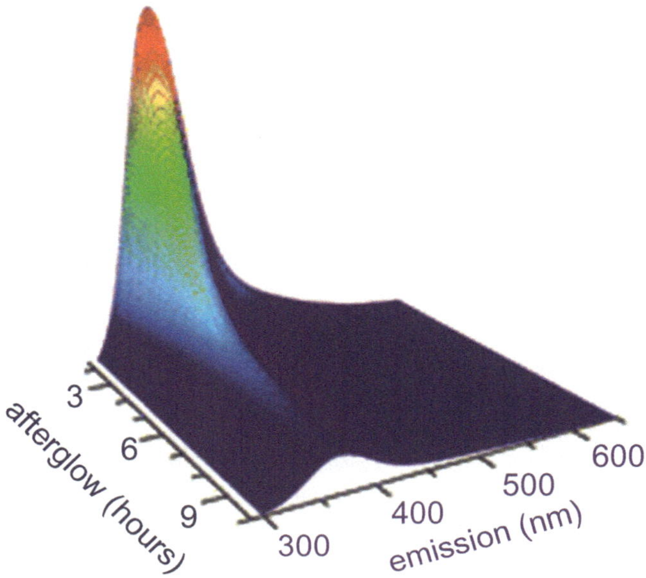

Fig. 5.43 3-D photoluminescence afterglow decay spectrum of $Sr_2MgGe_2O_7$:Pb^{2+} phosphor. Reproduced with permission from Y. Liang, F. Liu, Y. Chen, K. Sun, Z. Pan, Long persistent luminescence in the ultraviolet in Pb^{2+}-doped $Sr_2MgGe_2O_7$ persistent phosphor. Dalton Trans., 2016, 45, 1322–1326, Copyright 2016, American Chemical Society.

transitions of Eu^{3+} was observed after samples were excited at 254 nm, and the duration of the optimal sample can last for more than 35 min in dark. In another study a series of Eu^{3+} singly doped $CaWO_4$ phosphors were synthesized by solid-state reaction [113]. Red long afterglow was observed after samples were excited at 254 nm. The main emission peaks were ascribed to the $^5D_0 \rightarrow {}^7F_J$ ($J=0, 1, 2, 3, 4$) transitions of Eu^{3+}, and the duration of afterglow emission for the optimal sample could last for nearly 40 min in the dark seen with naked eyes.

5.5.1.9 *Titanates and zirconates*

Similar as wolframites the number of reports that focused on titanate- and zirconate-based persistent phosphors isn't abundant. Sun et al. observed red-orange long afterglow in $BaZrO_3$ powder phosphors doped with Eu^{3+} and Ti^{4+} prepared by the solid-state reaction method [69]. In this study the luminescence of the Eu^{3+} becomes weaker with the increase of Ti^{4+} concentration, and the optimum concentration of Ti^{4+} ions is 1.5%. Compared with the $BaZrO_3$:$0.025Eu^{3+}$ phosphors, the afterglow of the $BaZrO_3$: Eu^{3+}, Ti^{4+} phosphors exhibits longer decay time of up to 460 s. The results imply that Ti doping can be used to modulate the afterglow properties of $BaZrO_3$-based phosphors. Wang et al. studied the thermoluminescence, photostimulated luminescence, and photoluminescence of Pr^{3+}-doped $Ca_4Ti_3O_{10}$ to explore the optical storage property of this material [155]. The trivalent lanthanide (such as Y^{3+}, La^{3+}, Ce^{3+}, Gd^{3+}, and Dy^{3+}) codoping did not obviously alter the position of the emission peak, and no characteristic emissions of trivalent lanthanide ions are observed. This result indicates that Pr^{3+} ions act as emission center and trivalent lanthanide ions act as trap center in the codoped samples. Nevertheless, significant modification was found in the decay curves, especially for the Pr^{3+}-/Y^{3+}-codoped samples. They concluded that the photoluminescence and luminescence decay times of the Pr^{3+}-/Y^{3+}-codoped sample increase by 156% and 245% compared with the samples without doping of Y^{3+} ions under the same condition. To reveal the influence of excitation wavelengths on the persistent phosphorescence, the relationship between excitation energy and persistent intensity was investigated. Evidently the persistent phosphorescence spectrum is similar at different excitation wavelengths (ultraviolet blue), which means the persistent phosphorescence of the present samples can be efficiently excited by a broad spectral range of ultraviolet and blue light.

5.5.1.10 Niobates

Alkaline niobates, such as lithium niobate ($LiNbO_3$), sodium niobate ($NaNbO_3$), and potassium niobate ($KNbO_3$) crystals, are attractive candidates for host materials given their outstanding ferroelectric, piezoelectric, electrooptical, acousto-optical, and nonlinear optical characteristics, as well as their excellent mechanical, thermal, and chemical stability. Lin et al. observed persistent phosphorescence in Mg^{2+}- and Pr^{3+}-codoped lithium niobate perovskite-like single crystal, where Mg^{2+} acts as trap centers [154]. As shown in Fig. 5.44, the excitation peaks (inset of Fig. 5.44A) at 462, 487, and 501 nm match the energies of the transitions from $^3H_4$ to ^3P_J ($J = 0$, 1, and 2) levels of Pr^{3+} ions, indicating the negligible influence of Mg^{2+} dopants. By contrast, Mg^{2+} dopants play an important role in modifying the broad excitation bands ranging from 300 to 430 nm. From the emission spectra of $LiNbO_3$:Pr^{3+}:Mg^{2+} and $LiNbO_3$:Pr^{3+} (Fig. 5.44A) recorded at excitation wavelength of 365 nm, the $^1D_2 \rightarrow {}^3H_4$ emission at 619 nm is dominating. Both $LiNbO^3$:Pr^{3+}:Mg^{2+} and $LiNbO_3$:Pr^{3+} emit bright red light visible to naked eye when exposed to the ultraviolet light-emitting diode. The afterglow decay curves of $LiNbO_3$:Pr^{3+}:Mg^{2+} and $LiNbO_3$:Pr^{3+} are recorded at 619 nm after irradiation by ultraviolet light-emitting diode for one minute as shown in Fig. 5.44B. The data have been recorded as a function of luminescence intensity (I) versus time (t) for duration of 10 minutes. The luminescence of $LiNbO_3$:Pr^{3+} suddenly falls to zero at the withdrawal of the excitation source, indicating that no afterglow is present in $LiNbO_3$:Pr^{3+}. The afterglow intensity of $LiNbO_3$:Pr^{3+}:Mg^{2+} decreases quickly in the first dozens of seconds and then decays slowly, which is visualized by the series of images taken in a dark room using a digital camera at the time interval of 30 s after the removal of ultraviolet light-emitting diode (inset of Fig. 5.44B). The persistent phosphorescence intensity of $LiNbO_3$:Pr^{3+}:Mg^{2+} decays to 10.2 mcd m^{-2} after 100 s after ceasing ultraviolet light-emitting diode irradiation, which is much higher than the observable limit of human eyes (0.32 mcd m^{-2}).

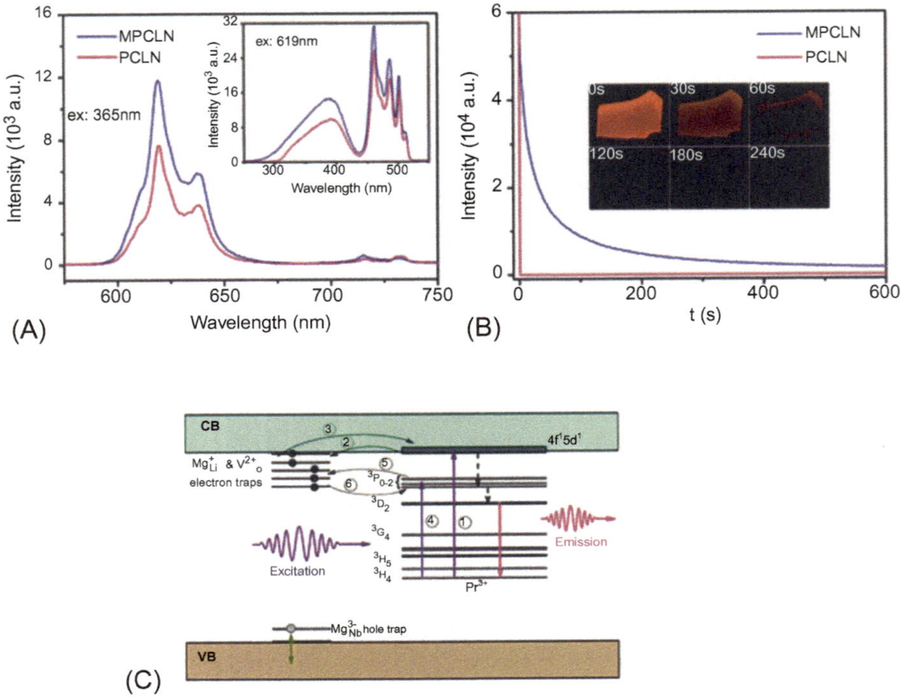

Fig. 5.44 (A) Photoluminescence and photoluminescence excitation (inset) spectra of $LiNbO_3$:Pr^{3+},Mg^{2+} (MPCLN) and $LiNbO_3$:Pr^{3+} (PCLN) recorded at room temperature. (B) Persistent phosphorescence decay curves for MPCLN and PCLN excited by the ultraviolet light-emitting diode at the region of 365–370 nm for 60 s, with the photographs of MPCLN shot in a dark room shown in the inset. (C) Schematic illustration of the luminescence and the persistent phosphorescence mechanism of the MPCLN single crystal phosphor.
Reproduced with permission from S. Lin, C. Xiong, D. Ma, H. Li, S. Long, B. Wang, Persistent luminescence found in Mg^{2+} and Pr^{3+} co-doped $LiNbO_3$ single crystal. J. Mater. Chem. C, 2018, 6, 10067–10072, Copyright 2018, The Royal Society of Chemistry.

Based on the lithium-vacancy model, the codopant Mg^{2+} would first substitute the antisite niobium Nb_{Li}^{4+}, forming Mg_{Li}^{+} and reducing V_{Li}^{-} until the threshold concentration (4.6 mol%) is reached, at which Nb_{Li}^{4+} is completely removed. At high concentration (>4.6 mol%) of Mg^{2+}, Mg^{2+} would enter into normal lithium and niobium sites, forming defects of Mg_{Li}^{+} and Mg_{Nb}^{3-}. Both types of defects tend to increase the Fermi energy, with the Mg_{Nb}^{3-} defect proven to be the major contributor. In consideration of charge compensation for Mg_{Nb}^{3-} and elimination of the conflicting Nb_{Li}^{4+}, oxygen vacancies V_O^{2+} would be possible at an excess concentration of Mg^{2+}. Thus the defects in $LiNbO_3$:Pr^{3+}:Mg^{2+} include Mg_{Nb}^{3-}, Mg_{Li}^{+}, V_{Li}^{-}, Pr_{Li}^{2+}, Pr_{Nb}^{2-}, and V_O^{2+} (Fig. 5.44C). Another report by Zhang et al. studied the persistent phosphorescence properties of a series of Pr^{3+}-doped perovskite-type $(Ca_{1-x}Na_x)[Ti_{1-x}Nb_x]O_3$:$0.001Pr^{3+}$ compounds, with $x = 0–1$ [142]. The authors found that the emission intensity shows the highest value at $x = 0.9$; however, the longest persistent time was only 1 min for the sample with $x = 0.1$.

5.5.2 Sulfides and oxysulfides

As known, sulfides are the oldest hosts that have been used to generate persistent phosphorescence. In 2018, Carvello et al. carried out a systematic investigation on the persistent phosphorescence properties of $RE_2O_2S:Ti^{3+}:Mg^{2+}$, RE = La, Gd, and Y [238]. As shown in Fig. 5.45, the emission spectra of the $RE_2O_2S:Ti^{3+},Mg^{2+}$ phosphors show a broad emission band originated from Ti^{3+} ions, and the emission peaks are located at 590, 615, and 625 nm for La, Gd, and Y oxysulfides, respectively (Fig. 5.45B). Ti^{3+} ions serve as the luminescent centers, and the broad emission bands for the oxysulfides cover a large part of the visible spectra and extend to the near-infrared region, resulting in an orange-red emission color. The position of emission band of Ti ions is strongly dependent on its oxidation state and the ligand field strength. The decay curves (Fig. 5.45A) for the $RE_2O_2S:Ti^{3+},Mg^{2+}$ materials show a very fast decay rate for $La_2O_2S:Ti^{3+},Mg^{2+}$, indicating the poor energy storage ability of this material. On the other hand, $RE_2O_2S:Ti^{3+},Mg^{2+}$ (RE: Y or Gd) phosphors show a persistent phosphorescence decay time, which can be observed for 5 h after cease of excitation by naked eye. These optical features are in good agreement with the trap structure from the normalized thermoluminescence glow curves (Fig. 5.45C). The thermoluminescence glow curves for $RE_2O_2S:Ti^{3+},Mg^{2+}$(RE: Y or Gd) are very similar, showing a high density of traps at 370 K, while $La_2O_2S:Ti^{3+},Mg^{2+}$ reveals strong thermoluminescence peak at 450 K, which reflects that the high density of traps was shifted to deeper level. These results suggest that the thermoluminescence peak at 350 K is responsible for the persistent phosphorescence in Gd_2O_2S and Y_2O_2S rather than La_2O_2S host doped with Ti^{3+} and Mg^{2+}. Consequently, white light-emitting diode devices can be fabricated by combining a single blue light-emitting diode with a proper conversion phosphor of $RE_2O_2S:Ti^{3+},Mg^{2+}$ (RE: Y or Gd) as shown in Fig. 5.45D. The design of white light-emitting diode is similar to the design of the blue-emitting InGaN light-emitting diode (450 nm) covered by the yellow-emitting $YAG:Ce^{3+}$ phosphor (550 nm). Application of $RE_2O_2S:Ti^{3+},Mg^{2+}$ as the spectral converting laser in solid-state lighting offers an additional feature of persistent phosphorescence after turning off the white light-emitting diode.

In the case of sulfide hosts, Wei et al. prepared a series of $Ba_{1-x}Ca_xS$: 2%Yb^{2+} ($x =$ 0, 0.1, 0.3, 0.5, 0.7, 0.9, 1) persistent phosphors via a conventional solid-state reaction method under H_2 atmosphere [186]. Wei investigated the photoluminescence and persistent phosphorescence properties of $Ba_{1-x}Ca_xS$: 2%Yb^{2+}. They observed that the $Ba_{1-x}Ca_xS$: 2%Yb^{2+} shows tunable broad emission from 586 to 650 nm as x value increases from 0 to 1. The obtained data indicated that the orange-red persistent phosphorescence can be lengthened up to 250 s by introducing a second metal cation (i.e., Ca^{2+}). Moreover the persistent phosphorescence can be effectively released under the excitation of a 980-nm laser diode. Due to the large penetration depth of the red and near-infrared light in tissues, $Ba_{1-x}Ca_xS$: 2%Yb^{2+} phosphors may find potential applications in fluorescence bioimaging.

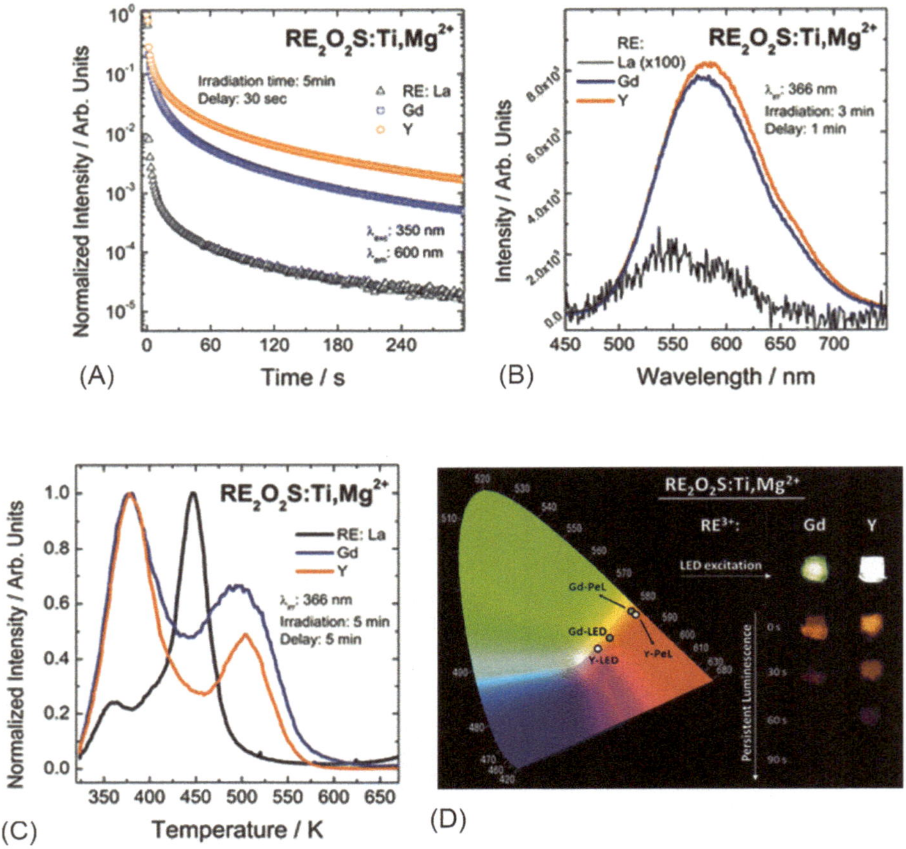

Fig. 5.45 (A) Persistent phosphorescence decay curves. (B) Emission spectra. (C) Thermoluminescence glow curves of RE_2O_2S:Ti,Mg^{2+} phosphors. (D) Persistent phosphorescence warm light-emitting diode based on a 280-nm light-emitting diode coated with polymer matrix containing RE_2O_2S:Ti,Mg^{2+} phosphorus particles (RE: Gd and Y). The international commission on illumination diagram shows the emission color of the light-emitting diode after excitation and the persistent phosphorescence after ceasing the excitation. Reproduced with permission from J. M. Carvalho, C. C. S. Pedroso, I. P. Machado, J. Hölsä, L. C. V. Rodrigues, P. Głuchowski, M. Lastusaari, H. F. Brito, Persistent luminescence warm-light LEDs based on Ti-doped RE_2O_2S materials prepared by rapid and energy-saving microwave-assisted synthesis. J. Mater. Chem. C, 2018, 6, 8897–8905, Copyright 2018, The Royal Society of Chemistry.

5.5.3 Nitrides and oxynitrides

Recently, nitrides have received growing interest as hosts to generate efficient persistent phosphorescence. Lv et al. reported Yb-doped $(Ba_{1-x}Sr_x)AlSi_5O_2N_7$ (x =0, 0.1, 0.2, 0.3) phosphors that showed intense persistent phosphorescence in the visible and near-infrared regions simultaneously [183]. As shown in Fig. 5.46C, under excitation of 437 nm, both Yb^{2+} and Yb^{3+} were found to be persistent phosphorescence centers, which presented a strong broad emission band centered at 664 nm ($4f^{13}5d \rightarrow 4f^{14}$) and a weak sharp emission at 980 nm (${}^2F_{5/2} \rightarrow {}^2F_{7/2}$), respectively. The long emission wavelength of Yb^{2+} is originated from the high N/O ratio in $Ba_{0.8}Sr_{0.2}AlSi_5O_2N_7$, which significantly lowers the emission energy level of Yb^{2+} ($4f^{13}5d^1$) owing to the large nephelauxetic effect, leading to longer wavelength emission. The excitation spectrum (monitored at 664 nm) exhibits two bands at 300 and 437 nm, while the

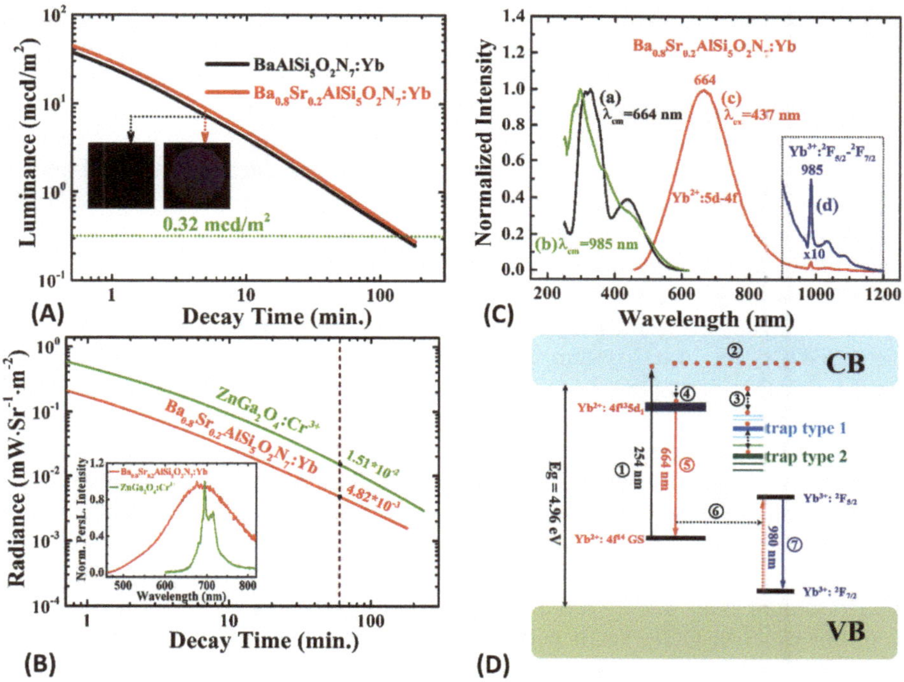

Fig. 5.46 (A) Decay curves of $BaAlSi_5O_2N_7$:Yb and $Ba_{0.8}Sr_{0.2}AlSi_5O_2N_7$:Yb (in units of mcd m^{-2}). The inset shows the digital photographs of the $BaAlSi_5O_2N_7$:Yb and $Ba_{0.8}Sr_{0.2}AlSi_5O_2N_7$:Yb phosphors taken 5 min later after ceasing 254-nm irradiation stops. (B) Decay curves of the $Ba_{0.8}Sr_{0.2}AlSi_5O_2N_7$:Yb and $ZnGa_2O_4:Cr^{3+}$ phosphors (in units of mW Sr^{-1} m^{-2}). Normalized persistent phosphorescence spectra of $Ba_{0.8}Sr_{0.2}AlSi_5O_2N_7$:Yb and $ZnGa_2O_4:Cr^{3+}$ are presented in the inset. (C) Normalized excitation (a, $\lambda_{em} = 664$ nm; b, $\lambda_{em} = 985$ nm) and emission (c, $\lambda_{ex} = 437$ nm) spectra of the $Ba_{0.8}Sr_{0.2}AlSi_5O_2N_7$:Yb phosphors. The blue solid line (D) is the 10-fold enlarged luminescence spectrum in the 900- to 1200-nm wavelength range. (E) Proposed persistent phosphorescence mechanism for the $Ba_{0.8}Sr_{0.2}AlSi_5O_2N_7$:Yb phosphors.

Reproduced with permission from Y. Lv, L. Wang, Y. Zhuang, T. L. Zhou, R. J. Xie, Discovery of the Yb^{2+}–Yb^{3+} couple as red-to-NIR persistent luminescence emitters in Yb-activated $(Ba_{1-x}Sr_x)AlSi_5O_2N_7$ phosphors. J. Mater. Chem. C, 2017, 5, 7095–7101, Copyright 2017, The Royal Society of Chemistry.

excitation spectrum monitored at 985 nm shows a shoulder around 446 nm, which matches well with the low-energy 4f–5d transitions of Yb^{2+}, confirming the energy transfer from Yb^{2+} to Yb^{3+} (Fig. 5.46D). However, they mentioned that all of the prepared Yb-activated phosphors show persistent phosphorescence and $Ba_{0.8}Sr_{0.2}AlSi_5O_2N_7$:Yb reveals the best persistent phosphor among $(Ba_{1-x}Sr_x)AlSi_5O_2N_7$:Yb phosphors. Fig. 5.46A demonstrates the persistent phosphorescence decay curves of $BaAlSi_5O_2N_7$:Yb and $Ba_{0.8}Sr_{0.2}AlSi_5O_2N_7$:Yb phosphors. We can observe that the emission of $BaAlSi_5O_2N_7$:Yb and $Ba_{0.8}Sr_{0.2}AlSi_5O_2N_7$:Yb can last for more than 2 h before reaching to 0.32 mcd m^{-2}. The $BaAlSi_5O_2N_7$:Yb and $Ba_{0.8}Sr_{0.2}AlSi_5O_2N_7$:Yb illustrate bright orange-red persistent phosphorescence owing to the broad band nature (inset of Fig. 5.46A). The prolonged persistent phosphorescence for $Ba_{0.8}Sr_{0.2}AlSi_5O_2N_7$:Yb with the Sr substitution may be caused by a suitable energy distance between the conduction band and the Yb^{2+} $(4f^{13}5d^1)$ energy level. Importantly the radiance of persistent phosphorescence in the $Ba_{0.8}Sr_{0.2}AlSi_5O_2N_7$:Yb phosphor at 60 min after ceasing ultraviolet irradiation was 4.82×10^{-3} mW Sr^{-1} m^{-2}, approximately one-third of the well-known $ZnGa_2O_4:Cr^{3+}$ phosphor as shown in Fig. 5.46B. In addition, $Ba_{0.8}Sr_{0.2}AlSi_5O_2N_7$:Yb shows a similar persistent phosphorescence peak compared with $ZnGa_2O_4:Cr^{3+}$ (inset of Fig. 5.46B). These results indicate that $Ba_{0.8}Sr_{0.2}AlSi_5O_2N_7$:Yb has potential applications for in vivo bioimaging.

Zhuang et al. successfully developed a series of deep trap persistent phosphors in the $SrSi_2O_2N_2:Ln^{2+}/Ln^{3+}$ system (Ln^{2+} = Yb, Eu; Ln^{3+} = Dy, Ho, Er), where Ln^{3+} served as trap centers [184]. As shown in Fig. 5.47A–H, the $SrSi_2O_2N_2:Yb^{2+}:Ln^{3+}$ and $SrSi_2O_2N_2:Yb^{2+}:Ln^{3+}$ phosphors show strong red (620 nm) and green (540 nm) luminescence intensity under excitation by 395- and 254-nm ultraviolet light, respectively. After switching off the excitation light, the samples codoped with various trivalent rare earth ions show large differences in the persistent phosphorescence property. The persistent phosphorescence in the $SrSi_2O_2N_2$:Yb,Dy phosphor after switching off the excitation is weak at room temperature. However, the $SrSi_2O_2N_2$:Yb,Ho sample gives intense red-orange persistent phosphorescence for a long duration. For the $SrSi_2O_2N_2$:Yb,Er pellet, extremely bright persistent phosphorescence is observed in the first several seconds, but it quickly decays in a short duration. Also, in the group of Eu^{2+}-activated phosphors, they find the similar effect of codopant on the brightness of green persistent phosphorescence. After being irradiated by a Xe lamp for 20 s (Fig. 5.47I and J), the initial intensity of the persistent phosphorescence in the $SrSi_2O_2N_2$:Yb,Dy phosphor is less than 10 mcd/m^2 and then drops below 0.32 mcd/m^2 in several minutes. The $SrSi_2O_2N_2$:Yb,Ho sample shows excellent red-orange persistent phosphorescence performance at room temperature, giving an afterglow time of approximately 200 min, which is two times longer than the decay time of the commercial red persistent phosphors (Ca,Sr)S:Eu. For the $SrSi_2O_2N_2$:Yb,Er sample, the initial intensity is more than 1000 mcd/m^2 (the highest value in this group), but the afterglow time is less than 50 min because of a high decay rate. In the case of Eu^{2+}-activated phosphors, the recorded afterglow time in the $SrSi_2O_2N_2$:Eu,Ho pellet reaches 360 min after irradiated by a 254-nm lamp, suggesting its potential for glow-in-the-dark applications. Finally the initial intensity of the persistent

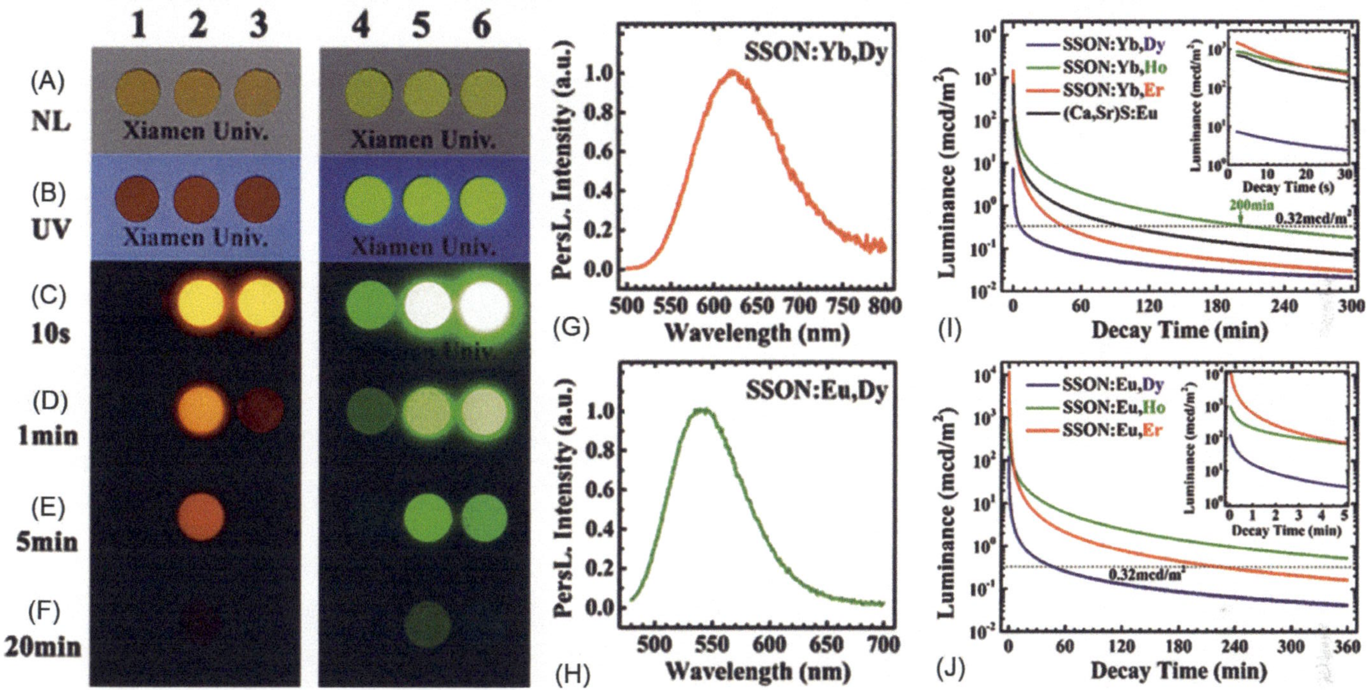

Fig. 5.47 (A–F) Photographic images of the pellet samples under natural light (A); under ultraviolet excitation (B); and 10 s (C), 1 min (D), 5 min (E), and 20 min (F) after removal of the excitation source at room temperature. The samples from [1, 2, 3] and [4, 5, 6] are [SSON:Yb,Ln_B, SSON = $SrSi_2O_2N_2$, Ln_B = Dy, Ho, Er for 1, 2, 3] and [SSON:Eu,Ln_B, SSON = $SrSi_2O_2N_2$, Ln_B = Dy, Ho, Er for 4, 5, 6], respectively. The ultraviolet excitation wavelengths for SSON:Yb,Ln_B and SSON:Eu,Ln_B are 395 and 254 nm, respectively. (G, H) Persistent phosphorescence spectra of SSON:Yb,Ho (G) and SSON:Eu,Ho (H). Exposure time of the digital camera is 1/200 s for (A, B) and 2 s for (C–F). (I, J) Persistent phosphorescence decay curves in SSON:Yb,Ln_B (I) and SSON:Eu,Ln_B (J) (Ln_B = Dy, Ho, and Er). The samples are irradiated by a Xe lamp (I) or a Hg lamp dominantly at 254 nm (J) for 20 s before recording the decay curves. The insets show the decay curves in the initial 5 min.

Reproduced with permission from Y. Zhuang, Y. Lv, L. Wang, W. Chen, T. L. Zhou, T. Takeda, N. Hirosaki, R. J. Xie, Trap depth engineering of $SrSi_2O_2N_2$: Ln^{2+}, Ln^{3+} (Ln^{2+}=Yb, Eu; Ln^{3+}=Dy, Ho, Er) persistent luminescence materials for information storage applications. ACS Appl. Mater. Interfaces, 2018, 10, 1854–1864, Copyright 2018, American Chemical Society.

phosphorescence in the $SrSi_2O_2N_2$:Eu,Dy sample is approximately 100 mcd/m^2 (1/100 of that in $SrSi_2O_2N_2$,Eu,Er), and the intensity after stopping irradiation is always much lower than that of the $SrSi_2O_2N_2$:Eu,Ho and $SrSi_2O_2N_2$:Eu,Er phosphors.

Xu et al. investigated the persistent phosphorescence properties of Mn^{2+}-doped AlN (AlN:Mn) red phosphor together with a commercial $SrAl_2O_4$:Eu^{2+}, Dy^{3+} green phosphor as a reference [208]. As shown in Fig. 5.48D, the AlN:Mn phosphor exhibits a red emission band centered at 600 nm owing to the Mn^{2+}: $^4T_1(^4G)\rightarrow{}^6A_1(^6S)$ transition with a relatively narrow full width at half maximum of 43 nm. In comparison with $SrAl_2O_4$:Eu^{2+},Dy^{3+}, $ZnGa_2O_4$:Cr^{3+},Bi^{3+}, and $MgGeO_3$:Mn^{2+},Bi^{3+} as shown in Fig. 5.48A–C, the emission of AlN:Mn powders was 0.65 mcd/m^2 at 60 min after ceasing ultraviolet illumination, and its afterglow duration above 0.32 mcd/m^2 could reach over 110 min, which is comparable with $ZnGa_2O_4$:Cr^{3+},Bi^{3+} phosphor. This result suggests that the prepared AlN:Mn phosphor possesses longer persistent phosphorescence duration than the commercial $MgGeO_3$:Mn^{2+},Bi^{3+} phosphor and shorter afterglow time compared with that of $SrAl_2O_4$:Eu^{2+},Dy^{3+} phosphor. However, Li et al. prepared persistent phosphors based on nitrides of $Ca_2Si_5N_8$:0.5%Eu^{2+},x%Tm^{3+} (x = 0, 0.5, 1, 2, 4) [264]. As shown in Fig. 5.49A, the emission spectrum of $Ca_2Si_5N_8$:0.5%Eu^{2+} exhibits strong orange emission bands in the wavelength range of 500–700 nm located at 582 nm that attributed to the allowed $4f^65d^1\rightarrow 4f^7$ transition

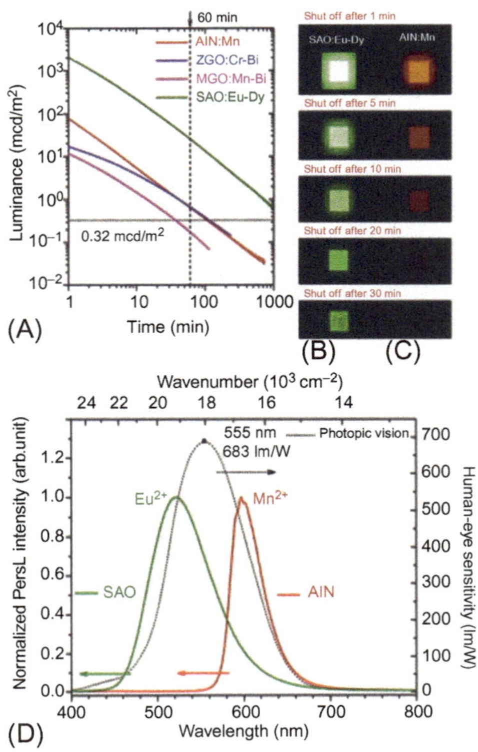

Fig. 5.48 (A) Persistent phosphorescence decay curves of the $SrAl_2O_4$:Eu^{2+}, Dy^{3+}(SAO:Eu-Dy), $ZnGa_2O_4$:Cr^{3+}, Bi^{3+}(ZGO:Cr-Bi), $MgGeO_3$:Mn^{2+}, Bi^{3+}(MGO:Mn-Bi), and AlN:Mn^{2+} phosphors as well as the photographs of the (B) $SrAl_2O_4$:Eu^{2+}, Dy^{3+} (C) AlN:Mn^{2+} powders with the same weight (0.2 g).(D) Normalized persistent phosphorescence spectra of the AlN:Mn^{2+} and $SrAl_2O_4$:Eu^{2+}, Dy^{3+} phosphors at 1 min after ceasing 254 and 365 nm excitation, respectively (the response curve of human-eye photopic vision plotted as a reference).
Reproduced with permission from J. Xu, N. J. Cherepy, J. Ueda, S. Tanabe, Red persistent luminescence in rare earth-free AlN:Mn^{2+} phosphor. Mater. Lett., 2017, 206, 175–177, Copyright 2017, Elsevier.

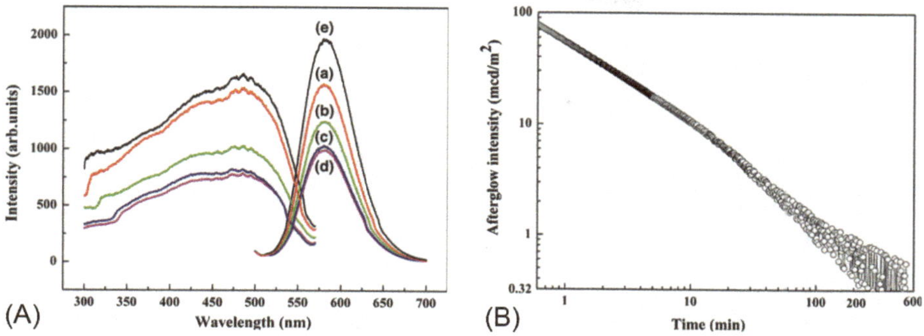

Fig. 5.49 (A) Excitation and emission spectra of (a) $Ca_2Si_5N_8$:0.5%Eu^{2+},1%Tm^{3+}, (b) $Ca_2Si_5N_8$:0.5%Eu^{2+},0.5%Tm^{3+}, (c) $Ca_2Si_5N_8$:0.5%Eu^{2+},2%Tm^{3+}, (d) $Ca_2Si_5N_8$:0.5% Eu^{2+},4%Tm^{3+}, and (e) $Ca_2Si_5N_8$:0.5%Eu^{2+} samples. (B) Afterglow decay curve of $Ca_2Si_5N_8$:0.5%Eu^{2+},1%Tm^{3+} after excitation for 15 min by a Xe lamp.
Reproduced with permission from J. Li, B. Lei, J. Qin, Y. Liu, X. Liu, Temperature-dependent emission spectra of $Ca_2Si_5N_8$:Eu^{2+}, Tm^{3+} phosphor and its afterglow properties. J. Am. Ceram. Soc., 2013, 96, 873–878, Copyright 2013, Wiley.

of Eu^{2+} ions. The substitution of Tm^{3+} ions into the $Ca_2Si_5N_8$:0.5%Eu^{2+} phosphor slightly decreases the luminescence intensity, and the optimal concentration is at 1% Tm^{3+}. Moreover the excitation spectrum monitored at 582 nm illustrates that this phosphor can be effectively excited in the range of 300–525 nm. After being excited by visible or ultraviolet light (Fig. 5.49B), the prepared phosphor (i.e., $Ca_2Si_5N_8$:0.5% Eu^{2+},1%Tm^{3+}) demonstrates orange afterglow time of more than 200 min until reach to 0.32 mcd/m^2. According to the general principles of the persistent phosphorescence phenomenon, if certain concentrations and depths of defects and traps can be introduced into the matrix to capture electrons or holes and then released them by thermal disturbance in ambient temperature, obvious afterglow emission can be observed.

5.5.4 Halides and oxyhalides

Owing to the low phonon energy of halide hosts, some scientists have focused on the synthesis of persistent phosphor-based halides to produce efficient persistent phosphorescence. Zheng et al. prepared a yellow-emitting persistent phosphor, Ca_2BO_3Cl:Eu^{2+},Dy^{3+} [50]. Ca_2BO_3Cl crystal has monoclinic structure with space group P_{121}/c_1 and provides two different sites for Ca^{2+} ions. The coordination number of Ca^{2+} is seven in which one site (Ca_1) is surrounded by 4O^{2-} ions and 3Cl^- ions and the other (Ca_2) by 5O^{2-} ions and 2Cl^- ions. As shown in Fig. 5.50A, the photoluminescence spectrum of Ca_2BO_3Cl:Eu^{2+} reveals only one asymmetric broad emission band located at 580 nm, ascribed to the 5d–4f transitions of Eu^{2+} ions, suggesting the doping Eu^{2+} ions could occupy both of the two Ca^{2+} sites in the Ca_2BO_3Cl crystal. However, the excitation spectrum shows that the prepared phosphor can be effectively excited with ultraviolet and/or blue light-emitting diode. As shown in Fig. 5.50B, the incorporation of Dy^{3+} ions that acts as trap centers doesn't

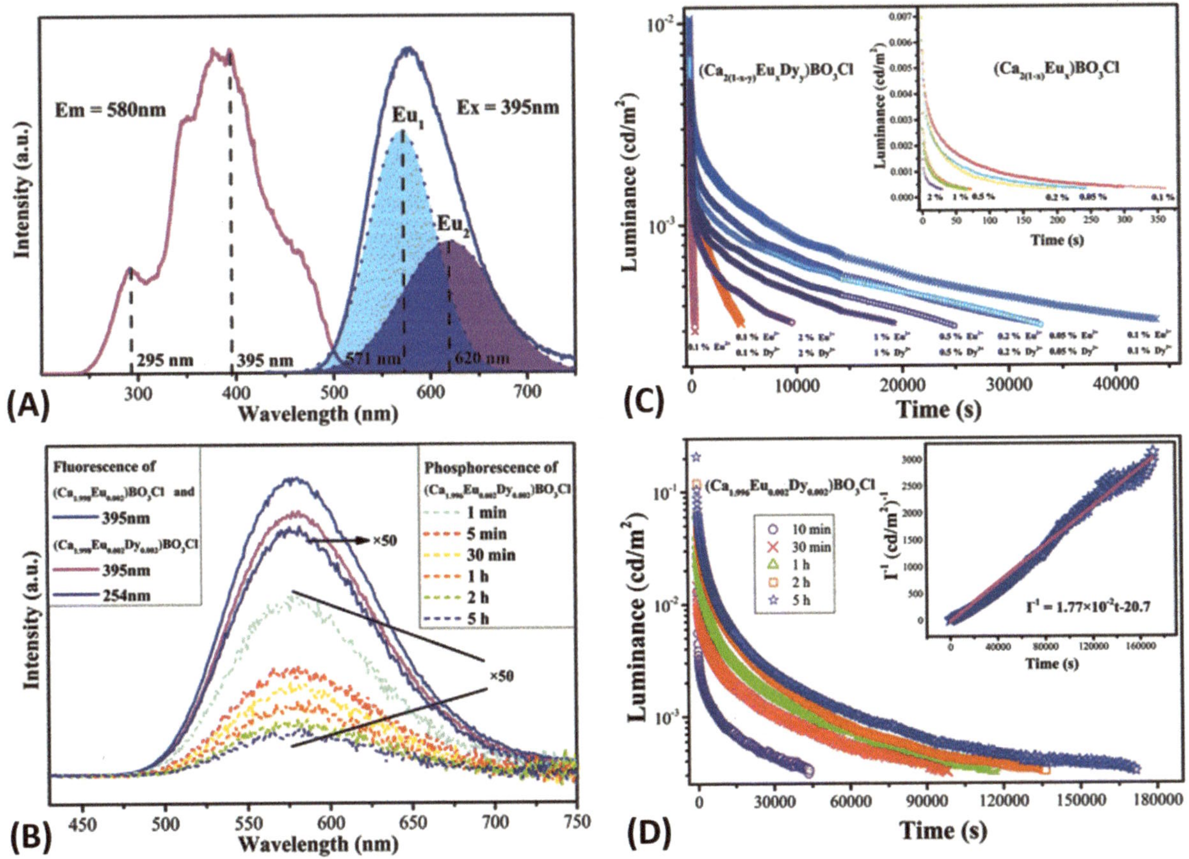

Fig. 5.50 (A) Excitation and emission spectra of the $(Ca_{1.998}Eu_{0.002})BO_3Cl$ phosphors. (B) Emission spectra of the $(Ca_{1.998}Eu_{0.002})BO_3Cl$ and $(Ca_{1.996}Eu_{0.002}Dy_{0.002})BO_3Cl$ phosphors and persistent phosphorescence spectra of the $(Ca_{1.996}Eu_{0.002}Dy_{0.002})BO_3Cl$ phosphors. (C) Afterglow decay curves of the $(Ca_{2(1-x-y)}Eu_xDy_y)BO_3Cl$ samples; (inset) afterglow decay curves of the series $(Ca_{2(1-x)}Eu_x)BO_3Cl$ samples. (D) Afterglow decay curve of the $(Ca_{1.996}Eu_{0.002}Dy_{0.002})BO_3Cl$ phosphors excited at different times. Inset: plot of reciprocal afterglow intensity (I^{-1}) versus time (t) of $(Ca_{1.996}Eu_{0.002}Dy_{0.002})BO_3Cl$ excited for 5 h.
Reproduced with permission from W. Zeng, Y. H. Wang, S. C. Han, W. B. Chen, G. Li, Y. Z. Wang, Y. Wen, Design, synthesis and characterization of a novel yellow long-persistent phosphor: Ca_2BO_3Cl: Eu^{2+}, Dy^{3+}, J. Mater. Chem. C. 2013, 1, 3004–3011, Copyright 2013, The Royal Society of Chemistry.

affect the photoluminescence spectrum and evidently enhances the persistent phosphorescence behavior of the phosphor that lasts up to 5 h after the 254-nm excitation source was switched off. To determine the optimal persistent phosphorescence duration, a series of samples of Eu^{2+} singly doped and Eu^{2+},Dy^{3+}-codoped Ca_2BO_3Cl with different concentrations were investigated as shown in Fig. 5.50C. The decay curves show that the optimal afterglow duration was obtained at 0.1% concentration of Eu^{2+} and Dy^{3+} ions (i.e., $Ca_{1.996}Eu_{0.002}Dy_{0.002}BO_3Cl$). As known the traps of most persistent phosphors could be filled within 10–20 min; thus extra excitation will not affect the afterglow properties. However, in $(Ca_{1.996}Eu_{0.002}Dy_{0.002})BO_3Cl$, they observed that the longer excitation time enhances the persistent phosphorescence time, until the excitation time reaches 5 h. As shown in Fig. 5.50D, the initial persistent phosphorescence of $(Ca_{1.996}Eu_{0.002}Dy_{0.002})BO_3Cl$ at 5 h is 20 times stronger than the 10 min one, and its afterglow can last for up to 48 h above the recognizable intensity level ($\geq$0.32 mcd m^{-2}). This result implies that Dy^{3+} ions should play a dominant trapping role in Ca_2BO_3Cl host.

Norrbo et al. prepared a heavy metal and rare earth–free low-cost phosphor, (Li, Na)$_8Al_6Si_6O_{24}(Cl,S)_2$:Ti, which can gives white persistent phosphorescence for 7 h above the 0.3 mcd m^{-2} limit and is observable for more than 100 h with a spectrometer [240]. Fig. 5.51A shows the photoluminescence emission spectrum of the synthetic hackmanite materials with the general formula $(Li,Na)_8Al_6Si_6O_{24}(Cl,S)_2$:2%Ti at room temperature. When excited by 310 nm, the $Na_8Al_6Si_6O_{24}(Cl,S)_2$:Ti demonstrates a broad blue emission band centered at 450 nm, assigned to Ti^{3+} ions that occupies the Al^{3+} tetrahedral sites. However, by replacing Na partly with Li in the $LiNa_7Al_6Si_6O_{24}(Cl,S)_2$:Ti, the PL peak (under excitation of 254 nm) remains unchanged, but the spectrum becomes more broader, resulting in white emission. The asymmetry of the emission bands suggests that the existence of two overlapping peaks at 452 and 510 nm is related to two different environments around Ti^{3+}. On the other hand the excitation spectrum of the blue emission in $Na_8Al_6Si_6O_{24}(Cl,S)_2$:Ti shows a broad spectrum (250–350 nm) centered at 310 nm, which is shifted to 275 nm after replacing Na with Li as shown in Fig. 5.51B. They suggest that the turning of emission color from blue to white and the shifting of excitation peak from 310 to 275 nm are because of the lack of absorption by the $(S_2)^{2-}$ species. In addition to the white photoluminescence, Fig. 5.51C shows the persistent phosphorescence decay curve of the $LiNa_7Al_6Si_6O_{24}(Cl,S)_2$:2%Ti phosphor after 30 min of excitation with a 254-nm ultraviolet lamp. It is typical decay curve for persistent phosphorescence behavior that the emission intensity decreases very fast in the beginning and then slows down for over 100 h. As shown in Fig. 5.51D, the decay curves for the commercial Y_2O_2S: Eu^{2+}, Ti^{4+}, Mg^{2+}, $CaAl_2O_4$: Eu^{2+}, Nd^{3+}, $SrAl_2O_4$: Eu^{2+},Dy^{3+}, and $Sr_2MgSi_2O_7$: Eu^{2+}, Dy^{3+} are plotted for comparison. The results indicate that the $Na_8Al_6Si_6O_{24}(Cl,S)_2$:Ti has clearly shorter persistent phosphorescence time than all of the plotted commercial phosphors. However, with 7 h of persistent phosphorescence above the 0.3 mcd m^{-2} limit, the performance of $LiNa_7Al_6Si_6O_{24}(Cl,S)_2$:Ti is comparable with the commercial $CaAl_2O_4$:Eu^{2+},Nd^{3+} and better than the performance ofY_2O_2S: Eu^{2+}, Ti^{4+}, Mg^{2+}. In addition, $LiNa_7Al_6Si_6O_{24}(Cl,S)_2$:Ti possesses

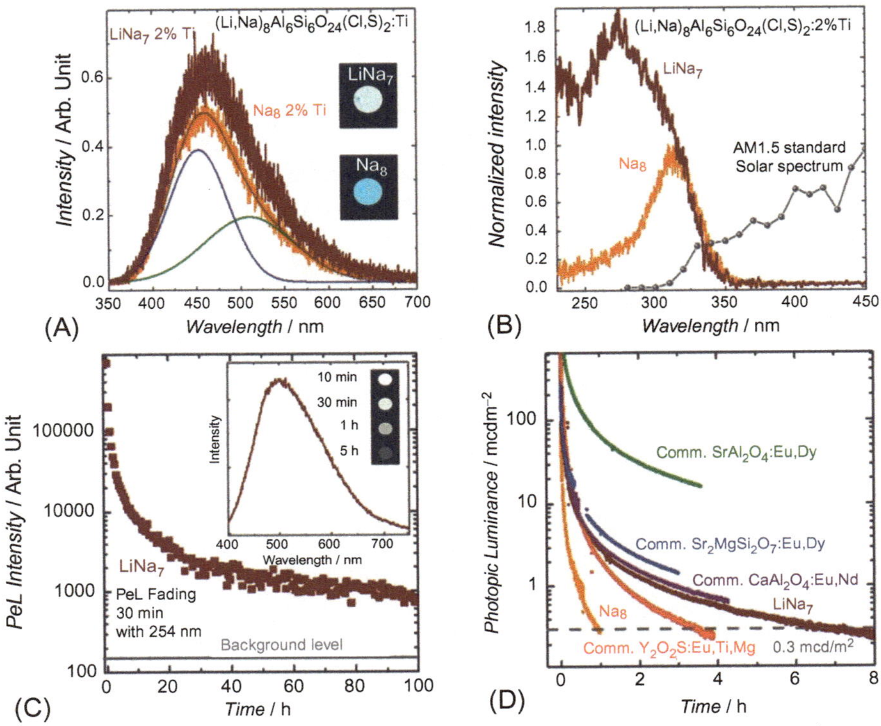

Fig. 5.51 Photoluminescence properties of the $(Li_xNa_{1-x})_8Al_6Si_6O_{24}(Cl_{1-2y}S_y)_2{:}2Ti^{3+}$ phosphors. (A) Emission spectra and photographs of the materials under 254 nm excitation. A deconvolution of the emission band of the Na_8 material is shown as well. (B) Afterglow excitation spectra for the $(Li_xNa_{1-x})_8Al_6Si_6O_{24}(Cl_{1-2y}S_y)_2{:}2\%Ti^{3+}$ samples and standard solar spectrum. (C) Afterglow decay curve for the $LiNa_7Al_6Si_6O_{24}(Cl_{0.8}S_{0.05\text{-}0.1})_2{:}2\%Ti^{3+}$ phosphor after irradiation for 30 min by 254 nm ultraviolet light. The inset shows the persistent phosphorescence spectrum and photographs of the material taken after removal of excitation light for different durations. (D) Comparison of the persistent phosphorescence decay curves of commercial materials and hackmanite measured as luminance.
Reproduced with permission from I. Norrbo, J. M. Carvalho, P. Laukkanen, J. Mäkelä, F. Mamedov, M. Peurla, H. Helminen, S. Pihlasalo, H. Härmä, J. Sinkkonen, M. Lastusaari, Lanthanide and heavy metal free long white persistent luminescence from Ti doped Li–hackmanite: a versatile, low-cost material. Adv. Funct. Mater., 2017, 27, 1606547, Copyright 2017, Wiley.

shorter persistent phosphorescence duration compared with $SrAl_2O_4$: Eu^{2+},Dy^{3+} and $Sr_2MgSi_2O_7$: Eu^{2+}, Dy^{3+} phosphors.

On the other hand, Yang et al. demonstrated a strategy for creating a new generation of persistent phosphors that exhibits strong ultraviolet C emission by judiciously selecting defect bearing fluoride elpasolite (i.e., Cs_2NaYF_6 as a host and Pr^{3+} ions as emitters) [158]. The resulting phosphor shows bright ultraviolet C emissions that can last for over 2 h after x-ray irradiation. As shown in Fig. 5.52A, after x-ray irradiation,

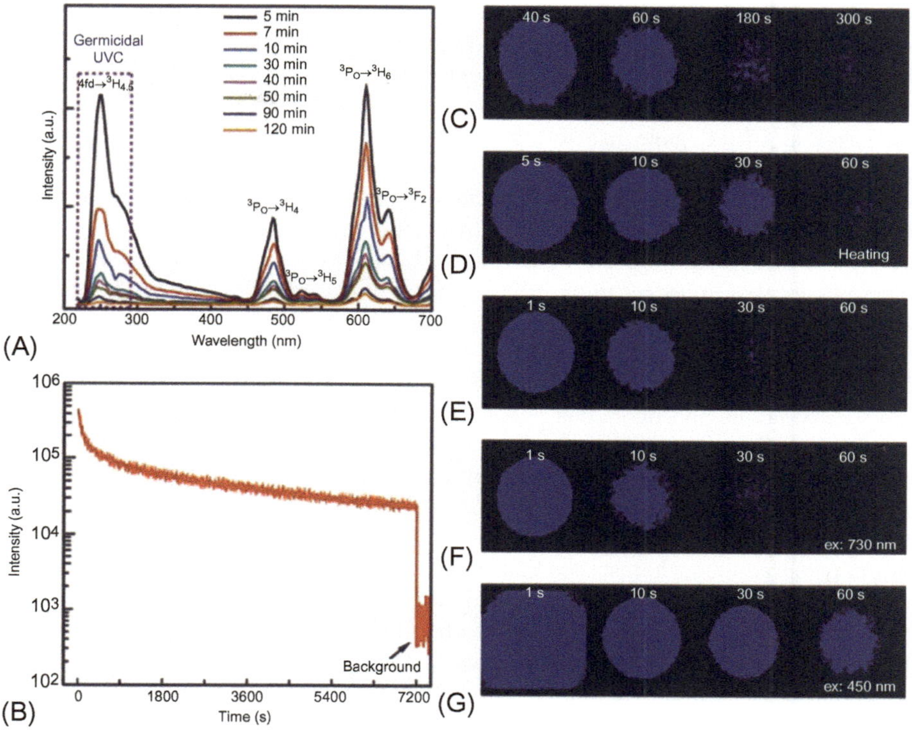

Fig. 5.52 Ultraviolet persistent phosphorescence is observed in $Cs_2NaY_{0.99}F_6$:$0.01Pr^{3+}$.
(A) Afterglow spectra recorded at different times after ceasing x-ray irradiation. (B) Afterglow decay detected at 250 nm as a function of time. The data were taken 5 min after stopping the x-ray irradiation. (C) Ultraviolet C images of phosphors taken at different afterglow times. (D) Ultraviolet C images of the 24-h-decayed phosphors heated to 200°C on a hot plate. (E–G) Ultraviolet C images of the 24-h decayed phosphors under laser irradiation with different wavelengths of (E) 793, (F) 730, and (G) 450 nm. The excitation power density is 1.77 W/cm^2 for the 793-, 730-, and 450-nm excitation wavelengths.
Reproduced with permission from Y. M. Yang, Z. Y. Li, J. Y. Zhang, Y. Lu, S. Q. Guo, Q. Zhao, X. Wang, Z. J. Yong, H. Li, J. P. Ma, Y. Kuroiwa, C. Moriyoshi, L. L. Hu, L. Y. Zhang, L. R. Zheng, H. T. Sun, X-ray-activated long persistent phosphors featuring strong UVC afterglow emissions. Light. Sci. Appl., 2018, 7, 88, Copyright 2018, Nature Publishing Group.

the emission band peaking at 250 nm and the emission shoulder at 270 nm can be assigned to the transitions of 4f5d → $^3H_4$ and 4f5d → $^3H_5$ of Pr^{3+}, respectively. In addition to the ultraviolet C emissions, visible emission bands are also observed. Fig. 5.52B displays the afterglow decay of $Cs_2NaY_{0.99}F_6$:$0.01Pr^{3+}$ detected at 250 nm following irradiation with an x-ray source for 1000 s with a dose of 20 Gy. Clearly the persistent phosphorescence in $Cs_2NaY_{0.99}F_6$:$0.01Pr^{3+}$ can last for over 2 h, and after 2 h, the intensity is still over one order of magnitude stronger than the background signal of the detection system. They further confirmed the ultraviolet

C emission by using ultraviolet C imager. Fig. 5.52C displays the dependence of the ultraviolet C emission intensity on the decay time for $Cs_2NaY_{0.99}F_6$:0.01Pr^{3+} after being irradiated by x-ray for 1000 s. Evidently the decay decreases much faster during the first several minutes, and then it slows down gradually. Heating the 24 h decayed phosphor at 200 °C gave rise to strong ultraviolet C luminescence again, with a glowing time of over 5 min (Fig. 5.52D). Interestingly, they find that these trapped electrons can also be photostimulated by laser irradiation with various photon energies (Fig. 5.52E–G). Specifically, irradiation of 450-nm light results in stronger emissions than those of 730- and 793-nm light, suggesting that there are more residual trapped electrons located in deeper traps.

5.5.5 Glasses

Unlike crystalline phosphors that possess persistent phosphorescence properties, the monitoring of persistent phosphorescence phenomenon in glasses has been investigated in some reports. In 1998, Qiu et al. reported a novel phenomenon in calcium aluminosilicate glasses doped with Ce^{3+}, Tb^{3+}, and Pr^{3+} [265]. After irradiation by an 800-nm femtosecond pulsed laser, the focal area of the laser in the glasses emits bright persistent phosphorescence that can be clearly seen with naked eye in the dark even 1 h after the removal of laser irradiation. Moreover, by selecting appropriate glass compositions and species of rare earth, optional three-dimensional image patterns emitting persistent phosphorescence in various colors, including blue, green, and red, can be formed within glass samples by moving the focal point of the laser. Based on absorption spectra, they concluded that the persistent phosphorescence is considered to be associated with thermostimulated recombination of holes and electrons at traps induced by the laser irradiation, which leaves holes and electrons in a metastable excited state at room temperature. Luo et al. fabricated a new type of Mn^{2+}-doped calcium aluminum germanate glasses ($(89-x)GeO_2$–$11Al_2O_3$–xCaO–1MnO, $x = 10, 15, 20, 25, 30, 35$) that demonstrated red to near-infrared persistent phosphorescence [261]. After switching off the 254-nm ultraviolet lamp, all Mn^{2+}-doped glasses shows red afterglow (Fig. 5.53C), where their afterglow intensity and duration decrease with increasing x values. As shown in Fig. 5.53A, the persistent phosphorescence decay curves are in good agreement with the photographs, suggesting the persistent phosphorescence time strongly depends on the glass composition. After 30 s of stopping ultraviolet lamp (Fig. 5.53B), the persistent phosphorescence spectrum at $x = 35$ shows broad red emission centered at 670 nm, and after additional 5 min, it shifts to 694 nm. However, the green persistent phosphorescence tail at 530–580 nm almost disappeared, and the remaining spectrum (i.e., deep-red region) gradually remains stable for 30 min. This result implies that Mn ions are located in both octahedral and tetrahedral coordination sites. To determine for how long the prepared glass can emit red emission, they measured the afterglow decay spectra at $x = 10$ as shown in Fig. 5.53D. Interestingly the persistent phosphorescence peak position at 664 nm within the first 30 s decay shifts to 684 nm with lengthened time, which shows a blueshift compared with the glass at $x = 35$. If the time of measurement is extended up to 24 h, the red band tail continues in expanding up to 820 nm.

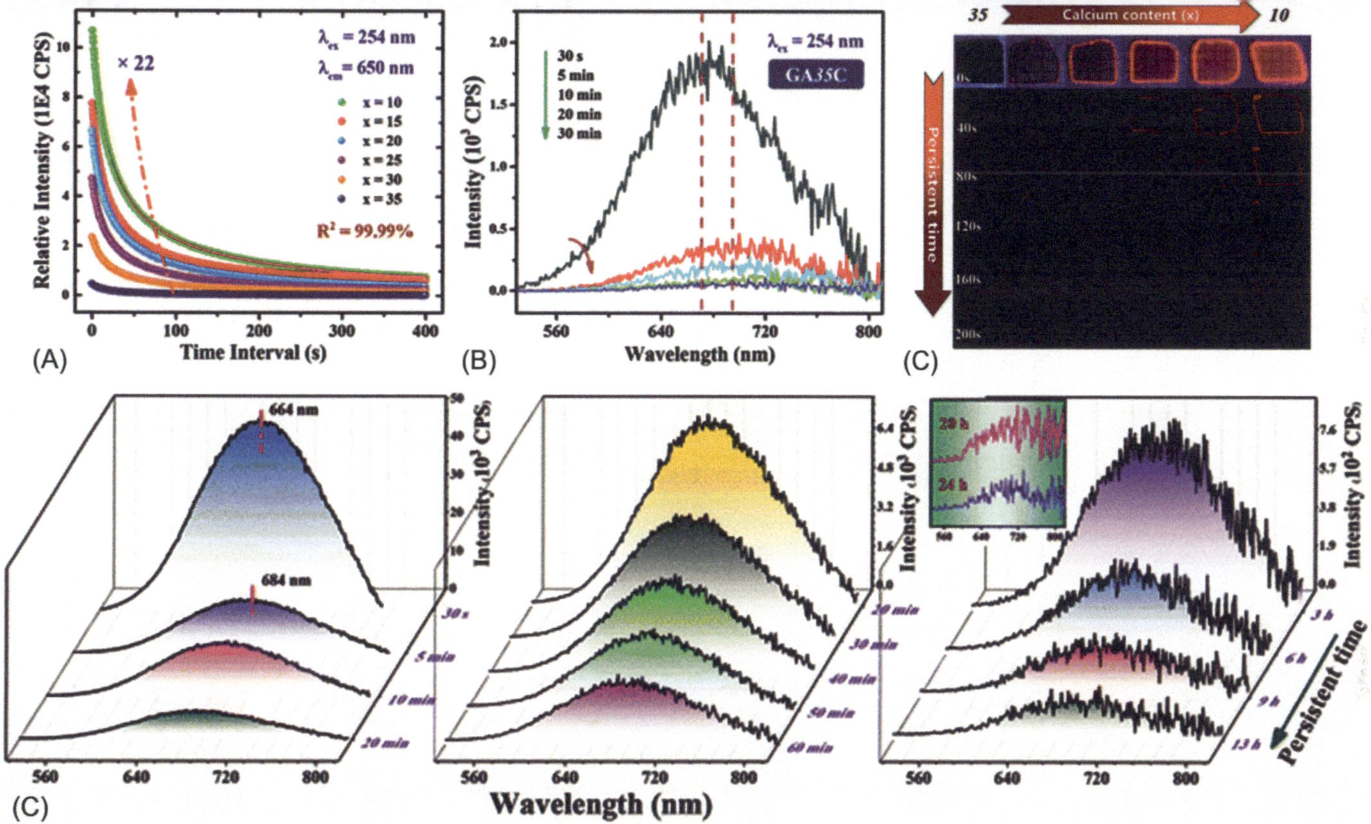

Fig. 5.53 (A) Afterglow decay curves of the $(89-x)GeO_2$–$11Al_2O_3$–$xCaO$–$1MnO$ glasses, $x = 10, 15, 20, 25, 30, 35$. (B) Afterglow spectra of the glass sample at $x = 35$. (C) Photographs of glass samples with different calcium contents in the presence (0 s) and absence (40–200 s) of the 254-nm irradiation source. (D) Afterglow spectra of sample at $x = 10$. All samples were preirradiated with a 254-nm lamp for 10 min and measured at different delay times (30 s to 24 h) after removing the lamp.

Reproduced with permission from H. Luo, J. Cao, X. Li, X. Wang, M. Peng, Topological tailoring of structure and defects to enhance red to near-infrared afterglow from Mn^{2+}-doped germanate photonic glasses. J. Mater. Chem. C, 2018, 6, 11525–11535, Copyright 2018, The Royal Society of Chemistry.

They finally concluded that the origin of persistent phosphorescence properties in these glasses may correlated with several reasons such as oxygen defects on Ge^{4+} entities, oxygen vacancies that created by occupying the octahedral sites of Al^{3+} by Mn^{2+}, and the photooxidized reaction of Mn^{2+} ions that occurs under ultraviolet irradiation, producing photooxidized $(Mn^{2+})^{+}$ ions. In addition, the decrease of Ca content in the glass composition leads to the appearance of several distorted octahedral sites that can be occupied by Mn ions and the increase of band gap of glass, and thus the persistent phosphorescence peak is significantly shifted at $x = 35$ rather than at $x = 35$, and the persistent phosphorescence time is extended from 30 min at $x = 35$ to longer than 24 h at $x = 10$.

Hu et al. reported a transparent oxyfluoride glass and glass ceramic containing α-Zn_2SiO_4:Mn^{2+} nanocrystals with multicolor persistent phosphorescence lasting for 1 h observable by naked eyes [266]. By adjusting the crystallization temperature and duration, the precipitation of α-Zn_2SiO_4 nanocrystals doped with Mn^{2+} within the glass matrix is achieved. As shown in Fig. 5.54A, the as-melt glass shows broad red emission centered at 620, which is assigned to $^4T_{1g}(G) \rightarrow {}^6A_{1g}(S)$ transitions of Mn^{2+} ions in octahedral sites. After annealing the as-melt glass at 575 °C for different periods of time, an additional green emission band at 524 nm appears, which is attributed to $^4T_1(G) \rightarrow {}^6A_1(S)$ transitions of Mn^{2+} ions in tetrahedral sites of the precipitated α-Zn_2SiO_4. Upon increasing the annealing time from 3 h to 9 h, the green emission becomes dominant, while the red emission almost disappears. This result implies that the Mn^{2+} ions gradually enter into the precipitated α-Zn_2SiO_4 nanocrystal as the glass crystallization rate increases. As shown in Fig. 5.54B, it is noteworthy that both of

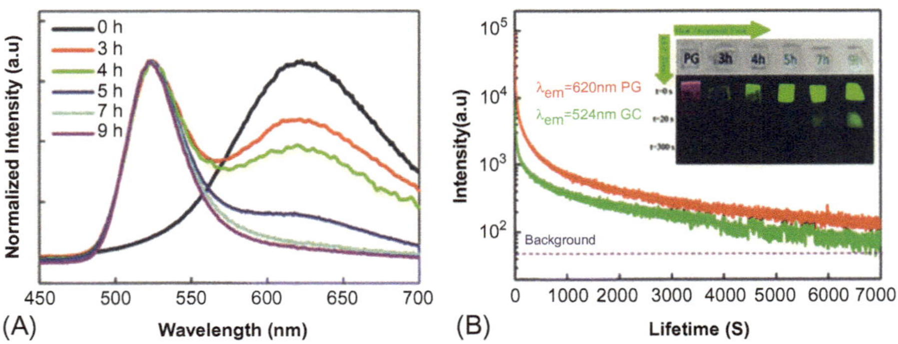

Fig. 5.54 (A) Photoluminescence spectra of the glass and glass ceramics after annealing at 575 °C for different durations. (B) Persistent phosphorescence decay curves of the glass and glass ceramics annealed for 9 h by monitoring at 620 and 524 nm, respectively. Before recording the decay curves the samples were irradiated at 260 nm for 5 min. The inset presents the photographs of the as-prepared samples and the luminescent ones under ultraviolet light and the afterglow taken after switching off the irradiation source.
Reproduced with permission from T. Hu, H. Lin, J. Xu, B. Wang, J. Wang, Y. Wang, Color-tunable persistent luminescence in oxyfluoride glass and glass ceramic containing Mn^{2+}:α-Zn_2SiO_4 nanocrystals. J. Mater. Chem. C, 2017, 5, 1479–1487, Copyright 2017, The Royal Society of Chemistry.

glass and glass ceramics exhibit red and green persistent phosphorescence, respectively. It can be clearly observed that the afterglow emission is measured by the spectrometer for more than 2 h and the afterglow duration of glass ceramics is shorter than that of glass, indicating that the trap depth distribution tends to be shallower in the glass ceramics rather than glass.

5.5.6 Others

However, all the aforementioned phosphors rely on expensive rare earth ions, which serve as activators or trapping centers. The search for a cheaper and green phosphor has stimulated much effort in the development of new luminescent materials with low cost and toxicity. Among them, boron-carbon oxynitride, which is simply consisted of a combination of graphite and hexagonal boron nitride, has demonstrated a great promise as an efficient phosphor for a white light-emitting diode with composition-tunable emission that covers full visible range [263]. In addition, theoretical calculations have also confirmed the tunability of the band gap of boron-carbon oxynitride through the adjusting of carbon to boron-nitride ratio. Liu et al. reported boron-carbon oxynitride as green persistent phosphorescence [262]. Boron-carbon oxynitride showed green emission band $\sim$ 520 nm with afterglow time of 2 h. However, there were no thermoluminescence glow curves reported on this boron-carbon oxynitride phosphor. By means of electron spin resonance, a signal at $g = 2.0$ was found to decrease synchronously with the afterglow decay [262, 263]. These results suggest the traps were assumed to consist of nitrogen vacancies, stabilized by the carbon impurities. The excitation spectra for the steady-state photoluminescence and the persistent phosphorescence were remarkably similar, producing persistent phosphorescence upon charging with visible light up to 500 nm.

On the other hand, cesium lead halide ($CsPbX_3$, X=Cl, Br, I) perovskite quantum dots are promising materials in the fields of solar cells, light-emitting diodes, and lasing, due to their excellent optoelectronic properties such as large absorption coefficient, narrow emission band, high photoluminescence quantum efficiency, and tunable band gap and photoluminescence emission. Gong et al. reported a convenient approach to tune the persistent phosphorescence by using $CsPbX_3$ quantum dots as efficient light conversion materials [267]. As shown Fig. 5.55, $CaAl_2O_4$:1%Eu^{2+}, 0.5%Nd^{3+} persistent phosphor is embedded in a polydimethylsiloxane substrate and spin-coated onto a quartz plate to serve as a light storage layer of ultraviolet light, which is then slowly released blue afterglow emission. They selected $CaAl_2O_4$:1% Eu^{2+},0.5%Nd^{3+} phosphor as the sensitizer because of its excellent persistent phosphorescence in the violet and blue regions, which overlaps well with the absorption of most $CsPbX_3$ quantum dots. The emitted blue light from $CaAl_2O_4$:1%Eu^{2+},0.5% Nd^{3+} is then absorbed by $CsPbX_3$ quantum dots in the light conversion layer of polydimethylsiloxane/$CsPbX_3$ quantum dots, which was spin-coated onto the surface of $CaAl_2O_4$:1%Eu^{2+},0.5%Nd^{3+}/polydimethylsiloxane, resulting in a series of deeply colored $CaAl_2O_4$:1%Eu^{2+},0.5%Nd^{3+}/$CsPbX_3$ quantum dots/polydimethylsiloxane composite films as shown in Fig. 5.56A. As a consequence, tunable persistent phosphorescence with wavelengths covering the entire visible spectral region could be

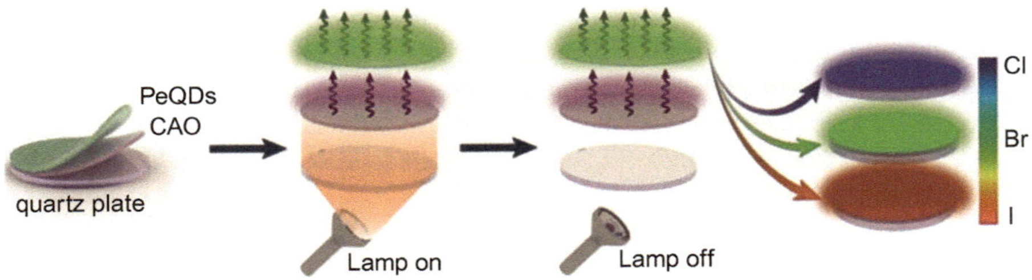

Fig. 5.55 Illustration for full-spectrum persistent phosphorescence tuning using all-inorganic $CsPbX_3$ quantum dots.
Reproduced with permission from Z. Gong, W. Zheng, Y. Gao, P. Huang, D. Tu, R. Li, J. Wei, W. Zhang, Y. Zhang, X. Chen, Full-spectrum persistent luminescence tuning using all inorganic perovskite quantum dots. Angew. Chem., 2019, 131, 7017–7021, Copyright 2019, Wiley.

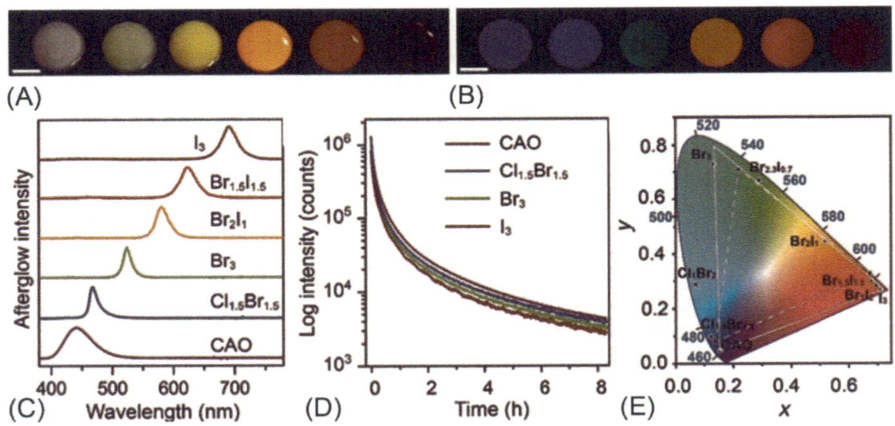

Fig. 5.56 (A) Photographs of the $CaAl_2O_4$:1%Eu^{2+},0.5%Nd^{3+}/$CsPbX_3$ quantum dots/polydimethylsiloxane films with varying halide compositions of $CsPbX_3$ quantum dots under daylight. (B) Afterglow images of the films in (a) recorded at a delay time of 10 min after 5 min of illumination with a 365-nm ultraviolet lamp. Scale bars: 1 cm. (C) Normalized afterglow emission spectra and (D) afterglow decay curves of the $CaAl_2O_4$:1%Eu^{2+},0.5%Nd^{3+}/$CsPbX_3$ quantum dots/polydimethylsiloxane films with varying $CsPbX_3$ quantum dots compositions. (E) The international commission on illumination diagram showing the color gamut of the afterglow emission colors (solid white triangle) from $CaAl_2O_4$:1%Eu^{2+},0.5%Nd^{3+}/$CsPbX_3$ quantum dots/polydimethylsiloxane films, compared with the color gamut (dashed white triangle) defined in the national television system committee color standard.
Reproduced with permission from Z. Gong, W. Zheng, Y. Gao, P. Huang, D. Tu, R. Li, J. Wei, W. Zhang, Y. Zhang, X. Chen, Full-spectrum persistent luminescence tuning using all inorganic perovskite quantum dots. Angew. Chem., 2019, 131, 7017–7021, Copyright 2019, Wiley.

realized via band gap tailoring of $CsPbX_3$ quantum dots, in addition to the benefits of narrow bandwidth of $CsPbX_3$ quantum dots and high synchronization in afterglow decays owing to the single energy storage source of $CaAl_2O_4$:1%Eu^{2+},0.5%Nd^{3+}. After irradiation with a 365-nm ultraviolet lamp for 5 min, these films displayed bright persistent phosphorescence with tunable emission colors from purple to green and deep red as the halide composition of the quantum dots changed from Cl to Br and I (Fig. 5.56B). Afterglow emission spectra of the films showed tunable emission bands from 440 to 694 nm (Fig. 5.56C). Further afterglow decay measurement showed that the afterglow intensities of the films decreased slowly over a long time of 8 h (Fig. 5.56D). Moreover, it was found that the afterglow intensities of $CsPbX_3$ quantum dots in the $CaAl_2O_4$:1%Eu^{2+},0.5%Nd^{3+}/$CsPbX_3$ quantum dots/polydimethylsiloxane composite films are similar to that of $CaAl_2O_4$:1%Eu^{2+},0.5%Nd^{3+} in the $CaAl_2O_4$:1% Eu^{2+},0.5%Nd^{3+}/polydimethylsiloxane composite films. This suggests that the persistent phosphorescence of the $CaAl_2O_4$:1%Eu^{2+},0.5%Nd^{3+}/$CsPbX_3$ quantum dots/polydimethylsiloxane composite films originates from the efficient energy transfer from $CaAl_2O_4$:1%Eu^{2+},0.5%Nd^{3+} to $CsPbX_3$ quantum dots. Importantly, because of the narrow emission bandwidth of $CsPbX_3$ quantum dots, highly saturated red-green-blue afterglow colors can be generated, leading to a wide color gamut encompassing 130% of that defined in the National Television System Committee color standard (Fig. 5.56E). The combination of afterglow feature with narrow bandwidth, wide color gamut, and highly synchronized afterglow decay in the $CaAl_2O_4$:1%Eu^{2+},0.5%Nd^{3+}/$CsPbX_3$ quantum dots/polydimethylsiloxane composite film is unreachable by traditional afterglow phosphors, which is highly attractive for optical coding and dark-light multicolor displays.

5.5.7 Discussions

In this chapter, we presented a comprehensive review about persistent phosphors. The most important purpose of this chapter is to provide the recent advances in the field of persistent phosphors. While significant and rapid progress has been made in this area, there are still many areas remain largely unexplored, including but not limited to (a) tailoring of luminescence efficiency, initial persistent phosphorescence intensity, and duration; (b) discovering of new activators and hosts especially in the second and third windows of near infrared; and (c) opening up novel multifunctional application fields. The investigations and analysis toward the species, concentration, and depth of traps and the probability and rate of electrons being captured and released should be evaluated as a whole by developing advanced characterization techniques. For fundamental physical and chemical insights into the afterglow mechanism, much detailed information needs to be revealed and extracted, that is, what are the species of defects, how to evaluate and then accurately control the trap concentration, how do the defects work in the host, and what is the process of electrons being captured and released. It is still necessary to study how defects are involved in the persistent phosphorescence process. Persistent phosphors should be given more consideration in advanced scientific applications. One of the key areas will be the integration of persistent phosphors with other peculiarities or special characteristics (such as magnetization and photothermal response) to open the door to novel multifunctional applications, that is, tracing and therapy and tracing and monitoring cancer cells in the fields of biomedicine and life science. On the other hand, persistent phosphors over a wide range of wavelengths and long persistent phosphorescence time offer more possibilities for incorporation in solar photovoltaic and optical information storage and displays, environmental sensing and monitoring, and photocatalysis on flexible platforms. The emission spectral region of near-infrared persistent phosphors should be further improved and extended to the second near-infrared window to increase their detection sensitivity for diverse disease biomarkers. The strategies to expand the spectral region to the long wavelength region include the selection of new emission centers, tailoring the crystal field surrounding the activator, and utilization of ET from sensitizers to acceptors. Although near-infrared persistent phosphors have been studied, more fundamental investigations on the role of activators (transition metal ions or rare earth ions) should be carried out because the known number of near-infrared phosphorescent centers is admittedly small. Cr^{3+} and gallate are commonly used as the optimized emission centers and hosts for near-infrared persistent phosphors. The predominance of Cr^{3+}-activated gallates might suggest that only gallates can be used as the hosts in Cr^{3+}-doped near-infrared persistent phosphors. It is challenging to identify the nature of defects, and it is difficult to predict the actual effectiveness and impact of intrinsic and substituted defects in gallates because comparison with Cr^{3+}-doped nongallate phosphors has yet to be performed. In addition, can we expect the discovery of new and even better persistent phosphors, apart from the gallates, aluminates, and silicates? In brief a great number of new concepts and novel materials are still in the research stage. Some of them may lead to much better afterglow properties in the coming decades. A long way is still ahead to achieving the goal from a "top-down" strategy, which is a concept of guiding the material design based on a clear mechanism and model. The energy band structures of hosts and activators, valence state and redox equilibrium of dopants, crystal structures of hosts, and even the volatility of raw materials have to be considered as the important and considerable factors in such a strategy.

References

[1] W. Hoogenstraaten, H.A. Klasens, Some properties of zinc sulfide activated with copper and cobalt, J. Electrochem. Soc. 100 (1953) 366–375.

[2] W.M. Yen, H. Yamamoto, Phosphor Handbook, CRC Press, Boca Raton, 2006.

[3] H. Yamamoto, T. Matsuzawa, Mechanism of long phosphorescence of $SrAl_2O_4:Eu^{2+}$, Dy^{3+} and $CaAl_2O_4:Eu^{2+}$, Nd^{3+}, J. Lumin. 72–74 (1997) 287–289.

[4] C.K. Chang, D.L. Mao, Long lasting phosphorescence of $Sr_4Al_{14}O_{25}:Eu^{2+},Dy^{3+}$ thin films by magnetron sputtering, Thin Solid Films 460 (2004) 48–52.

[5] B.C. Cheng, L.T. Fang, Z.D. Zhang, Y.H. Xiao, S.J. Lei, $BaAl_2O_4:Eu^{2+}$, Dy^{3+} nanotube synthesis by heating conversion of homogeneous coprecipitates and afterglow characteristics, J. Phys. Chem. C 115 (2011) 1708–1713.

[6] R.X. Zhong, J.H. Zhang, X. Zhang, S.Z. Lu, X.J. Wang, Red phosphorescence in $Sr_4Al_{14}O_{25}:Cr^{3+},Eu^{2+},Dy^{3+}$ through persistent energy transfer, Appl. Phys. Lett. 88 (2006) 201916.

[7] N.Y. Yu, F. Liu, X.F. Li, Z.W. Pan, Near infrared long-persistent phosphorescence in $SrAl_2O_4$:Eu^{2+},Dy^{3+},Er^{3+} phosphors based on persistent energy transfer, Appl. Phys. Lett. 95 (2009) 231110.

[8] Y. Teng, J.J. Zhou, Z.J. Ma, M.M. Smedskjaer, J.R. Qiu, Persistent near infrared phosphorescence from rare earth ions co-doped strontium aluminate phosphors, J. Electrochem. Soc. 158 (2011) K17–K19.

[9] W.B. Dai, Investigation of the luminescent properties of Ce^{3+} doped and Ce^{3+}/Mn^{2+} co-doped $CaAl_2Si_2O_8$ phosphors, RSC Adv. 4 (2014) 11206–11215.

[10] J. Liu, G. Li, H. Guo, D. Liu, P. Feng, Y. Wang, Design, synthesis and characterization of a novel bluish-green long-lasting phosphorescence phosphor $BaLu_2Si_3O_{10}$:Eu^{2+},Nd^{3+}, RSC Adv. 8 (2018) 10246–10254.

[11] Y. Gong, Y.H. Wang, X.H. Xu, Y.Q. Li, S.Y. Xin, L.R. Shi, The persistent energy transfer of Eu^{2+} and Mn^{2+} and the thermoluminescence properties of long-lasting phosphor $Sr_3MgSi_2O_8$:Eu^{2+}, Mn^{2+}, Dy^{3+}, Opt. Mater. 33 (2011) 1781–1785.

[12] M. Lastusaari, A.J.J. Bos, P. Dorenbos, T. Laamanen, M. Malkamaki, L.C.V. Rodrigues, J. Holsa, Wavelength-sensitive energy storage in $Sr_3MgSi_2O_8$:Eu^{2+},Dy^{3+}, J. Therm. Anal. Calorim. 121 (2015) 29–35.

[13] J.Y. Kuang, Y.L. Liu, J.X. Zhang, Effects of RE^{3+} as a co-dopant in blue-emitting long-lasting phosphors, $Sr_3Al_{10}SiO_{20}$:Eu^{2+}, J. Mater. Sci. 41 (2006) 5500–5503.

[14] M. Zheng, X. Chen, B. Lei, Y. Xiao, R. Liu, H. Zhang, H. Dong, Y. Liu, X. Liu, Synthesis and luminescence properties of flower-like $Sr_2MgSi_2O_7$:Eu^{2+},Dy^{3+} phosphor via hydrothermal-homogeneous coprecipitation route, ECS Solid State Lett. 2 (2013) R19–R22.

[15] Y.H. Lin, Z.T. Zhang, Z.L. Tang, X.X. Wang, J.Y. Zhang, Z.S. Zheng, Luminescent properties of a new long afterglow Eu^{2+} and Dy^{3+} activated $Ca_3MgSi_2O_8$ phosphor, J. Eur. Ceram. Soc. 21 (2001) 683–685.

[16] Y.H. Wang, Y. Gong, X.H. Xu, Y.Q. Li, Recent progress in multicolor long persistent phosphors, J. Lumin. 133 (2013) 25–29.

[17] Y. Gong, Y.H. Wang, Y.Q. Li, X.H. Xu, Ce^{3+}, Dy^{3+} co-doped white-light long-lasting phosphor: $Sr_2Al_2SiO_7$ through energy transfer, J. Electrochem. Soc. 157 (2010) J208–J211.

[18] Y. Yao, Z. Zhou, F. Ye, Properties of a novel $Ba_5Si_8O_{21}$:Eu^{2+}, Nd^{3+} phosphor: bulk and 1D nanostructure with PVP synthesized by sol-gel and electrospinning, J. Alloys Compd. 712 (2017) 213–218.

[19] Z.F. Yang, Y.H. Hu, L. Chen, X.J. Wang, Fluorescence and phosphorescence properties of $Ba_4(Si_3O_8)_2$:Eu^{2+}, Re^{3+} (Re:Dy,Ho), Opt. Mater. 35 (2013) 1264–1267.

[20] J. Yan, L.X. Ning, Y.C. Huang, C.M. Liu, D.J. Hou, B.B. Zhang, Y. Huang, Y. Tao, H.B. Liang, Luminescence and electronic properties of $Ba_2MgSi_2O_7$:Eu^{2+}: a combined experimental and hybrid density functional theory study, J. Mater. Chem. C 2 (2014) 8328–8332.

[21] H. Guo, Y. Wang, G. Li, J. Liu, P. Feng, D. Liu, Cyan emissive super-persistent luminescence and thermoluminescence in $BaZrSi_3O_9$:Eu^{2+},Pr^{3+} phosphors, J. Mater. Chem. C 5 (2017) 2844–2851.

[22] Y. Li, Y. Wang, Y. Gong, X. Xu, F. Zhang, Photoionization behavior of Eu^{2+}-doped $BaMgSiO_4$ long-persisting phosphor upon UV irradiation, Acta Mater. 59 (2011) 3174–3183.

[23] H. He, R.L. Fu, X.F. Song, R. Li, Z.W. Pan, X.R. Zhao, Z.H. Deng, Y.G. Cao, Observation of fluorescence and phosphorescence in $Ca_2MgSi_2O_7$: Eu^{2+}, Dy^{3+} phosphors, J. Electrochem. Soc. 157 (2010) J69–J73.

[24] Y. Gong, Y.H. Wang, Y.Q. Li, X.H. Xu, W. Zeng, Fluorescence and phosphorescence properties of new long-lasting phosphor $Ba_4(Si_3O_8)_2$:Eu^{2+},Dy^{3+}, Opt. Express 19 (2011) 4310–4315.
[25] G.F. Ju, Y.H. Hu, L. Chen, X.J. Wang, Z.F. Mu, Persistent luminescence in $CaAl_2Si_2O_8$: Eu^{2+},R^{3+} (R = Pr, Nd, Dy, Ho and Er), J. Lumin. 146 (2014) 102–108.
[26] W.B. Dai, M. Zhou, Z.Y. Xian, L.K. Zeng, Structure and photoluminescence characteristics of europium(iii) doped in $CaAl_2Si_2O_8$ phosphors, RSC Adv. 4 (2014) 25470–25478.
[27] A. Madej, E. Zych, Controlled synthesis of the monoclinic and orthorhombic polymorphs of Sr_2SiO_4 activated with Ce^{3+} or Eu^{2+}, RSC Adv. 5 (2015) 104441.
[28] K. Szczodrowski, A. Chruscinska, J. Barzowska, K. Przegietka, K. Anders, R. Piramidowicz, M. Grinberg, Influence of Ti^{4+} on the long lasting luminescence of Sr_2SiO_4:Eu^{2+}, RSC Adv. 5 (2015) 65236–65244.
[29] Y. Li, B.H. Li, C.C. Ni, S.X. Yuan, J. Wang, Q. Tang, Q. Su, Synthesis, persistent luminescence, and thermoluminescence properties of yellow Sr_3SiO_5:Eu^{2+},RE^{3+} (RE = Ce, Nd, Dy, Ho, Er, Tm, Yb) and orange–Red $Sr_{3-x}Ba_xSiO_5$:Eu^{2+}, Dy^{3+} phosphor, Chem. Asian J. 9 (2014) 494–499.
[30] X. Liu, J.H. Zhang, X. Zhang, Z.D. Hao, J. Qiao, X.L. Dong, Strongly enhancing photostimulated luminescence by doping Tm^{3+} in Sr_3SiO_5: Eu^{2+}, Opt. Lett. 38 (2013) 148–150.
[31] W.H. Shen, X.H. Xu, B.H. Zhang, T. Wang, J.B. Qiu, X. Yu, Distribution of luminescence centers and trapping centers in $Ba_{13}Al_{22}Si_{10}O_{66}$: Eu^{2+}, Tm^{3+}, ECS Solid State Lett. 4 (2015) R31–R33.
[32] L.L. Jiang, S.G. Xiao, X.L. Yang, X.A. Zhang, X.H. Liu, B.Y. Zhou, X.L. Jin, Preparation and luminescence properties of yellow long-lasting phosphor $Ca_2ZnSi_2O_7$:Eu^{2+}, Dy^{3+}, Mater. Sci. Eng. B 178 (2013) 123–126.
[33] L. Lin, C.S. Shi, Z.F. Wang, W.P. Zhang, M. Yin, A kinetics model of red long-lasting phosphorescence in $MgSiO_3$:Eu^{2+}, Dy^{3+}, Mn^{2+}, J. Alloys Compd. 466 (2008) 546–550.
[34] Y. Jia, W. Sun, R. Pang, T. Ma, D. Li, H. Li, S. Zhang, J. Fu, L. Jiang, C. Li, Sunlight activated new long persistent luminescence phosphor $BaSiO_3$:Eu^{2+},Nd^{3+},Tm^{3+}:optical properties and mechanism, Mater. Des. 90 (2016) 218–224.
[35] Q. le Masne de Chermont, C. Chaneac, J. Seguin, F. Pelle, S. Maitrejean, J.P. Jolivet, D. Gourier, M. Bessodes, D. Scherman, Nanoprobes with near-infrared persistent luminescence for in vivo imaging, Proc. Natl. Acad. Sci. U. S. A. 104 (2007) 9266–9271.
[36] G.F. Ju, Y.H. Hu, L. Chen, X.J. Wang, L.H. Hung, Persistent luminescence properties of $SrMg_2(PO_4)_2$:Eu^{2+},Tb^{3+}, Appl. Phys. A Mater. Sci. Process. 114 (2014) 867–874.
[37] R. Pang, C.Y. Li, S. Zhang, Q. Su, Luminescent properties of a new blue long-lasting phosphor $Ca_2P_2O_7$:Eu^{2+}, Y^{3+}, Mater. Chem. Phys. 113 (2009) 215–218.
[38] R. Pang, C.Y. Li, L.L. Shi, Q. Su, A novel blue-emitting long-lasting proyphosphate phosphor $Sr_2P_2O_7$:Eu^{2+},Y^{3+}, J. Phys. Chem. Solids 70 (2009) 303–306.
[39] G.F. Ju, Y.H. Hu, L. Chen, X.J. Wang, Z.F. Mu, Blue persistent luminescence in Eu^{2+} doped $Ca_3Mg_3(PO_4)_4$, Opt. Mater. 36 (2014) 1183–1188.
[40] K. Li, M.M. Shang, Y. Zhang, J. Fan, H.Z. Lian, J. Lin, Photoluminescence properties of single-component white-emitting $Ca_9Bi(PO_4)_7$:Ce^{3+},Tb^{3+},Mn^{2+} phosphors for UV LEDs, J. Mater. Chem. C 3 (2015) 7096–7104.
[41] H. Guo, W. Chen, W. Zeng, G. Li, Y. Wang, Y. Li, Y. Li, X. Ding, Structure and luminescence properties of a novel yellow super long-lasting phosphate phosphor $Ca_6BaP_4O_{17}$:Eu^{2+},Ho^{3+}, J. Mater. Chem. C 3 (2015) 5844–5850.

[42] H.R. Zhang, M.T. Zheng, B.F. Lei, Y. Xiao, H.W. Dong, Y.L. Liu, X.T. Liu, J.J. Deng, J. W. Deng, Z.L. Huang, Preparation and long-lasting phosphorescence properties of $BaAlSi_5N_7O_2:Eu^{2+}$, ECS Solid State Lett. 2 (2013) R16–R18.
[43] H.R. Zhang, B.F. Lei, J.L. Qin, J.F. Li, Y.L. Liu, X.T. Liu, M.T. Zheng, Y. Xiao, H.W. Dong, Preparation of $SrSi_2O_2N_2:Eu^{2+}$ phosphor by SrSi alloy precursor and its long-lasting phosphorescence properties, J. Am. Ceram. Soc. 96 (2013) 1810–1814.
[44] P.F. Smet, J. Botterman, K. Van den Eeckhout, K. Korthout, D. Poelman, Persistent luminescence in nitride and oxynitride phosphors: A review, Opt. Mater. 36 (2014) 1913–1919.
[45] X. Liu, Z. Song, S. Wang, Q. Liu, The red persistent luminescence of $(Sr,Ca)AlSiN_3:Eu^{2+}$ and mechanism different to $SrAl_2O_4:Eu^{2+},Dy^{3+}$, J. Lumin. 218 (2019) 313–321.
[46] J. Wang, H.R. Zhang, B.F. Lei, H.W. Dong, H.M. Zhang, Y.L. Liu, N.L. Lai, Y. Fang, Z. J. Chen, Red persistent and photo-stimulated luminescence properties of $SrCaSi_5N_8$: Eu^{2+}, Tm^{3+} solid solution, Opt. Mater. 36 (2014) 1855–1858.
[47] G.F. Ju, Y.H. Hu, L. Chen, X.J. Wang, Persistent luminescence and its mechanism of $Ba_5(PO_4)_3Cl:Ce^{3+},Eu^{2+}$, J. Appl. Phys. 111 (2012) 113508.
[48] C.Q. Wu, J.C. Zhang, P.F. Feng, Y.M. Duan, Z.Y. Zhang, Y.H. Wang, Blue photoluminescence and long lasting phosphorescence properties of a novel chloride phosphate phosphor: $Sr_5(PO_4)_3Cl:Eu^{2+}$, J. Lumin. 147 (2014) 229–234.
[49] B.F. Lei, L. Sha, H.R. Zhang, Y.L. Liu, S.Q. Man, S. Yue, Preparation and luminescence properties of green-light-emitting afterglow phosphor $Ca_8Mg(SiO_4)_4Cl_2:Eu^{2+}$, Solid State Sci. 12 (2010) 2177–2180.
[50] W. Zeng, Y.H. Wang, S.C. Han, W.B. Chen, G. Li, Y.Z. Wang, Y. Wen, Design, synthesis and characterization of a novel yellow long-persistent phosphor: $Ca_2BO_3Cl:Eu^{2+},Dy^{3+}$, J. Mater. Chem. C 1 (2013) 3004–3011.
[51] Y.H. Jin, Y.H. Hu, L. Chen, X.J. Wang, Synthesis and persistent luminescence mechanism of a novel orange emitting persistent phosphor $Sr_5(BO_3)_3Cl:Eu^{2+}$, J. Am. Ceram. Soc. 97 (2014) 2573–2579.
[52] Y.Q. Li, Y.H. Wang, Y. Gong, X.H. Xu, M.J. Zhou, Design, synthesis and characterization of an orange-yellow long persistent phosphor: $Sr_3Al_2O_5Cl_2:Eu^{2+},Tm^{3+}$, Opt. Express 18 (2010) 24853–24858.
[53] P. Feng, G. Li, H. Guo, D. Liu, Q. Ye, Y. Wang, Identifying a cyan ultralong persistent phosphorescence (Ba, Li) (Si, Ge, P)$_2$O$_5$:Eu^{2+}, Pr^{3+} via solid solution strategy, J. Phys. Chem. C 123 (5) (2019) 3102–3109.
[54] H.D. Luo, A.J.J. Bos, A. Dobrowolska, P. Dorenbos, Low-temperature VUV photoluminescence and thermoluminescence of UV excited afterglow phosphor $Sr_3Al_xSi_{1-x}O_5:Ce^{3+},Ln^{3+}$ (Ln = Er, Nd, Sm, Dy and Tm), Phys. Chem. Chem. Phys. 17 (2015) 15419–15427.
[55] K. Van den Eeckhout, P.F. Smet, D. Poelman, Persistent luminescence in Eu^{2+}-doped compounds: A Review, Materials 3 (2010) 2536–2566.
[56] R. Pang, C. Li, L. Jiang, Q. Su, Blue long lasting phosphorescence of Tm^{3+} in zinc pyrophosphate phosphor, J. Alloys Compd. 471 (2009) 364–367.
[57] X.G. Zhang, L.Y. Zhou, M.L. Gong, Thermal quenching of luminescence of $LiSr_4(BO_3)_3:Eu^{2+}$ orange-emitting phosphor, Luminescence 29 (2014) 104–108.
[58] L. Wu, X.L. Chen, H. Li, M. He, Y.P. Xu, X.Z. Li, Structure Determination and Relative Properties of Novel Cubic Borates $MM'_4(BO_3)_3$ (M = Li, M' = Sr; M = Na, M' = Sr, Ba), Inorg. Chem. 44 (2005) 6409–6414.

[59] D.C. Rodriguez Burbano, S.K. Sharma, P. Dorenbos, B. Viana, J.A. Capobianco, Persistent and photostimulated red emission in CaS:Eu^{2+},Dy^{3+} nanophosphors, Adv. Opt. Mater. 3 (2015) 551–557.

[60] M.A. Ali, X. Liu, J. Qiu, A comparative investigation on upconversion luminescence in glass–ceramics containing LaF_3 and CaF_2 nanocrystals, J. Mater. Sci. Mater. Electron. 29 (2018) 8701–8709.

[61] M.A. Ali, J. Ren, X. Liu, J. Qiu, Effect of ligand field symmetry on upconversion luminescence in heat treated $LaBGeO_5$:Yb^{3+},Er^{3+} glass, J. Am. Ceram. Soc. 101 (2018) 4387–4396.

[62] M.A. Ali, J. Ren, X. Liu, X. Qiao, J. Qiu, Understanding enhanced upconversion luminescence in oxyfluoride glass-ceramics based on local structure characterizations and molecular dynamics simulations, J. Phys. Chem. C 121 (2017) 15384–15391.

[63] J. Xu, Z. Ju, X. Yang, X. Tang, W. Liu, Considerable improvement of $Na_2CaSn_2Ge_3O_{12}$: Eu^{3+} reddish orange long-persistent phosphorescence with co-doping of Dy^{3+}, Dyes Pigments 162 (2019) 951–958.

[64] J. Kuang, Y. Liu, B. Lei, Effect of RE^{3+} as a co-dopant in long-lasting phosphorescence $CdSiO_3$:Mn^{2+} (RE=Y, La, Gd, Lu), J. Lumin. 118 (2006) 33–38.

[65] J. Fu, ChemInform abstract: orange and violet-emitting long-lasting phosphors, ChemInform 33 (2002) 255–257.

[66] K. Feng-Wen, H. Yi-Hua, W. Hao-Yi, J. Gui-Fang, Red afterglow properties of Eu^{3+} in $CaMoO_4$ phosphor, Chin. Phys. Lett. 28 (2011) 107201–107204.

[67] F. Kang, Y. Hu, H. Wu, Z. Mu, G. Ju, C. Fu, N. Li, Luminescence and red long afterglow investigation of Eu^{3+}–Sm^{3+} co-doped $CaWO_4$ phosphor, J. Lumin. 132 (2012) 887–894.

[68] F. Kang, Y. Hu, L. Chen, X. Wang, Z. Mu, H. Wu, G. Ju, Eu^{3+} doped $CaWO_4$—a potential red long afterglow phosphor, Appl. Phys. B Lasers Opt. 107 (2012) 833–837.

[69] D. Sun, D. Li, Z. Zhu, J. Xiao, Z. Tao, W. Liu, Photoluminescence properties of europium and titanium Co-doped $BaZrO_3$ phosphors powders synthesized by the solid-state reaction method, Opt. Mater. 34 (2012) 1890–1896.

[70] Y. Lin, C.W. Nan, N. Cai, X. Zhou, H. Wang, D. Chen, Anomalous afterglow from Y_2O_3-based phosphor, J. Alloys Compd. 361 (2003) 92–95.

[71] J. Fu, Orange and red emitting long-lasting phosphors MO:Eu^{3+} (M = Ca, Sr, Ba), Electrochem. Solid-State Lett. 3 (2000) 350–351.

[72] J. Trojan-Piegza, E. Zych, Afterglow luminescence of Lu_2O_3:Eu ceramics synthesized at different atmospheres, J. Phys. Chem. C 114 (2010) 4215–4220.

[73] X.G. Wang, F.P. Du, D.L. Wei, Y.L. Huang, H.J. Seo, A new long-lasting phosphor Zr^{4+} and Eu^{3+} co-doped $SrMg_2(PO_4)_2$, Sensors Actuators B Chem. 158 (2011) 171–175.

[74] X.P. Gao, Z.Y. Zhang, C. Wang, J. Xu, Z.H. Ju, Y.Q. An, W.S. Liu, The persistent energy transfer and effect of oxygen vacancies on red long-persistent phosphorescence phosphors Ca_2SnO_4: Gd^{3+}, Eu^{3+}, J. Electrochem. Soc. 158 (2011) J405–J408.

[75] J. Kong, W. Zheng, Y. Liu, R. Li, E. Ma, H. Zhu, X. Chen, Persistent luminescence from Eu^{3+} in SnO_2 nanoparticles, Nanoscale 7 (2015) 11048–11054.

[76] G. Li, Y. Wang, W. Zeng, W. Chen, S. Han, H. Guo, Luminescent and magnetic properties of the afterglow phosphors $GdSr_2AlO_5$:RE^{3+} ($RE^{3+} = Eu^{3+}$, Sm^{3+}, Pr^{3+} and Dy^{3+}), RSC Adv. 5 (2015) 20884–20889.

[77] C.E. Cui, G.W. Jiang, P. Huang, L. Wang, D. Liu, Synthesis and characterization of Y_2O_2S:Eu^{3+}, Mg^{2+}, Ti^{4+} nanotubes via hydrothermal method, J. Lumin. 145 (2014) 665–668.

[78] T. Hang, Q. Liu, D.L. Mao, C.K. Chang, Long lasting behavior of Gd_2O_2S:Eu^{3+} phosphor synthesized by hydrothermal routine, Mater. Chem. Phys. 107 (2008) 142–147.

[79] I.P. Sahu, D.P. Bisen, N. Brahme, Dysprosium doped di-strontium magnesium di-silicate white light emitting phosphor by solid state reaction method, Displays 35 (2014) 279–286.
[80] Y. Chen, X. Cheng, M. Liu, Z. Qi, C. Shi, Comparison study of the luminescent properties of the white-light long afterglow phosphors: $Ca_xMgSi_2O_{5+x}$:Dy^{3+} (x = 1, 2, 3), J. Lumin. 129 (2009) 531–535.
[81] X. Xu, Y. Wang, W. Zeng, Y. Gong, B. Liu, Luminescent properties of the multicolor afterglow phosphors $Ca_3SnSi_2O_9$:Re^{3+} (Re = Pr, Tb, Sm), J. Am. Ceram. Soc. 94 (2011) 3632–3635.
[82] B. Liu, L. Kong, C. Shi, White-light long-lasting phosphor $Sr_2MgSi_2O_7$:Dy^{3+}, J. Lumin. 122–123 (2007) 121–124.
[83] J. Kuang, Y. Liu, J. Zhang, White-light-emitting long-lasting phosphorescence in Dy^{3+}-doped $SrSiO_3$, J. Solid State Chem. 179 (2006) 266–269.
[84] J. Kuang, Y. Liu, White-emitting long-lasting phosphor Sr_2SiO_4:Dy^{3+}, Chem. Lett. 34 (2005) 598–599.
[85] L. Zhang, P. Han, K. Wang, Z. Lu, L. Wang, Y. Zhu, Q. Zhang, Enhanced luminescence of Sr_2SiO_4:Dy^{3+} by sensitization (Ce^{3+}/Bi^{3+}) and its composition-induced phase transition, J. Alloys Compd. 541 (2012) 54–59.
[86] L. Fu, L. Liang, Y. Ren, A novel white light emitting long-lasting phosphor, Chin. Chem. Lett. 15 (2004) 335–338.
[87] Y. Liu, B. Lei, C. Shi, Luminescent properties of a white afterglow phosphor $CdSiO_3$:Dy^{3+}, Chem. Mater. 17 (2005) 2108–2113.
[88] P. Dewangan, D.P. Bisen, N. Brahme, S. Sharma, Structural characterization and luminescence properties of Dy^{3+} doped $Ca_3MgSi_2O_8$ phosphors, J. Alloys Compd. 777 (2019) 423–433.
[89] G.B. Che, C.B. Liu, Q.W. Wang, Z.L. Xu, White-light-emission afterglow phosphor $CaZnGe_2O_6$:Dy^{3+}, Chem. Lett. 37 (2008) 136–137.
[90] L.V. Ermakova, I.I. Leonidov, Towards a new group of olivine-type afterglow phosphors: the case of Ca_2GeO_4:Dy^{3+}, Mater. Lett. 233 (2018) 39–41.
[91] M. Shi, D. Zhang, C. Chang, Dy^{3+}:Ca_2SnO_4, a new yellow phosphor with afterglow behavior, J. Alloys Compd. 639 (2015) 168–172.
[92] J. Lai, Z. Long, J. Qiu, D. Zhou, J. Zhou, C. Zhu, S. Hu, K. Zhang, Q. Wang, Warm white light emitting from single composition $SrGa_{12}O_{19}$:Dy^{3+} phosphors for AC-LED, J. Am. Ceram. Soc. 103 (2020) 335–345.
[93] B. Liu, C.S. Shi, Z.M. Qi, Potential white-light long-lasting phosphor: Dy^{3+}-doped aluminate, Appl. Phys. Lett. 86 (2005) 191111.
[94] P. Huang, F. Yang, C. Cui, L. Wang, X. Lei, Luminescence properties of white-light long-lasting phosphor Y_2O_2S:Dy^{3+}, Mg^{2+}, Si^{4+}, Ceram. Int. 39 (2013) 7193–7197.
[95] P. Huang, X. He, C. Cui, L. Wang, The effect of concentration of Dy^{3+} ions on luminescence properties of Y_2O_2S:Dy^{3+}, Mg^{2+}, Ti^{4+} phosphors, Ceram. Int. 40 (2014) 6435–6440.
[96] J. Zeler, P. Bolek, D. Kulesza, E. Zych, On thermoluminescence mechanism and energy leakage in Lu_2O_3:Tb,V storage phosphor, Opt. Mater.:X 1 (2019) 100001.
[97] J. Trojan-Piegza, E. Zych, J. Holsa, J. Niittykoski, Spectroscopic properties of persistent luminescence phosphors: Lu_2O_3:Tb^{3+},M^{2+} (M= Ca, Sr, Ba), J. Phys. Chem. C 113 (2009) 20493–20498.
[98] T. Takayama, T. Katsumata, S. Komuro, T. Morikawa, Growth and characteristics of a new long afterglow Phosphorescent yttrium tantalate crystal, J. Cryst. Growth 275 (2005) e2013–e2017.

[99] F. Xue, Y. Hu, G. Ju, L. Chen, M. He, T. Wang, Y. Jin, S. Zhang, J. Lin, Photoluminescence and long persistent luminescence properties of a novel green emitting phosphor $Sr_3TaAl_3Si_2O_{14}:Tb^{3+}$, Appl. Phys. A Mater. Sci. Process. 122 (2016) 612.
[100] G.F.C. Bispo, A.B. Andrade, I.R.F.A. da Silva, Z.S. Macedo, M.E.G. Valerio, Thermoluminescence and persistent luminescence of Tb^{3+} activated $CaYAl_3O_7$, Opt. Mater. 91 (2019) 413–418.
[101] X. Fan, X. Xu, X. Yu, W. Chen, D. Zhou, J. Qiu, Wide band long persistent luminescence of $Ca_3Ga_2Ge_3O_{12}:Tb^{3+},Tm^{3+}$ phosphor with synergistic effect of different traps, Mater. Res. Bull. 99 (2018) 398–402.
[102] X. Li, C. Liang, S. Guo, Y. Xiao, C. Chang, Photoluminescence and afterglow behavior of Tb^{3+} activated Li_2SrSiO_4 phosphor, J. Lumin. 188 (2017) 199–203.
[103] X. Yu, S. Wang, Y. Zhu, J. Liang, J. Qiu, X. Xu, W. Lu, High-temperature long persistent and photo-stimulated luminescence in Tb^{3+}doped gallate phosphor, J. Alloys. Compd. 701 (2017) 774–779.
[104] S. Zhang, Z. Mu, Y. Lv, L. Fan, Y. Li, G. Ju, Y. Hu, White-light long persistent luminescence of Tb^{3+}-doped $Y_3Al_2Ga_3O_{12}$ phosphor, J. Alloys. Compd. 729 (2017) 418–425.
[105] X. Fu, S. Zheng, J. Shi, Y. Li, H. Zhang, Long persistent luminescence property of a novel green emitting $SrLaGaO_4:Tb^{3+}$ phosphor, J. Lumin. 184 (2017) 199–204.
[106] J. Zhou, X. Yu, T. Wang, D. Zhou, J. Qiu, Improved optical storage properties of $NaAlSiO_4:Tb^{3+}$ induced by Bi^{3+}, Opt. Mater. 57 (2016) 140–145.
[107] S. Zhang, Y. Hu, L. Chen, G. Ju, Z. Wang, J. Lin, Luminescent properties of a green emitting persistent phosphor $CdGeO_3:Tb^{3+}$, Opt. Mater. 47 (2015) 203–210.
[108] J. Trojan-Piegza, J. Niittykoski, J. Holsa, E. Zych, Thermoluminescence and kinetics of persistent luminescence of vacuum-sintered Tb^{3+}-doped and Tb^{3+},Ca^{2+}-codoped Lu_2O_3 materials, Chem. Mater. 20 (2008) 2252–2261.
[109] K.K. Satapathy, G.C. Mishra, R.S. Kher, S.J. Dhoble, Mechanoluminescence and thermoluminescence characterization of Tb^{3+} doped $CaAl_2O_4$: a theoretical and experimental study, RSC Adv. 5 (2015) 79391–79396.
[110] K. Van den Eeckhout, D. Poelman, P.F. Smet, Persistent luminescence in non-Eu^{2+}-doped compounds: A Review, Materials 6 (2013) 2789–2818.
[111] Z.W. Liu, Y.L. Liu, Synthesis and luminescent properties of a new green afterglow phosphor $CaSnO_3$:Tb, Mater. Chem. Phys. 93 (2005) 129–132.
[112] Y.H. Jin, Y.H. Hu, L. Chen, X.J. Wang, G.F. Ju, Z.F. Mu, Luminescent properties of Tb^{3+}-doped Ca_2SnO_4 phosphor, J. Lumin. 138 (2013) 83–88.
[113] Y.G. Su, L.P. Li, G.S. Li, Synthesis and optimum luminescence of $CaWO_4$-based red phosphors with codoping of Eu^{3+} and Na^+, Chem. Mater. 20 (2008) 6060–6067.
[114] L.C. Rodrigues, H.F. Brito, J. Holsa, M. Lastusaari, Persistent luminescence behavior of materials doped with Eu^{2+} and Tb^{3+}, Opt. Mater. Express. 2 (2012) 382–390.
[115] A. Bessière, R.A.T. Benhamou, G. Wallez, A. Lecointre, B. Viana, Site occupancy and mechanisms of thermally stimulated luminescence in $Ca_9Ln(PO_4)_7$ (Ln = lanthanide), Acta Mater. 60 (2012) 6641–6649.
[116] J.C. Zhang, X.L. Ma, Q.S. Qin, L.R. Shi, J.Y. Sun, M.J. Zhou, B.T. Liu, Y.H. Wang, The synthesis and afterglow luminescence properties of a novel red afterglow phosphor: $SnO_2:Sm^{3+},Zr^{4+}$, Mater. Chem. Phys. 136 (2012) 320–324.
[117] X. Zhou, G. Ju, T. Dai, Y. Li, H. Wu, Y. Jin, Y. Hu, Strontium substitution enhancing a novel Sm^{3+}-doped barium gallate phosphor with bright and red long persistent luminescence, J. Lumin. 218 (2020) 116820.
[118] Z. Zhao, Y. Wang, The synthesis and afterglow luminescence properties of a novel red afterglow phosphor: $ZrO_2:Sm^{3+},Sn^{4+}$, J. Lumin. 132 (2012) 2842–2846.

[119] G. Li, W. Chen, Y. Wang, B. Duhan, Electronic structure, photoluminescence and phosphorescence properties in $Sr_2ScGaO_5:Sm^{3+}$, Dyes Pigments 157 (2018) 259–266.
[120] G. Li, Y.H. Wang, S.C. Han, W. Zeng, W.B. Chen, Y. Gong, A red long-lasting phosphorescence material $Gd_{9.33}(SiO_4)_6O_2:Sm^{3+}$ and effect of oxygen vacancies on its performance, ECS J. Solid State Sci. Technol. 2 (2013) R161–R164.
[121] Y.Y. Zhang, R. Pang, C.Y. Li, C.Y. Zang, Q.A. Su, Reddish orange long lasting phosphorescence of Sm^{3+} in $Sr_2ZnSi_2O_7:Sm^{3+}$ phosphors, J. Rare Earths 28 (2010) 705–708.
[122] J. Xu, Z.H. Ju, X.P. Gao, Y.Q. An, X.L. Tang, W.S. Liu, $Na_2CaSn_2Ge_3O_{12}$: A novel host lattice for Sm^{3+}-doped long-persistent phosphorescence materials emitting reddish orange light, Inorg. Chem. 52 (2013) 13875–13881.
[123] T. Wang, Y. Hu, L. Chen, X. Wang, G. Ju, Luminescent properties of a reddish orange long afterglow phosphor $SrSnO_3:Sm^{3+}$, Radiat. Meas. 73 (2015) 7–13.
[124] Z.H. Ju, R.P. Wei, J.R. Zheng, X.P. Gao, S.H. Zhang, W.S. Liu, Synthesis and phosphorescence mechanism of a reddish orange emissive long afterglow phosphor Sm^{3+}-doped Ca_2SnO_4, Appl. Phys. Lett. 98 (2011) 1906.
[125] X.H. Xu, Y.H. Wang, Y. Gong, W. Zeng, Y.Q. Li, Effect of oxygen vacancies on the red phosphorescence of $Sr_2SnO_4:Sm^{3+}$ phosphor, Opt. Express 18 (2010) 16989–16994.
[126] J. Yahong, H. Yihua, C. Li, F. Yinrong, M. Zhongfei, J. Guifang, L. Jun, W. Zhonghua, X. Feihong, A novel orange emitting long persistent phosphor $CdGeO_3:Sm^{3+}$, Sci. Adv. Mater. 9 (2017) 386–391.
[127] B.F. Lei, S.Q. Man, Y.L. Du, S. Yue, Luminescence properties of Sm^{3+}-doped $Sr_3Sn_2O_7$ phosphor, Mater. Chem. Phys. 124 (2010) 912–915.
[128] X. Xu, Y. Wang, Z. Wei, G. Yu, Luminescence and storage properties of Sm-doped alkaline-earth atannates, J. Electrochem. Soc. 158 (2011) J305–J309.
[129] R. Pang, W.Z. Sun, J.P. Fu, H.F. Li, Y.L. Jia, D. Li, L.H. Jiang, S. Zhang, C.Y. Li, Luminescence properties of a novel reddish orange long-lasting phosphorescence phosphor $Zn_2P_2O_7:Sm^{3+},Li^+$, RSC Adv. 5 (2015) 82704–82710.
[130] Z.M. Shao, S. Saitzek, P. Roussel, R. Desfeux, Stability limit of the layered-perovskite structure in $Ln_2Ti_2O_7$ (Ln = lanthanide) thin films grown on (110)-oriented $SrTiO_3$ substrates by the sol–gel route, J. Mater. Chem. 22 (2012) 24894–24901.
[131] Y.J. Li, M.W. Wang, L.D. Zhang, D. Gao, S.X. Liu, Soft chemical synthesis and luminescence properties of red long-lasting phosphors $Y_2O_2S:Sm^{3+}$, Int. J. Miner. Metall. Mater. 20 (2013) 972–977.
[132] G.D. Liu, Q.H. Zhang, H.Z. Wang, Y.G. Li, A reddish La_2O_2S-based long-afterglow phosphor with effective absorption in the visible light region, Mater. Sci. Eng. B 177 (2012) 316–320.
[133] R. Pang, Y.L. Jia, R. Zhao, H.F. Li, J.P. Fu, W.Z. Sun, L.H. Jiang, S. Zhang, C.Y. Li, Q. Su, Tunable long lasting phosphorescence due to the selective energy transfer from defects to luminescent centres via tunnelling in Mn^{2+} and Tm^{3+} co-doped zinc pyrophosphate, Dalton Trans. 43 (2014) 9661–9668.
[134] R. Panga, C. Lia, L. Jianga, Q. Su, Blue long lasting phosphorescence of Tm^{3+} in zinc pyrophosphate phosphor, J. Alloys Compd. 471 (2009) 364–367.
[135] Y. Liang, F. Liu, Y. Chen, X. Wang, K. Sun, Z. Pan, Extending the applications for lanthanide ions: efficient emitters in short-wave infrared persistent luminescence, J. Mater. Chem. C 5 (2017) 6488–6492.
[136] P. Xiong, M. Peng, Visible to near-infrared persistent luminescence from Tm^{3+}-doped two-dimensional layered perovskite Sr_2SnO_4, J. Mater. Chem. C 7 (2019) 8303–8309.
[137] J. Ueda, J.L. Leaño, C. Richard, K. Asami, S. Tanabe, R.S. Liu, Broadband near-infrared persistent luminescence of $Ba[Mg_2Al_2N_4]$ with Eu^{2+} and Tm^{3+} after red light charging, J. Mater. Chem. C 7 (2019) 1705–1712.

[138] D. Haranath, A.F. Khan, H. Chander, Bright red luminescence and energy transfer of Pr^{3+}-doped (Ca,Zn)TiO_3 phosphor for long decay applications, J. Phys. D. Appl. Phys. 39 (2006) 4956–4960.

[139] Y. Liang, F. Liu, Y. Chen, X. Wang, K. Sun, Z. Pan, Red/near-infrared/short-wave infrared multi-band persistent luminescence in Pr^{3+}-doped persistent phosphors, Dalton Trans. 46 (2017) 11149–11153.

[140] P. Xiong, M. Peng, K. Qin, F. Xu, X. Xu, Visible to near-infrared persistent luminescence and mechanoluminescence from Pr^{3+}-doped $LiGa_5O_8$ for energy storage and bioimaging, Adv. Opt. Mater. 7 (2019) 1901107.

[141] D. Haranath, A.F. Khan, H. Chander, Luminescence enhancement of (Ca,Zn)TiO_3:Pr^{3+} phosphor using nanosized silica powder, Appl. Phys. Lett. 89 (2006), 091903.

[142] R. Zhang, Z. Song, L. He, Z. Xia, Q. Liu, Improvement of red-emitting afterglow properties via tuning electronic structure in perovskite-type $(Ca_{1-x}Na_x)[Ti_{1-x}Nb_x]O_3$:$Pr^{3+}$ compounds, J. Alloys. Compd. 729 (2017) 663–670.

[143] Z. Zou, C. Cao, T. Zhang, L. Feng, J. Zhang, Z. Ci, Z. Zhang, Y. Wang, Structure, long persistent luminescent properties and mechanism of a novel efficient red emitting $Ca_2Ga_2GeO_7$:Pr^{3+} phosphor, J. Alloys. Compd. 680 (2016) 397–405.

[144] S. Tian, B. Liu, L. Zhao, Z. Liu, X. Fan, L. Yang, Q. Min, H. Zhang, X. Yu, J. Qiu, X. Xu, Red photo-stimulated luminescence from deep traps of $BaZrGe_3O_9$:Pr^{3+} for optical imaging application, J. Alloys. Compd. 800 (2019) 224–230.

[145] C. Minghui, J. Dapeng, Z. Chengjiu, L. Bin, Long-lasting phosphorescence properties of pyrochlore $La_2Ti_2O_7$:Pr^{3+} phosphor, Chin. Phys. Lett. 27 (2004) 208–211.

[146] A. Lecointre, A. Bessière, A.J.J. Bos, P. Dorenbos, B. Viana, S. Jacquart, Designing a red persistent luminescence phosphor: the example of YPO_4:Pr^{3+},Ln^{3+} (Ln = Nd, Er, Ho, Dy), J. Phys. Chem. C 115 (2011) 4217–4227.

[147] A. Wiatrowska, E. Zych, Lu_2O_3:Pr,Hf storage phosphor: compositional and technological issues, Materials 7 (2014) 157–169.

[148] S. Zhang, C.Y. Li, R. Pang, L.H. Jiang, L.L. Shi, Q. Su, Energy transfer and excitation wavelength dependent long-lasting phosphorescence in Pr^{3+} activated $Y_3Al_5O_{12}$, J. Lumin. 131 (2011) 2730–2734.

[149] H. Li, R. Pang, Y. Luo, H. Wu, S. Zhang, L. Jiang, D. Li, C. Li, J. Feng, H. Zhang, Commendable Pr^{3+}-activated $Ba_2Ga_2GeO_7$ phosphor with high-brightness white long-persistent luminescence, J. Mater. Chem. C 7 (2019) 6698–6705.

[150] Y.H. Jin, Y.H. Hu, L. Chen, Y.R. Fu, Z.F. Mu, T. Wang, J. Lin, Photoluminescence, reddish orange long persistent luminescence and photostimulated luminescence properties of praseodymium doped $CdGeO_3$ phosphor, J. Alloys Compd. 616 (2014) 159–165.

[151] Y.H. Jin, Y.H. Hu, L. Chen, X.J. Wang, G.F. Ju, Luminescent properties of a red afterglow phosphor Ca_2SnO_4:Pr^{3+}, Opt. Mater. 35 (2013) 1378–1384.

[152] T. Nakamura, M. Shima, M. Yasukawa, K. Ueda, Synthesis of Pr^{3+} doped or Tb^{3+}–Mg codoped $CaSnO_3$ perovskite phosphor by the polymerized complex method, J. Sol-Gel Sci. Technol. 61 (2012) 362–366.

[153] P. Boutinaud, L. Sarakha, R. Mahiou, $NaNbO_3$:Pr^{3+}: a new red phosphor showing persistent luminescence, J. Phys. Condens. Matter 21 (2009), 025901.

[154] S. Lin, C. Xiong, D. Ma, H. Li, S. Long, B. Wang, Persistent luminescence found in Mg^{2+} and Pr^{3+} co-doped $LiNbO_3$ single crystal, J. Mater. Chem. C 6 (2018) 10067–10072.

[155] B. Wang, X. Li, Y. Chen, Y. Chen, J. Zhou, Q. Zeng, Long persistent and photo-stimulated luminescence in Pr^{3+}-doped layered perovskite phosphor for optical data storage, J. Am. Ceram. Soc. 101 (2018) 4598–4607.

[156] L.L. Noto, M.Y.A. Yagoub, O.M. Ntwaeaborwa, H.C. Swart, Persistent photoluminescence emission from $SrTa_2O_6:Pr^{3+}$ phosphor prepared at different temperatures, Ceram. Int. 41 (2015) 8828–8836.
[157] L.L. Noto, M.L. Chitambo, O.M. Ntwaeaborwa, H.C. Swart, Photoluminescence and thermoluminescence properties of Pr^{3+} doped $ZnTa_2O_6$ phosphor, Powder Technol. 247 (2013) 147–150.
[158] Y.M. Yang, Z.Y. Li, J.Y. Zhang, Y. Lu, S.Q. Guo, Q. Zhao, X. Wang, Z.J. Yong, H. Li, J. P. Ma, Y. Kuroiwa, C. Moriyoshi, L.L. Hu, L.Y. Zhang, L.R. Zheng, H.T. Sun, X-ray-activated long persistent phosphors featuring strong UVC afterglow emissions, Light. Sci. Appl. 7 (2018) 88.
[159] H. Shi, Z. An, Ultraviolet afterglow, Nat. Photonics 13 (2019) 74–75.
[160] M. Yamaga, Y. Ohsumi, T. Nakayama, T.P.J. Han, Persistent phosphorescence in Ce-doped Lu_2SiO_5, Opt. Mater. Express 2 (2012) 413–419.
[161] S.K. Sharma, M. Bettinelli, I. Carrasco, J. Beyer, R. Gloaguen, J. Heitmann, Dynamics of charges in super long black light-emitting $CaB_2O_4:Ce^{3+}$ persistent phosphor, J. Phys. Chem. C 123 (23) (2019) 14639–14646.
[162] K. Van den Eeckhout, P.F. Smet, D. Poelman, Persistent luminescence in Eu^{2+}-doped compounds: a review, Materials 3 (2010) 2536–2566.
[163] D.D. Jia, X.J. Wang, W.M. Yen, Delocalization, thermal ionization, and energy transfer in singly doped and codoped $CaAl_4O_7$ and Y_2O_3, Phys. Rev. B: Condens. Matter Mater. Phys. 69 (2004) 235113.
[164] J. Ueda, S. Miyano, S. Tanabe, Formation of deep electron traps by Yb^{3+} codoping leads to super-long persistent luminescence in Ce^{3+}-doped yttrium aluminum gallium garnet phosphors, ACS Appl. Mater. Interfaces 10 (24) (2018) 20652–20660.
[165] L.X. Ning, C.C. Zhou, W.P. Chen, Y.C. Huang, C.K. Duan, P. Dorenbos, Y. Tao, H.B. Liang, Electronic properties of Ce^{3+}-doped $Sr_3Al_2O_5Cl_2$: a combined spectroscopic and theoretical study, J. Phys. Chem. C 119 (2015) 6785–6792.
[166] J. Ueda, K. Aishima, S. Nishiura, S. Tanabe, Afterglow luminescence in Ce^{3+}-Doped $Y_3Sc_2Ga_3O_{12}$ Ceramics, Appl. Phys. Express 4 (2011), 042602.
[167] J. Xu, J. Ueda, Y.X. Zhuang, B. Viana, S. Tanabe, $Y_3Al_{5-x}Ga_xO_{12}:Cr^{3+}$: A novel red persistent phosphor with high brightness, Appl. Phys. Express 8 (2015), 042602.
[168] J. Ueda, K. Kuroishi, S. Tanabe, Yellow persistent luminescence in Ce^{3+}–Cr^{3+}-codoped gadolinium aluminum gallium garnet transparent ceramics after blue-light excitation, Appl. Phys. Express 7 (2014), 062201.
[169] X.J. Wang, D.D. Jia, W.M. Yen, Mn^{2+} activated green, yellow, and red long persistent phosphors, J. Lumin. 102 (2003) 34–37.
[170] W.B. Dai, Investigation of the luminescent properties of Ce^{3+} doped and Ce^{3+}/Mn^{2+} co-doped $CaAl_2Si_2O_8$ phosphors, RSC Adv. 4 (2014) 11206–11215.
[171] X.H. Xu, Y.H. Wang, X. Yu, Y.Q. Li, Y. Gong, Electric field-assisted sintering in comparison with the hot pressing of yttria-stabilized zirconia, J. Am. Ceram. Soc. 94 (2011) 24–27.
[172] D. Jia, R.S. Meltzer, W.M. Yen, W. Jia, X. Wang, Green phosphorescence of $CaAl_2O_4$: Tb^{3+},Ce^{3+} through persistence energy transfer, Appl. Phys. Lett. 80 (2002) 1535–1537.
[173] Y. Luo, Z.G. Xia, Effect of Al/Ga substitution on photoluminescence and phosphorescence properties of garnet-type $Y_3Sc_2Ga_{3-x}Al_xO_{12}:Ce^{3+}$ phosphor, J. Phys. Chem. C 118 (2014) 23297–23305.
[174] Y. Gong, Y. Wang, Y. Li, X. Xu, Ce^{3+}, Dy^{3+} co-doped white-light long-lasting phosphor: $Sr_2Al_2SiO_7$ through energy transfer, J. Electrochem. Soc. 157 (2010) J208–J211.

[175] Y. Gong, X. Xu, W. Zeng, C. Wu, Y. Wang, Ce^{3+}, Mn^{2+} co-doped red–light long–lasting phosphor: $BaMg_2Si_2O_7$ through energy transfer, Phys. Procedia 29 (2012) 86–90.
[176] Y. Ito, A. Komeno, K. Uematsu, K. Toda, M. Sato, Luminescence properties of long-persistence silicate phosphors, J. Alloys Compd. 408–412 (2006) 907–910.
[177] S.K. Sharma, M. Bettinelli, M. Karlsson, Green persistent luminescence excitable by multiple wavelengths in the $CaSc_2O_4$:Ce^{3+} phosphor co-doped with Mg^{2+}, J. Lumin. 196 (2018) 437–441.
[178] K. Van den Eeckhout, D. Poelman, P.F. Smet, Persistent luminescence in non-Eu^{2+}-doped compounds: a review, Materials 6 (2013) 2789–2818.
[179] M. Zhong-Fei, W. Yin-Hai, H. Yi-Hua, The afterglow and thermoluminescence properties of $Y_3Al_5O_{12}$:Ce^{3+}, Acta Phys. Sin. 60 (2011) 012110–013201.
[180] H. Kanno, K. Noda, K. Matsui, New long-afterglow phosphors:Yb^{2+}-doped strontium aluminates, Chem. Phys. Lett. 580 (2013) 103–107.
[181] Z. Ma, S. Tian, H. Wu, J. Zhang, H. Li, Self-reduction of Yb^{3+} to Yb^{2+} and red persistent luminescence of $Na_2CaGe_2O_6$:Yb^{2+} phosphor, Ceram. Int. 44 (2018) 14582–14586.
[182] F. Yu, Y. Yang, X. Su, C. Mi, H.J. Seo, Novel long persistent luminescence phosphors: Yb^{2+} codoped MAl_2O_4 (M = Ba, Sr), Opt. Mater. Express 5 (2015) 585–595.
[183] Y. Lv, L. Wang, Y. Zhuang, T.L. Zhou, R.J. Xie, Discovery of the Yb^{2+}–Yb^{3+} couple as red-to-NIR persistent luminescence emitters in Yb-activated $(Ba_{1-x}Sr_x)AlSi_5O_2N_7$ phosphors, J. Mater. Chem. C 5 (2017) 7095–7101.
[184] Y. Zhuang, Y. Lv, L. Wang, W. Chen, T.L. Zhou, T. Takeda, N. Hirosaki, R.J. Xie, Trap depth engineering of $SrSi_2O_2N_2$:Ln^{2+},Ln^{3+} (Ln^{2+}=Yb, Eu; Ln^{3+}=Dy, Ho, Er) persistent luminescence materials for information storage applications, ACS Appl. Mater. Interfaces 10 (2018) 1854–1864.
[185] Y.M. Yang, X.D. Li, X.Y. Su, Z.Q. Li, L.L. Liu, Y.Z. Liu, J. Zhang, C. Mi, F. Yu, A novel Yb^{2+} doped SrS long persistent luminescence phosphors, Opt. Mater. 36 (2014) 1822–1825.
[186] J. Wei, L. Chen, W. Zhang, Y. Yang, Z. Li, The orange–red persistent luminescence of $Ba_{1-x}Ca_xS$: Yb^{2+}, J. Lumin. 181 (2017) 427–432.
[187] S. Kamimura, C.N. Xu, H. Yamada, N. Terasaki, M. Fujihala, Long-persistent luminescence in the near-infrared from Nd^{3+}-doped Sr_2SnO_4 for in vivo optical imaging, Jpn. J. Appl. Phys. 53 (2014), 062403.
[188] J. Xu, S. Tanabe, A.D. Sontakke, J. Ueda, Near-infrared multi-wavelengths long persistent luminescence of Nd^{3+} ion through persistent energy transfer in Ce^{3+}, Cr^{3+} co-doped $Y_3Al_2Ga_3O_{12}$ for the first and second bio-imaging windows, Appl. Phys. Lett. 107 (2015), 081903.
[189] L. Pihlgren, T. Laihinen, L.C.V. Rodrigues, S. Carlson, K.O. Eskola, A. Kotlov, M. Lastusaari, T. Soukka, H.F. Brito, J. Holsa, On the mechanism of persistent up-conversion luminescence in the ZrO_2:Yb^{3+},Er^{3+} nanomaterials, Opt. Mater. 36 (2014) 1698–1704.
[190] D. Zhang, M. Shi, Y. Sun, Y. Guo, C. Chang, Long afterglow property of Er^{3+} doped Ca_2SnO_4 phosphor, J. Alloys. Compd. 667 (2016) 235–239.
[191] J. Xu, D. Murata, J. Ueda, S. Tanabe, Near-infrared long persistent luminescence of Er^{3+} in garnet for the third bio-imaging window, J. Mater. Chem. C 4 (2016) 11096–11103.
[192] X.F. Qu, L.X. Cao, W. Liu, G. Su, P.P. Wang, I. Schultz, Sol–gel synthesis of long-lasting phosphors $CdSiO_3$: Mn^{2+}, RE^{3+} (RE = Tb, Eu, Nd) and luminescence mechanism research, Mater. Res. Bull. 47 (2012) 1598–1603.
[193] T. Maldiney, A. Lecointre, B. Viana, A.L. Bessière, M. Bessodes, D. Gourier, C. Richard, D. Scherman, Controlling electron trap depth to enhance optical properties of persistent

luminescence nanoparticles for in vivo imaging, J. Am. Chem. Soc. 133 (2011) 11810–11815.

[194] L. Lin, M. Yin, C. Shi, W. Zhang, Luminescence properties of a new red long-lasting phosphor: $Mg_2SiO_4:Dy^{3+},Mn^{2+}$, J. Alloys Compd. 455 (2008) 327–330.

[195] X.J. Wang, D. Jia, W.M. Yen, Mn^{2+} activated green, yellow, and red long persistent phosphors, J. Lumin. 102 (2003) 34–37.

[196] Z.F. Mu, Y.H. Hu, Y.H. Wang, H.Y. Wu, C.J. Fu, F.W. Kang, The structure and luminescence properties of long afterglow phosphor $Y_{3-x}Mn_xAl_{5-x}Si_xO_{12}$, J. Lumin. 131 (2011) 676–681.

[197] Y.X. Zhuang, J. Ueda, S. Tanabe, Multi-color persistent luminescence in transparent glass ceramics containing spinel nano-crystals with Mn^{2+} ions, Appl. Phys. Lett. 105 (2014) 191904.

[198] X.Q. Xu, J. Ren, G.R. Chen, D.S. Kong, C.J. Gu, C.M. Chen, L.R. Kong, Bright green emission from the Mn^{2+}-doped zinc gallogermanate phosphors, Opt. Mater. Express 3 (2013) 1727–1732.

[199] Z.R. Liu, R.X. Zhong, Green and red long lasting phosphorescence (LLP) in γ-$Zn_3(PO_4)_2:Mn^{2+}/Ga^{3+}$, J. Alloys Compd. 556 (2013) 6–11.

[200] A. Bessiere, A. Lecointre, R.A. Benhamou, E. Suard, G. Wallez, B. Viana, How to induce red persistent luminescence in biocompatible $Ca_3(PO_4)_2$, J. Mater. Chem. C 1 (2013) 1252–1259.

[201] A. Lecointre, R. Ait Benhamou, A. Bessiére, G. Wallez, M. Elaatmani, B. Viana, Red long-lasting phosphorescence (LLP) in β-TCP type $Ca_{9.5}Mn(PO_4)_7$ compounds, Opt. Mater. 34 (2011) 376–380.

[202] Y.H. Jin, Y.H. Hu, H. Duan, L. Chen, X.J. Wang, The long persistent luminescence properties of phosphors: Li_2ZnGeO_4 and $Li_2ZnGeO_4:Mn^{2+}$, RSC Adv. 4 (2014) 11360–11366.

[203] Y. Takahashi, M. Ando, R. Ihara, T. Fujiwara, Green-emissive Mn-activated nanocrystallized glass with willemite-type Zn_2GeO_4, Opt. Mater. Express 1 (2011) 372–378.

[204] G.B. Che, X.Y. Li, C.B. Liu, H. Wang, Y. Liu, Z.L. Xu, Long-lasting phosphorescence properties of Mn^{2+}-doped $Cd_2Ge_7O_{16}$ orange light-emitting phosphor, *Phys*, Status Solidi A 205 (2008) 194–198.

[205] Y.X. Zhuang, Y. Katayama, J. Ueda, S. Tanabe, A brief review on red to near-infrared persistent luminescence in transition-metal-activated phosphors, Opt. Mater. 36 (2014) 1907–1912.

[206] G.B. Che, C.B. Liu, X.Y. Li, Z.L. Xu, Y. Liu, H. Wang, Luminescence properties of a new Mn^{2+}-activated red long-afterglow phosphor, J. Phys. Chem. Solids 69 (2008) 2091–2095.

[207] B.F. Lei, B. Li, X.J. Wang, W. Li, Green emitting long lasting phosphorescence (LLP) properties of $Mg_2SnO_4:Mn^{2+}$ phosphor, J. Lumin. 118 (2006) 173–178.

[208] J. Xu, N.J. Cherepy, J. Ueda, S. Tanabe, Red persistent luminescence in rare earth-free $AlN:Mn^{2+}$ phosphor, Mater. Lett. 206 (2017) 175–177.

[209] J. Xu, D.Q. Chen, Y.L. Yu, W.J. Zhu, J.C. Zhou, Y.S. Wang, $Cr^{3+}:SrGa_{12}O_{19}$: A broadband near-infrared long-persistent phosphor, Chem.-Asian J. 9 (2014) 1020–1025.

[210] F. Sun, R. Xie, L. Guan, C. Zhang, The near-infrared long-persistent phosphorescence of Cr^{3+}-activated non-gallate phosphor, Mater. Lett. 164 (2016) 39–43.

[211] Y. Zhang, R. Huang, H. Li, D. Hou, Z. Lin, J. Song, Y. Guob, H. Lin, C. Song, Z. Lin, J. Robertson, Germanium substitution endowing Cr^{3+}-doped zinc aluminate phosphors with bright and super-long near-infrared persistent luminescence, Acta Mater. 155 (2018) 214–221.

[212] F. Sun, R. Xie, L. Guan, C. Zhang, Cr^{3+}doped $Ca_{14}Zn_6Ga_{10}O_{35}$: A near-infrared long persistent luminescence phosphor, J. Lumin. 180 (2016) 251–257.
[213] Y.L. Wu, Y. Li, X.X. Qin, R.C. Chen, D.K. Wu, S.J. Liu, J.R. Qiu, Dual mode NIR long persistent phosphorescence and NIR-to-NIR Stokes luminescence in $La_3Ga_5GeO_{14}$:Cr^{3+}, Nd^{3+} phosphor, J. Alloys Compd. 649 (2015) 62–66.
[214] X. Yi, Z.T. Chen, S. Ye, Y. Li, E.H. Song, Q.Y. Zhang, Multifunctionalities of near-infrared upconversion luminescence, optical temperature sensing and long persistent luminescence in $La_3Ga_5GeO_{14}$:Cr^{3+},Yb^{3+},Er^{3+} and their potential coupling, RSC Adv. 5 (2015) 49680–49687.
[215] W.Z. Yan, F. Liu, Y.Y. Lu, X.J. Wang, M. Yin, Z.W. Pan, Near infrared long-persistent phosphorescence in $La_3Ga_5GeO_{14}$:Cr^{3+} phosphor, Opt. Express 18 (2010) 20215–20221.
[216] Y. Li, Y.Y. Li, R.C. Chen, K. Sharafudeen, S.F. Zhou, M. Gecevicius, H.H. Wang, G.P. Dong, Y.L. Wu, X.X. Qin, J.R. Qiu, Tailoring of the trap distribution and crystal field in Cr^{3+}-doped non-gallate phosphors with near-infrared long-persistence phosphorescence, NPG Asia Mater. 7 (2015), e180.
[217] D.Q. Chen, Near-infrared long-lasting phosphorescence in transparent glass ceramics embedding Cr^{3+}-doped $LiGa_5O_8$ nanocrystals, J. Eur. Ceram. Soc. 34 (2014) 4069–4075.
[218] Y.J. Chuang, Z.P. Zhen, F. Zhang, F. Liu, J.P. Mishra, W. Tang, H.M. Chen, X.L. Huang, L.C. Wang, X.Y. Chen, J. Xie, Z.W. Pan, Photostimulable near-infrared persistent luminescent nanoprobes for ultrasensitive and longitudinal deep-tissue bio-Imaging, Theranostics 4 (2014) 1112–1122.
[219] F. Liu, W.Z. Yan, Y.J. Chuang, Z.P. Zhen, J. Xie, Z.W. Pan, Photostimulated near-infrared persistent luminescence as a new optical read-out from Cr^{3+}-doped $LiGa_5O_8$, Sci. Rep. 3 (2013), 01554.
[220] Y. Li, S.F. Zhou, Y.Y. Li, K. Sharafudeen, Z.J. Ma, G.P. Dong, M.Y. Peng, J.R. Qiu, Long persistent and photo-stimulated luminescence in Cr^{3+}-doped Zn–Ga–Sn–O phosphors for deep and reproducible tissue imaging, J. Mater. Chem. C 2 (2014) 2657–2663.
[221] N. Basavaraju, K.R. Priolkar, D. Gourier, S.K. Sharma, A. Bessiere, B. Viana, The importance of inversion disorder in the visible light induced persistent luminescence in Cr^{3+} doped AB_2O_4 (A = Zn or Mg and B = Ga or Al), Phys. Chem. Chem. Phys. 17 (2015) 1790–1799.
[222] N. Basavaraju, K.R. Priolkar, D. Gourier, A. Bessiere, B. Viana, Order and disorder around Cr^{3+} in chromium doped persistent luminescent AB_2O_4 spinels, Phys. Chem. Chem. Phys. 17 (2015) 10993–10999.
[223] Z.W. Pan, Y.Y. Lu, F. Liu, Sunlight-activated long-persistent luminescence in the near-infrared from Cr^{3+}-doped zinc gallogermanates, Nat. Mater. 11 (2012) 58–63.
[224] Y. L. Wu, Y. Li, X. X. Qin, R. C. Chen, D. K. Wu, S. J. Liu, J. R. Qiu, Near-infrared long-persistent phosphor of $Zn_3Ga_2Ge_2O_{10}$:Cr^{3+} sintered in different atmosphere. Spectrochim. Acta, Part A, 2015, 151, 385–389.
[225] D.Q. Chen, Y. Chen, H.W. Lu, Z.G. Ji, A bifunctional Cr/Yb/Tm:$Ca_3Ga_2Ge_3O_{12}$ phosphor with near-infrared long-lasting phosphorescence and upconversion luminescence, Inorg. Chem. 53 (2014) 8638–8645.
[226] Y.X. Zhuang, J. Ueda, S. Tanabe, P. Dorenbos, Band-gap variation and a self-redox effect induced by compositional deviation in $Zn_xGa_2O_{3+x}$:Cr^{3+} persistent phosphors, J. Mater. Chem. C 2 (2014) 5502–5509.
[227] M. Allix, S. Chenu, E. Veron, T. Poumeyrol, E.A. Kouadri-Boudjelthia, S. Alahrache, F. Porcher, D. Massiot, F. Fayon, Considerable improvement of long-persistent luminescence in germanium and tin substituted $ZnGa_2O_4$, Chem. Mater. 25 (2013) 1600–1606.

[228] X.S. Wang, W.S. Li, J.Q. Situ, X.Y. Ying, H. Chen, Y. Jin, Y.Z. Du, Multi-functional mesoporous β-Ga_2O_3: Cr^{3+} nanorod with long lasting near infrared luminescence for in vivo imaging and drug delivery, RSC Adv. 5 (2015) 12886–12889.
[229] Y. Zhan, Y. Jin, H. Wu, L. Yuan, G. Ju, Y. Lv, Y. Hu, Cr^{3+}-doped $Mg_4Ga_4Ge_3O_{16}$ near-infrared phosphor membrane for optical information storage and recording, J. Alloys. Compd. 777 (2019) 991–1000.
[230] Y. Li, Y.Y. Li, K. Sharafudeen, G.P. Dong, S.F. Zhou, Z.J. Ma, M.Y. Peng, J.R. Qiu, A strategy for developing near infrared long-persistent phosphors: taking $MAlO_3$:Mn^{4+}, Ge^{4+} (M = La, Gd) as an example, J. Mater. Chem. C 2 (2014) 2019–2027.
[231] F. Xue, Y. Hu, L. Chen, H. Wu, G. Ju, T. Wang, L. Yang, A novel rare-earth free red long-persistent phosphor: Mg_2GeO_4:Mn^{4+}, Ceram. Int. 43 (2017) 15141–15145.
[232] X. Zhang, J. Nie, S. Liu, Y. Li, J. Qiu, Deep-red photoluminescence and long persistent luminescence in double perovskite-type La_2MgGeO_6:Mn^{4+}, J. Am. Ceram. Soc. 101 (2018) 1576–1584.
[233] Z.J. Chi, Q.Q. Song, Y.M. Hui, Z.H. Liang, Photoluminescence and persistent luminescence properties of non-doped and Ti^{4+}-doped Mg_2SnO_4 phosphors, Chin. Phys. B 20 (2011), 094211.
[234] B.P. Chandra, Persistent mechanoluminescence induced by elastic deformation of ZrO_2: Ti phosphors, J. Lumin. 130 (2010) 2218–2222.
[235] J. Qiu, K. Miyauchi, Y. Kawamoto, Long-lasting phosphorescence in Sn^{2+}–Cu^{2+} codoped silicate glass and its high-pressure treatment effect, Appl. Phys. Lett. 81 (2002) 394.
[236] J.M. Carvalho, L.C.V. Rodrigues, J. Hölsä, M. Lastusaari, L.A.O. Nunes, M.C.F.C. Felinto, O.L. Malta, H.F. Brito, Influence of titanium and lutetium on the persistent luminescence of ZrO_2, Opt. Mater. Express 2 (2012) 331–340.
[237] H. Zhang, Q. Yang, S. Lu, D. Huang, Y. Wang, Z. Wei, Q. Wang, Y. Zhang, Fabrication, spectral and laser performance of 5 at.% Yb^{3+} doped $(La_{0.10}Y_{0.90})_2O_3$ transparent ceramic, Opt. Mater. 35 (2013) 766–769.
[238] J.M. Carvalho, C.C.S. Pedroso, I.P. Machado, J. Hölsä, L.C.V. Rodrigues, P. Głuchowski, M. Lastusaari, H.F. Brito, Persistent luminescence warm-light LEDs based on Ti-doped RE_2O_2S materials prepared by rapid and energy-saving microwave-assisted synthesis, J. Mater. Chem. C 6 (2018) 8897–8905.
[239] L.X. Wang, L. Zhang, Y.D. Huang, D.Z. Jia, J.J. Lu, Effects of Gd^{3+} and Lu^{3+} co-doping on the long afterglow properties of yellowish-orange phosphor Y_2O_2S:Ti^{4+}, Mg^{2+}, J. Lumin. 129 (2009) 1032–1035.
[240] I. Norrbo, J.M. Carvalho, P. Laukkanen, J. Mäkelä, F. Mamedov, M. Peurla, H. Helminen, S. Pihlasalo, H. Härmä, J. Sinkkonen, M. Lastusaari, Lanthanide and heavy metal free long white persistent luminescence from Ti doped Li–hackmanite: a versatile, low-cost material, Adv. Funct. Mater. 27 (2017) 1606547.
[241] S. Lai, Z. Yang, J. Liao, J. Qiu, Z. Song, Y. Yang, D. Zhou, Investigation of persistent luminescence property of Bi^{3+}, Dy^{3+} co-doped $CdSiO_3$ phosphor, Mater. Res. Bull. 60 (2014) 714–718.
[242] Y.X. Zhuang, J. Ueda, S. Tanabe, Photochromism and white long-lasting persistent luminescence in Bi^{3+}-doped $ZnGa_2O_4$ ceramics, Opt. Mater. Express 2 (2012) 1378–1383.
[243] S. Zhang, Y. Hu, R. Chen, X. Wang, Z. Wang, Photoluminescence and persistent luminescence in Bi^{3+}-doped Zn_2GeO_4 phosphors, Opt. Mater. 36 (2014) 1830–1835.
[244] F.W. Kang, M.Y. Peng, A new study on the energy transfer in the color-tunable phosphor $CaWO_4$:Bi, Dalton Trans. 43 (2014) 277–284.

[245] X. Dou, H. Xiang, P. Wei, S. Zhang, G. Ju, Z. Meng, L. Chen, Y. Hu, Y. Li, A novel phosphor $CaZnGe_2O_6:Bi^{3+}$ with persistent luminescence and photo-stimulated luminescence, Mater. Res. Bull. 105 (2018) 226–230.
[246] Z. Liu, Y. Ding, C. Zheng, Y. Meng, J. Tang, Q. Qiu, An efficient UV converted blue-emitting $Lu_2CaGeO_6:Bi^{3+}$ persistent phosphor for potential application in photocatalysis, Ceram. Int. 44 (2018) 14712–14716.
[247] H. Li, J. Cai, R. Pang, G. Liu, S. Zhang, L. Jiang, D. Li, C. Li, J. Feng, H. Zhang, A strategy for developing thermal-quenching-resistant emission and super-long persistent luminescence in $BaGa_2O_4:Bi^{3+}$, J. Mater. Chem. C 7 (2019) 13088–13096.
[248] W. Sun, R. Pang, H. Li, D. Li, L. Jiang, S. Zhang, J. Fu, C. Li, Investigation of a novel color tunable long afterglow phosphor $KGaGeO_4:Bi^{3+}$: luminescence properties and mechanism, J. Mater. Chem. C 5 (2017) 1346–1355.
[249] X. Wang, P. Boutinaud, L. Li, J. Cao, P. Xiong, X. Li, H. Luo, M. Peng, Novel persistent and tribo-luminescence from bismuth ion pairs doped strontium gallate, J. Mater. Chem. C 6 (2018) 10367–10375.
[250] J. Shi, X. Sun, S. Zheng, X. Fu, Y. Yang, J. Wang, H. Zhang, Super-long persistent luminescence in the ultraviolet A region from a Bi^{3+}-doped $LiYGeO_4$ phosphor, Adv. Opt. Mater. 7 (2019) 1900526.
[251] S. Wang, W. Chen, D. Zhou, J. Qiu, X. Xu, X. Yu, Long persistent properties of $CaGa_2O_4:Bi^{3+}$ at different ambient temperature, J. Am. Ceram. Soc. 100 (2017) 3514–3521.
[252] Y.H. Jin, Y.H. Hu, L. Chen, X.J. Wang, G.F. Ju, Z.F. Mou, A novel NIR long phosphorescent phosphor:$SrSnO_3:Bi^{2+}$, RSC Adv. 5 (2015) 101347–101352.
[253] Y. Liang, F. Liu, Y. Chen, K. Sun, Z. Pan, Long persistent luminescence in the ultraviolet in Pb^{2+}-doped $Sr_2MgGe_2O_7$ persistent phosphor, Dalton Trans. 45 (2016) 1322–1326.
[254] L.C.V. Rodrigues, J. Holsa, M. Lastusaari, M.C.F.C. Felinto, H.F. Brito, Defect to R^{3+} energy transfer: colour tuning of persistent luminescence in $CdSiO_3$, J. Mater. Chem. C 2 (2014) 1612–1618.
[255] X.H. Xu, J.B. Qiu, Y. Xue, Multicolor long persistent luminescence phosphors $CaSnSiO_5:R^{3+}$ (R = Pr, Tb, Sm, Dy), ECS Solid State Lett. 2 (2013) R9–R11.
[256] X. Xu, Y.H. Wang, W. Zeng, Y. Gong, B.T. Liu, Luminescent properties of the multicolor afterglow phosphors $Ca_3SnSi_2O_9:Re^{3+}$ (Re = Pr, Tb, Sm), J. Am. Ceram. Soc. 94 (2011) 3632–3635.
[257] P.F. Feng, J.C. Zhang, C.Q. Wu, X. Liu, Y.H. Wang, Self-activated afterglow luminescence of un-doped $Ca_2ZrSi_4O_{12}$ material and explorations of new afterglow phosphors in a rare earth element-doped $Ca_2ZrSi_4O_{12}$ system, Mater. Chem. Phys. 141 (2013) 495–501.
[258] M.M. Shi, D.Y. Zhang, C.K. Chang, Dy^{3+}:Ca_2SnO_4, a new yellow phosphor with afterglow behavior, J. Alloys Compd. 639 (2015) 168–172.
[259] J. Zhang, M. Yu, Q. Qin, H. Zhou, M. Zhou, X. Xu, Y. Wang, The persistent luminescence and up conversion photostimulated luminescence properties of nondoped Mg_2SnO_4 Material, J. Appl. Phys. 108 (2010) 123518.
[260] X.L. Ma, J.C. Zhang, H.H. Li, B.C. Duan, L.N. Guo, M.D. Que, Y.H. Wang, Violet blue long-lasting phosphorescence properties of Mg-doped $BaZrO_3$ and its ability to assist photocatalysis, J. Alloys Compd. 580 (2013) 564–569.
[261] H. Luo, J. Cao, X. Li, X. Wang, M. Peng, Topological tailoring of structure and defects to enhance red to near-infrared afterglow from Mn^{2+}-doped germanate photonic glasses, J. Mater. Chem. C 6 (2018) 11525–11535.

[262] X. F. Liu, Y. B. Qiao, G. P. Dong, S. Ye, B. Zhu, Y. X. Zhuang, J. R. Qiu, BCNO-based long-persistent phosphor. J. Electrochem. Soc., 2009, 156, P81–P84.

[263] X.F. Liu, S. Ye, G.P. Dong, Y.B. Qiao, J. Ruan, Y.X. Zhuang, Q. Zhang, G. Lin, D.P. Chen, J.R. Qiu, Spectroscopic investigation on BCNO-based phosphor: photoluminescence and long persistent phosphorescence, J. Phys. D-Appl. Phys. 42 (2009) 215409.

[264] J. Li, B. Lei, J. Qin, Y. Liu, X. Liu, Temperature-dependent emission spectra of $Ca_2Si_5N_8$:Eu^{2+}, Tm^{3+} phosphor and its afterglow properties, J. Am. Ceram. Soc. 96 (2013) 873–878.

[265] J. Qiu, K. Miura, H. Inouye, Y. Kondo, T. Mitsuyu, K. Hirao, Femtosecond laser-induced three-dimensional bright and long-lasting phosphorescence inside calcium aluminosilicate glasses doped with rare earth ions, Appl. Phys. Lett. 73 (1998) 1763.

[266] T. Hu, H. Lin, J. Xu, B. Wang, J. Wang, Y. Wang, Color-tunable persistent luminescence in oxyfluoride glass and glass ceramic containing Mn^{2+}:α-Zn_2SiO_4 nanocrystals, J. Mater. Chem. C 5 (2017) 1479–1487.

[267] Z. Gong, W. Zheng, Y. Gao, P. Huang, D. Tu, R. Li, J. Wei, W. Zhang, Y. Zhang, X. Chen, Full-spectrum persistent luminescence tuning using all inorganic perovskite quantum dots, Angew. Chem. 131 (2019) 7017–7021.

Control of afterglow properties

6

In long persistent phosphors (PPs), dopant plays a leading role in their performances, while the host can be used to tune and optimize the optical properties. The main emitters in long persistent phosphors include transition metals and lanthanides. Lanthanides are divided into two groups: divalent rare earth ions and trivalent rare earth ions. Divalent rare earth ions, such as Eu^{2+} and Yb^{2+}, exhibit a broadband 4f-5d transition, while most trivalent rare earth ions are usually characterized by the line-shaped 4f-4f transitions [1].

As we know, very strong shielding effect of the 4f electrons by the 5 s and 5p orbitals and the weaker shielding of the 5d electron exist in lanthanide ions. Thus the optical performance involving 5d electrons is complicated, and it is usually determined by the factors including the bond lengths from the activator ion to the coordinating anions, the degree of covalency between the activator ion and its ligands, the coordination environment, and the symmetry of the activator site. Defects exert a strong influence of the aforementioned factors separately or synergistically and thus modulate the optical performance. Relatively the shielded 4f electrons show only slight disturbance by the local environment. However, the parity forbidden 4f-4f intraconfigurational transitions of lanthanides ions can be partially allowed by mixing certain odd-parity configurations when embedded in specific inorganic lattices, thus allowing for the tuning of optical properties of persistent phosphors by introducing defects [2–4].

Besides the largest group of elements in the periodical table is the transition metals. Of all the groups of elements, the transition metals are most confusing to specify because there are different definitions of which elements should be included. A transition metal is an element with a partially filled d electron subshell. Elements from Groups 3 to 12 on the periodic table, including the f-block elements (lanthanides and actinides, below the main body of the periodic table), are all classified as transition metals. Different from the common d-block elements, the lanthanides and actinides are usually called "inner transition metals." These elements are generally hard, with high melting points and boiling points. Moving from left to right across the periodic table, electron filling of the d (or f) orbitals is increased. The d electrons are loosely bound, which contributes to the high electrical conductivity and malleability of the transition elements. The transition elements have low ionization energies, and they can be stabilized in different positive oxidation states. The positive oxidation states allow transition elements to form many different ionic and partially ionic compounds. The formation of complexes causes the d orbitals to split into two energy sublevels, which enables many of the complexes to absorb specific frequencies of light. When dissolved in solution are crystallized into compounds, these complexes exhibit characteristic colors originating from the transitions involving d electrons. In addition, complexation reactions sometimes enhance the relatively low solubility of some compounds.

Persistent Phosphors. https://doi.org/10.1016/B978-0-12-818637-4.00006-9

6.1 Implantation of defect

Hereinafter, in situ generation of defect by chemical methods will be discussed. These methods include stoichiometry control, ion doping, annealing at high temperatures, high pressure, strong external field and oxidizing/reducing atmosphere, and surface modification of bonding chemistry.

6.1.1 High-temperature behaviors of crystals

In the previous section, we have discussed common crystal structures found in most inorganic compounds. Defects are also characterized by their atomic structure. Intrinsic defects, or native defects, are atoms that are located at the wrong places with respect to the perfect crystal. The simplest example of a native defect is a vacancy. Extrinsic defects, in comparison, are formed by introducing impurity atoms that are foreign to the crystal. Dopants usually refer to impurity atoms that are introduced intentionally. A complex defect involves a combination of two or more impurities or native defects.

Thermal expansion at high temperatures is a very natural thermodynamic process for materials and the modeling of the equation of state of the earth's interior. It is tempting to think about solids as if the particles were locked into position, and the way bricks are used to build a wall. To a certain extent, this is analogy could be inappropriate, because the particles in a solid are in more or less constant motion rocking back and forth and rotating with respect to their equilibrium positions in solids. This motion depends on two factors: the temperature of the system and the strength of the interactions that hold the particles together. The higher the temperature, the faster the particles are moving. The stronger the force of attraction between particles, the smaller separations of the particles will be. Because the van der Waals forces that hold molecules together are much weaker than the bonds between atoms in a metal or between positive and negative ions in an ionic compound, molecular crystals expand more when being heated than metals and ionic compounds.

According to Grüneisen's theory [5], the thermal expansion of crystals at high temperature should be a slowly varying function of temperature, especially above the Debye temperature. It is indicated that at high temperature, the thermal expansion is quite different from theoretically predicted. It was Lawson who first postulated that abnormal expansion at high temperature arises from thermal defects [6].

Theoretical calculations indicate that thermal expansion is one of the most sensitive thermodynamic properties to the presence of lattice defects. Merriam et al. recorded thermal expansion from x-ray diffraction and assumed that at high temperature there was a linear relationship between thermal expansion and temperature [7]. The deviation between their measurements and a linear relationship was attributed to the contribution of thermal vacancies. By assuming a linear relationship between the anomalous high-temperature thermal expansion and the concentration of thermal vacancies, they obtained the formation energy for rock salt structure. Pathak and coworkers employed the same approach to calculate the formation energies of defects for several alkali halides. Although their results are close to the theoretical prediction [8], later more physically rigorous derivations indicate that their assumption is not very sound (Fig. 6.1) [8].

Wang et al. used a semiempirical quasi-harmonic model (QHM) to describe the thermal expansion of perfect ionic crystals [9]. They used the thermodynamic theory of point defects to identify the difference of the thermal expansion between real and perfect crystals at high temperatures. They found that the formation energy for defects changes with temperature at low temperatures. While in the range where temperature is not low and also not close to the melting point of the crystal, the formation energy does not change with temperature. A linear trend is suggested between the formation enthalpies and the Debye temperatures of the alkali halides. The calculated smaller formation enthalpies of defects confirm the assumption that the vacancy distribution at high temperatures is vibrationally ordered into vacancy clusters like vacancy pairs and tetravacancies.

6.1.2 High-pressure effects

At elevated temperatures the formation of defects is an energetically favored process. A system in thermodynamic equilibrium, at a pressure P and temperature T, will minimize the Gibbs free energy:

$$G = H - TS \tag{6.1}$$

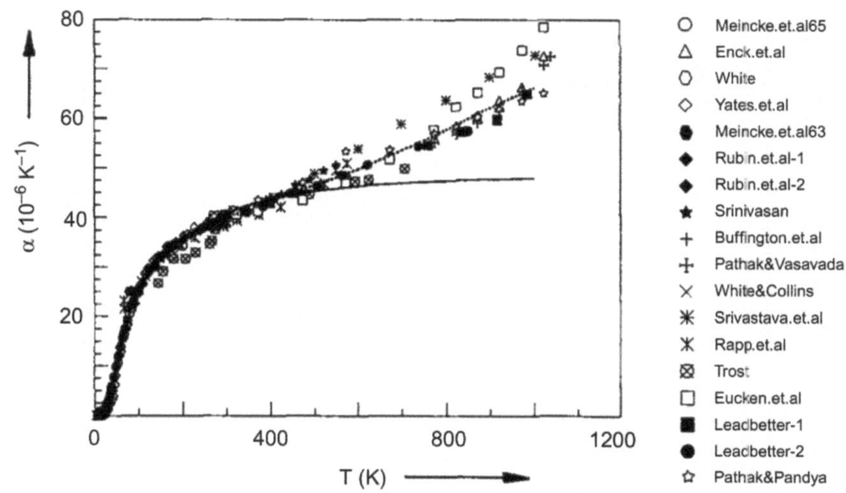

Fig. 6.1 Comparison of thermal expansion data, quasi-harmonic model, and modified quasi-harmonic model for NaCI. *Solid line* represents the quasi-harmonic model, and *dashed line* is the modified quasi-harmonic model.

Reproduced with permission from A.K. Pathak, T. Mukherjee, D.K. Maity, Theoretical studies on photoelectron and IR spectral properties of Br-2(center dot-)(H_2O)(n) clusters, J. Chem. Phys. 127 (2007), 8, Copyright 2007, American Institute of Physics.

where H is the enthalpy and S is the entropy. The enthalpy is defined as

$$H = E + PV \tag{6.2}$$

where E is the total energy and V is the volume of the sample. At ambient pressures, the PV term plays a minor role. However, under high-pressure environments, such as those produced by diamond-anvil cells, energy and enthalpy are essentially interchangeable in the solid phase.

A simple derivation of the dependence of the vacancy concentration on temperature has been given by Kittel (2005) [10], who considered a monoatomic crystal with vacancies. The enthalpy required to remove one atom from the interior of the crystal and to place it on the surface. To create the vacancies with an amount of enthalpy must be gained.

Another study about the dependence of the diffusivity, D; on pressure, p; and stress, σ, provides valuable information, which helps to elucidate the atomistic mechanism(s) of diffusion [11]. This model enables the understanding of the predominant mechanism related with point defects independent of the assumptions inherent in the currently used kinetic models. The only necessary assumption is that the experimental timescale is either very long or very short compared with that of point defect equilibration between the experimental diffusion zone and the nearest surface. For persistent phosphors discussed here, the timescale is assumed to be long. In addition, nonhydrostatic stresses in crystalline strained-layer epitaxial semiconductors can be very large. An understanding of their effects on diffusion and crystallization is therefore an important part of the study of the stability of these materials. Although many researchers have studied the effects of these stresses on the energetics of growth and other kinetic processes in strained epitaxial films, our emphasis has been on the study of the effects of stress on the point defect concentrations and mobilities. The atomic diffusivity in a material with cubic crystallography is given by this formula.

$$D = 1/6\lambda^2 \Gamma P_{\mathrm{m}} f \tag{6.3}$$

where λ is the jump distance, P_{m} is the fraction of the mobile atoms, Γ is the average jump rate of the mobile species, and f is the departures from a true random-walk are accounted for by the correlation factor. For example, for diffusion of an impurity "A" by a normal vacancy mechanism, the mobile species (which is called an A-V complex) is A on a lattice site adjacent to a vacancy. P_{m} and Γ are given, respectively, by [12].

$$P_{\mathrm{m}} = \exp\left(-G^{\mathrm{O}}{}_{\mathrm{f}}/k_{\mathrm{B}}T\right) \tag{6.4}$$

$$\Gamma = \nu \exp\left(-G_{\mathrm{m}}/k_{\mathrm{B}}T\right) \tag{6.5}$$

where $G_{\mathrm{f}}^{\mathrm{O}}$ is the standard free energy for the formation of the mobile species. The free energy of migration G_{m} is the additional change in free energy when the species moves to the saddle point of its migration path, and ν is the effective vibrational frequency. Substituting into Eq. (6.1), we obtain the familiar form for D:

$$D = 1/6\lambda^2\, \nu f\, \exp\left(\left(G^{\mathrm{O}}{}_{\mathrm{f}} + G^{\mathrm{m}}\right)/k_{\mathrm{B}}T\right) \tag{6.6}$$

The effect of pressure on D is characterized by the activation volume, V^*:

$$V^* \equiv -k_B T\left(\frac{\partial \ln D}{\partial p}\right)_T \tag{6.7}$$

V^* can be either positive or negative, depending upon whether D decreases or increases with p. As can be seen by differentiating of Eq. (6.4), when the pressure dependence of λ, ν, and f is neglected, V^* is the sum of two components:

$$V^* = V^{\mathrm{f}} + V^{\mathrm{m}} \tag{6.8}$$

where

$$V^{\mathrm{f}} = \left(\frac{\partial G_{\mathrm{f}}^{O}}{\partial p}\right)_T \tag{6.9}$$

and

$$V^{\mathrm{m}} = \left(\frac{\partial G^{\mathrm{m}}}{\partial p}\right)_T \tag{6.10}$$

Neglecting variations in ν and λ is practically always justified, but the pressure dependence may not be negligible, depending on whether there is a significant pressure dependence of the three interactions between the impurity and the defect. The formation volume, V^{f}, is the volume change of the system upon formation of a defect in its standard state; the migration volume, V^{m}, is the additional volume change when the defect reaches the saddle point in its migration path. V^{f} characterizes the pressure dependence of the equilibrium point defect concentration (e.g., A-V complex for the vacancy mechanism; A on an interstitial site for the interstitialcy mechanism); V^{m} characterizes the pressure dependence of the defect mobility. These volume changes are shown schematically in Fig. 6.2 [13].

A measurement of the activation volume can be a revealing test for distinguishing between potential diffusion mechanisms because, unlike activation energies, which are always positive, activation volumes are expected to have different signs for different types of point defect mechanisms. The formation volume is

$$V^{\mathrm{f}} = \pm\Omega + V^{\mathrm{r}} \tag{6.11}$$

where Ω is the lattice site volume, the plus sign is for vacancy formation, and the minus sign is for interstitial formation. The relaxation volume V^{r} is the amount of outward relaxation of the sample surfaces (if the relaxation is inward, V^{r} is negative) due to the newly created point defect. V^{r} is determined by site volume, but is not necessarily equal to the volume change of a shell of neighbors surrounding the point defect [14].

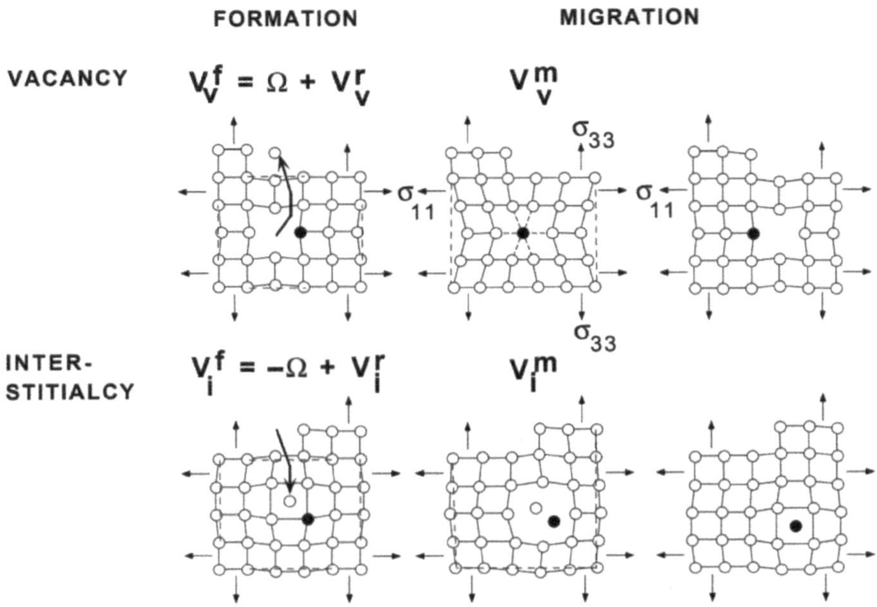

Fig. 6.2 Schematic illustration of volume changes associated with diffusion by vacancy (top) and the interstitialcy (bottom) mechanism. Pressure effect is determined by volume changes of sample upon defect formation (V^f) and migration (V^m). To create a vacancy, one moves an atom from the interior to a surface site; to create an interstitial atom, one moves an atom from a surface site to the interior. Stress effect is determined by shape change of sample upon defect formation and migration and, for formation, depends upon which surface the defect is formed from. (If point defect equilibration occurs at an internal dislocation instead of the surface, dislocation climb results in an identical volume change; however, the shape change depends on the direction of the Burgers vector.) Stress-strain work done by point defect formation and migration is indicated by interactions between shape changes and stresses (*thin arrows*). Reproduced with permission from Sugino and Oshiyama, Microscopic mechanism of atomic diffusion in Si under pressure, Phys. Rev. B, Condensed Matter 46 (1992), 12335–12341, Copyright 1992, American Physical Society.

If the shell is taken large enough that lattice distortions outside of it obey linear elasticity, then V^r can be calculated directly from the expansion of the shell itself. Because V^r and V^m are almost universally assumed to be much smaller than Ω in magnitude, the $\pm\ \Omega$ term is expected to dominate the pressure dependence of D, and a positive (negative) activation volume is taken as indication of a vacancy (interstitial)-based mechanism. The advent of ab initio atomistic calculations of point defect energies and configurations permits the aforementioned assumptions to be lifted and a more rigorous test of point defect mechanisms to be performed using application of high pressure.

Antonelli and Bernholc calculated relaxation volumes for several self-diffusion pathways for neutral and positively charged point defects [15]. For the neutral vacancy mechanism, they found $V^r = -0.24\ \Omega$. The migration pathway for the interstitialcy mechanism is still controversial, but for the tetrahedral interstitial ground state and bond-centered interstitial saddle point chosen by Antonelli and Bernholc, their results imply $V^{r} = +0.72\ \Omega$ and $V^{m} = -0.28\ \Omega$ for the neutral defect. For the direct interchange or "concerted exchange" mechanism (for which $V^{f} \equiv 0$ because no point defect is involved), they found $V^{m} = -0.04\ \Omega$. Sugino and Oshiyama calculated the relaxation volumes for dopant diffusion via vacancy and interstitialcy mechanisms and the relaxation volume for vacancy formation in self-diffusion [16].

For the latter, their results imply $V^{r} = -1.47\ \Omega$. Other volumetric calculations have also been reported. The disagreements among all of these calculations are not only due to differences in Hamiltonians (classical vs tight binding approximation vs local density approximation vs generalized gradient approximation). Significant differences also arise due to the implementation, for example, the number of relaxational degrees of freedom that are permitted, the size of supercell, and the method of deducing formation volumes from shell volume changes or from lattice parameter dependencies of formation energies. In some cases the $\pm\ \Omega$ term has been overlooked. Nevertheless, it is apparent that the assumption of negligible relaxation and migration volumes should be made with great caution pending the outcome of more definitive atomistic calculations [15].

The effect of biaxial strain on diffusion in semiconductor thin films is controversial at present and the reported strain effect varying from huge to barely discernible. To understand the various influences of pressure and stress on the diffusivity, we consider separately the effects on the point defect concentrations and on the point defect mobilities. The thermodynamics of point defect formation in heavily distorted crystals under hydrostatic or nonhydrostatic stress has been well understood for decades.

When experimental conditions are such that the point defect concentrations equilibrate rapidly with the point defect sources and sinks compared with the experimental timescale, the pressure dependence of the point defect concentration is characterized by the (scalar) formation volume. The new situation that gives the point defect formation volume in a cubic crystal a tensor character is that large single crystals entirely are free of extended defects, at which point defects can internally equilibrate.

Consequently, in a single crystal free of internal point defects, the volume change upon point defect formation does not tell the entire story: the crystal changes shape, the shape change depends upon on which surface(s) the defects are equilibrated, and nonhydrostatic stresses interact with the components of that shape change in the characteristic stress strain tensor manner. In the succeeding text is a synopsis of the derivation and of a more thorough explanation [17].

Let's suppose that rapid heating of an initially perfect tile-shaped "wafer" from absolute zero results in the introduction of only monovacancies, that their relaxation volume is zero, and that the lattice parameter is independent of T and p. The stress-strain relation has been examined for a wafer with compressive stress applied to each set of faces in turn. A gravity-driven device for applying different stresses to each set of faces is shown in Fig. 6.2. Suppose that, when applying compressive stress to the lateral faces, we can do this without buckling the wafer (alternatively, we could use tensile stress) [18].

It is assumed that there are enough steps on the crystal surfaces that vacancies can be formed without changing the surface area. The fraction of vacant lattice sites in equilibrium (neglecting local entropy effects for the sake of simplicity) is a Boltzmann factor in the vacancy formation-induced change in the total energy of the system (energy of the crystal plus gravitational potential energy of the weights). An equivalent statement is that the fraction of vacant lattice sites in equilibrium is a Boltzmann factor in the total energy change of the crystal itself, plus the work done against the agents imposing the stresses (in this case, gravity). If vacancies are formed at the (001) surfaces but not at the (100) surfaces (say, because the latter are coated with a film preventing them from serving as efficient sources/sinks), the work done against gravity is $p3V^f$. In case that vacancies are formed at the (100) surfaces but not at the (001) surfaces, the work done against gravity is $p1V^f$. The equilibrium defect concentrations depend on the location of the sources/sinks. A similar argument applies for the formation of self-interstitials [19].

6.1.3 Ionizing irradiation

Ionizing irradiation provides a flexible approach to control the luminescence by quantified atomic defect implantation in the targets. With regard to the dynamic process, the projectile energy is transferred to the target atoms through the collision between the nuclei/electrons of target and penetrating energetic particles—ions or electrons. Various atomic-scale defects may appear in the target as long as the target recoil atom acquires kinetic energy enough to leave its position in the atomic network. Essentially the frequently used sources of ionizing irradiations include x-ray, α-ray, and swift heavy ion (SHI), which give rise to the quantified defects with tunable parameters like energy, fluence, beam size, and position. Ionizing irradiation has been extensively employed in the control of defects in PPs during the past few decades. For example, the defects that are responsible for the persistent spectral hole burning of Eu^{3+}-doped glasses were formed by x-ray irradiation; the SHI like Ag, Ni, and Au ions have been used to create defects in PPs, so as to enhance the luminescence intensity [20].

Swift heavy ion (SHI) irradiation has found a wide range of applications in many fields, such as biology, medicine, physics, and materials sciences. The intense energy deposition from SHI irradiation onto the target material is several orders of magnitude higher than that of conventional ionizing radiation processes such as x-rays and electrons. The resultant dense electronic excitation leads to defect formation along the trajectory of the incident ions. The formation of defects has been observed in many materials, including semiconductors, insulators, and various metals. Therefore SHI irradiation is a strong candidate for improving the afterglow of persistent phosphors through the formation of lattice defects. Xu et al. developed a novel approach for improving the afterglow of persistent phosphors with swift heavy ion irradiation. The afterglow of the famous persistent phosphor $SrAl_2O_4$: Eu^{2+} was dramatically enhanced after swift heavy ion irradiation. Furthermore, it was found that the enhancement was dependent on the electronic stopping power and the irradiation influence [21].

Electron beam bombardment (EBB) is used to evaporate materials that are hard to evaporate by thermal evaporation due to high melting points of these materials. That means that it is a very efficient way to obtain high temperature, which we think can be used for sintering and calcinations of solid materials. A novel process of preparation of Eu- and Dy-codoped strontium aluminate based on EBB in vacuum chamber has been proposed by carefully controlling the *E*-beam power. Results show that this method is highly efficient and energy saving and induces very low contamination from the furnace or crucible. In addition, this method uses or produces no hazardous by-product, which is harmful to the environment [22].

Ji et al. have summarized the advantages of this process. First, it can achieve high temperature in a very short time, which is difficult for conventional sintering or calcination techniques. Second, it is a very effective way to reduce Eu^{3+} to Eu^{2+} since *E*-beam supplies electrons, which reduce Eu^{3+} to Eu^{2+} [23]. Third, it is carried out in high vacuum, where small amount of oxygen atoms might be liberated from the materials, resulting in the reduction of Eu^{3+} to Eu^{2+} due to loss of oxygen. Finally, it is a clean process that neither uses nor produces harmful by-product since no chemical agent is used in this process. Smet et al. also found that E-beam annealing is a viable option to reduce Eu^{3+} in $CaAl_2O_4$: Eu, Nd powders so as to produce a long and bright afterglow from the material. The differences between this method and standard thermal treatments in vacuum or in reducing atmosphere are a much faster heating rate, a higher temperature, a short dwelling time, and a fast cooling rate [24].

Among the techniques used to prepare persistent phosphors, pulsed laser ablation/deposition has several attractive features, including stoichiometric transfer of the target material, fast generation of plume of energetic species, hyperthermal reaction between the ablated cations and molecular O_2 in the ablation plasma, and compatibility with background pressures ranging from ultrahigh vacuum (UHV) to 100 Pa. The plasma generated during pulsed laser ablation is very energetic, and its mobility can be easily controlled by changing processing parameters. The presence of a background gas in the chamber has a strong influence on the quality of the plasma produced by the laser. The gas can modify the kinetic energy and the spatial distribution of the ejected species present in the plasma, and it may also induce compositional changes in the deposited films. Taking $SrAl_2O_4$: Eu^{2+}, Dy^{3+} as an example, to meet the demand of various industrial applications such as device fabrication and surface coatings, it is very important to prepare thin film with higher lateral smoothness with small grains, high thermal stability, reduced outgassing, and good adhesion to solid substrates by pulsed laser ablation/deposition technology [25–28].

Laser irradiation approaches using, for example, high-power UV rays from excimer lasers, have been employed to control defect formation in bulk materials, especially optical glasses for several decades. It's regarded as a facile way to form defects via two-photon excitation. Besides, with the rapid development of modern pulse lasers, especially femtosecond lasers, laser irradiation becomes a powerful tool for material processing with high precision. When a single high fluence femtosecond laser pulse irradiates onto the target, the laser energy is first absorbed by the target, and this induces generation of photoelectrons. These electrons then transfer their kinetic energy to the lattice over a picosecond timescale. Various microstructures are generated during the laser-matter interaction. For instance, point defects (vacancies, interstice atoms, and color centers) can be easily produced by femtosecond laser irradiation of lanthanide-doped glasses and crystals [29].

6.1.4 Stoichiometry control and doping

Nonstoichiometry refers to the deviation from ideal stoichiometry for crystalline compounds, which are usually dictated by the octet rule (8 or 18 electron rule). In nonstoichiometric materials the valence electrons from the anions or cations are not fully neutralized, and usually, defects are generated to stabilize the lattice. Taking the YAG host as an example, there are a variety of defect formation mechanisms that conserve site, charge, and mass balance while facilitating deviations from the stoichiometric in the YAG composition. Possible reactions could be expressed using the Kröger-Vink notation, which are listed as follows by Patel et al. [30].

Y_2O_3 excess: [31].

$$4Y_2O_3 + 5Al_{Al}^X \rightleftharpoons 5Y_{Al,16a}^X + Y_3Al_5O_{12},$$

$$4Y_2O_3 + 5Al_{Al}^X \rightleftharpoons 5Y_{Al,24d}^X + Y_3Al_5O_{12},$$

$$3Y_2O_3 + 5Al_{Al}^X + 3O_O^X \rightleftharpoons 3Y_{Al,16a}^X + 2V_{Al,16a}''' + 3V_O^{\bullet\bullet} + Y_3Al_5O_{12},$$

$$3Y_2O_3 + 5Al_{Al}^X + 3O_O^X \rightleftharpoons 3Y_{Al,16a}^X + 2V_{Al,24d}''' + 3V_O^{\bullet\bullet} + Y_3Al_5O_{12},$$

$$3Y_2O_3 + 5Al_{Al}^X + 3O_O^X \rightleftharpoons 3Y_{Al,24d}^X + 2V_{Al,16a}''' + 3V_O^{\bullet\bullet} + Y_3Al_5O_{12},$$

$$3Y_2O_3 + 5Al_{Al}^X + 3O_O^X \rightleftharpoons 3Y_{Al,24d}^X + 2V_{Al,24d}''' + 3V_O^{\bullet\bullet} + Y_3Al_5O_{12},$$

$$4Y_2O_3 + 5Al_{Al}^X \rightleftharpoons 5V_{Al,16a}''' + 5Y_i^{\bullet\bullet\bullet} + Y_3Al_5O_{12},$$

$$4Y_2O_3 + 5Al_{Al}^X \rightleftharpoons 5V_{Al,24d}''' + 5Y_i^{\bullet\bullet\bullet} + Y_3Al_5O_{12},$$

$$2Y_2O_3 + 5Al_{Al}^X + 6O_O^X \rightleftharpoons 5V_{Al,16a}''' + 6V_O^{\bullet\bullet} + Y_i^{\bullet\bullet\bullet} + Y_3Al_5O_{12},$$

$$2Y_2O_3 + 5Al_{Al}^X + 6O_O^X \rightleftharpoons 5V_{Al,24d}''' + 6V_O^{\bullet\bullet} + Y_i^{\bullet\bullet\bullet} + Y_3Al_5O_{12},$$

$$2Y_2O_3 + 5Al_{Al}^X + 6O_O^X \rightleftharpoons 4V_{Al,16a}''' + 6V_O^{\bullet\bullet} + Y_{Al,16a}^X + Y_3Al_5O_{12},$$

$$2Y_2O_3 + 5Al_{Al}^X + 6O_O^X \rightleftharpoons 4V_{Al,24d}''' + 6V_O^{\bullet\bullet} + Y_{Al,24d}^X + Y_3Al_5O_{12},$$

Al_2O_3 excess: [32]

$$4Al_2O_3 + 3Y_Y^X \rightleftharpoons 3Al_Y^X + Y_3Al_5O_{12},$$

$$3Al_2O_3 + 3Y_Y^X + 3O_O^X \rightleftharpoons Al_Y^X + 2V_Y''' + 3V_O^{\bullet\bullet} + Y_3Al_5O_{12},$$

$$4Al_2O_3 + 3Y_Y^X \rightleftharpoons 3Al_i^{\bullet\bullet\bullet} + 3V_Y''' + Y_3Al_5O_{12},$$

$$5Al_2O_3 + 6Y_Y^X + 9O_O^X \rightleftharpoons 6V_Y''' + 9V_O^{\bullet\bullet} + 2Y_3Al_5O_{12},$$

$$2Al_2O_3 + 3Y_Y^X + Al_{Al}^X + 6O_O^X \rightleftharpoons 3V_Y''' + V_{Al,24d}''' + 6V_O^{\bullet\bullet} + Y_3Al_5O_{12}.$$

By the earlier reactions considering site, charge, and mass balance, one can establish a nonstoichiometric structure with predicted defects for targeted optical performance.

Besides, adding a codopant into the host is another way to implant the traps. Intentionally introducing foreign ions into a pure host would create defects of different types, which has become an important approach to realize defect control in persistent phosphors. Dopants and defects are important in semiconductor and optoelectronic devices. Strategies for controlling doping and defects have been the focus of semiconductor physics research during the past decades and remain highly active even today. Codoping is a promising strategy that can be used for effectively tuning the dopant populations, electronic properties, and optical properties. It can enhance the solubility of dopants and improve the stability of desired defects. During the past 20 years, significant experimental and theoretical efforts have been devoted to study the characteristics of codopants in persistent phosphors [33].

In general, codoping can be efficient for increasing the dopant solubility, activation rate by lowering the ionization energy of acceptors and donors, and carrier mobility. Enhanced dopant incorporation is especially significant under nonequilibrium growth conditions owing to the introduction of new chemical species as codopants. Despite these fundamental investigations on codoping, to the best of our knowledge, "codope" (including "codoping" and "codoped") was introduced in 1963 and was later used in other areas. The research at that time aimed to determine the mechanisms of charge compensation for unintentional introduction of multiple dopants, and chemical mechanisms of growing and doping in materials such as alkaline earth halides (e.g., CaF_2 and BaF_2) doped with rare earths or uranium (U), for potential use in optically pumped lasers. In addition, optical energy transfer between codopants was also a particular concern in laser crystals like chromium (Cr) and neodymium (Nd) doubly doped into $Y_3Al_5O_{12}$ (YAG) and other Cr-doped materials [34, 35].

Although the discussion and organization of this chapter follows the logical order of different passivation mechanisms of codoping, some very important experimental and theoretical investigations on unintentional codoping and surface effects on codoping should be emphasized and will serve as an extension of the general concept of codoping. Unintentional codoping usually happens when unexpected dopants that are present in the growth atmosphere are incorporated into semiconductors and form defect complexes with original dopants or intrinsic defects. Background contaminations, such as H, C, or O, are often the source of such unexpected dopants. In addition, defect complexes of intrinsic defects and dopants can sometimes be regarded as resulting from unintentional codoping. In this sense the surface as a special type of defect plays an important role in the incorporation of dopants, and surfactant-enhanced doping or codoping becomes especially interesting [36].

6.1.5 Annealing in oxidizing/reducing atmosphere

Atmosphere-controlled annealing is a refined approach that introduces a balanced equilibrium between the interior oxygen vacancy concentrations and the oxygen partial pressure of the atmosphere. In principle the following defect equilibrium provides a guidance to control of defect concentration: [37].

$$O_O^X \rightarrow V_O^{\bullet\bullet} + \frac{1}{2}O_2 \uparrow + 2e' \tag{6.12}$$

It could be found that decreasing of oxygen partial pressure results in deprival of oxygen from the lattice and formation oxygen vacancies ($\boldsymbol{V_O^{\bullet\bullet}}$) and excess electrons ($\boldsymbol{e'}$). By calculating the temperature-dependent equilibrium constant (K_R) according to the following equation, [37].

$$K_R(T) = \left[V_O^{\bullet\bullet}\right] \cdot n^2 \cdot p_{O_2}^{1/2} \tag{6.13}$$

where n stands for the concentration of electrons, it can be predicted that the formation of oxygen vacancies is followed by thermally ionization to $\boldsymbol{V_O^{\bullet\bullet}}$ [37].

Surface-based methods hold special promise for defect manipulation in nanostructures where surface-to-volume ratios are high. In general, surface defects including dangling bonds and adsorbed molecules or ions are parasitic to the host structure. But in most cases these surface defects are detrimental to the optical quantum efficiency; as a result, great efforts have been devoted to eliminate these surface defects. For example, epitaxial growth of a shell layer on top of the synthesized nanoparticles is reported as an extremely effective way to mitigate surface quenching.

6.2 Control of emission band

6.2.1 Emission centers

Generation of point defects inside a rigid polyhedral structure is generally associated with charge imbalance or geometric distortion. Consequently, electrons (or holes) are transferred from the defects to the Ln^{3+} ions. Bivalent lanthanides are produced through reduction from Ln^{3+} to Ln^{2+} by, for example, heat treatment in a reducing atmosphere. Divalent ions (Sm^{2+}, Eu^{2+}, or Yb^{2+}) can generate broad and strong emissions from the 5d-4f transitions when being doped in different hosts. A plenty of hosts have been reported to provide a reducing environment that stabilizes divalent lanthanide ions, including borates (SrB_4O_7, SrB_6O_{10}, BaB_8O_{13}, or CaB_2O_4), haloborate ($Sr_2B_5O_9Cl$), aluminates ($Sr_4Al_{14}O_2$ or $BaAl_2O_4$), phosphates ($Ba_3(PO_4)_2$ or $NaCaPO_4$), borophosphates ($MBPO_5$, M = Ca, Sr, or Ba), silicate ($BaMgSiO_4$), sulfate ($BaSO_4$), pollucite ($CsAlSi_2O_6$), borate glass ($15Na_2O$–$5Al_2O_3$–$80B_2O_3$ (mol%)), borophosphate glasses (ZnO-B_2O_3-P_2O_5), Eu/Al-codoped high silica glasses $95.5SiO_2$-$4B_2O_3$-$0.5Na_2O$ (wt%), and aluminosilicate glass ($40CaO$-$30Al_2O_3$-$27SiO_2$-$3GeO_2$) [38–42].

Among these hosts the reduction of oxidation state are controlled by different mechanisms including (1) aliovalent doping, for example, replacing Ca^{2+}, Sr^{2+}, and Ba^{2+} by Eu^{3+}; (2) modifying the covalency and polarizability of activator-ligand bonds, for example, replacing Al^{3+}-F^- by Si^{4+}-O^{2-}; and (3) insertion of Li into the vacant sites, which is specific for a $NaZr_2(PO_4)_3$-type structure. For instance, it is difficult to reduce Eu^{3+} to Eu^{2+} when being doped in $CaYAlO_4$ even in reducing atmosphere. However, by employing the defect engineering approach to the $CaYAlO_4$ host through incorporation of Si^{4+}-Ca^{2+} to replace Al^{3+}-Y^{3+}, partial reduction of Eu^{3+} to Eu^{2+} can be realized (Fig. 6.3A). The resultant tunable luminescence from Eu^{2+} and Eu^{3+} under UV excitation makes this phosphor promising for white light–emitting diodes (WLED) [43–45].

The configuration of 4f electrons in lanthanides is characterized by the abundant energy levels, thus allowing for multitudinous transitions. Though the 4f-4f transitions are influenced by the shielding from the $5s^25p^6$ subshells, some of the transitions are known as hypersensitive and strongly affected by the surrounding environment. For example, the $^5D_0 \rightarrow {}^7F_{0,2,4}$ transitions that are of forced electric-dipole nature are strictly forbidden in a centrosymmetric octahedral (O_h symmetry) field. Breakdown of the crystallographic site symmetry through defect engineering, for example, substitutional doping, enables the forbidden transitions to become partially allowed (Fig. 6.3D) [42].

Recently, it has been reported that the radiative lifetime of lanthanides can be efficiently tuned by defects in nanomaterials with accurate control in a short time scale. Due to trapping by different defects, the radiative lifetime could be increased to dozens of hours. Here, we illustrate the mechanism for the increase in lifetime in a typical persistent phosphors $SrAl_2O_4$: Eu^{2+}, Dy^{3+}. As shown in Fig. 6.3E, upon UV excitation, 4f-5d transition of Eu^{2+} occurs; meanwhile, holes are produced simultaneously at the 4f ground state and are released thermally to the valence band. In this process, Eu^{2+} is assumed to be converted into Eu^+. The released holes migrate through the valence band and are trapped by Dy^{3+} ions, which produce a defect level that has very long decay time. After excitation stops, the hole is reexcited thermally and liberated to the valence band and finally captured by Eu^+. Consequently the recombination takes place, resulting in the increment in emission lifetime and the observation of persistent luminescence [42–45].

Energy transfer process can be used to tailor the emission intensity of different centers and thus to control the emission color. This enables us to generate multiband emission from a single material, which has advantages like low cost, simple fabrication, and high emission efficiency. As shown in Fig. 6.4A, three individual primary colors are generated from a piece of Er^{3+}–Eu^{3+} codoped oxygen-deficient germanate glass under UV light irradiation. Vividly blue, green, and red emissions originated from Ge-related oxygen-deficient defect centers (GODCs), Er^{3+} ions, and Eu^{3+} ions could be selectively excited by UV lights at wavelength of 340, 379, and 395 nm,

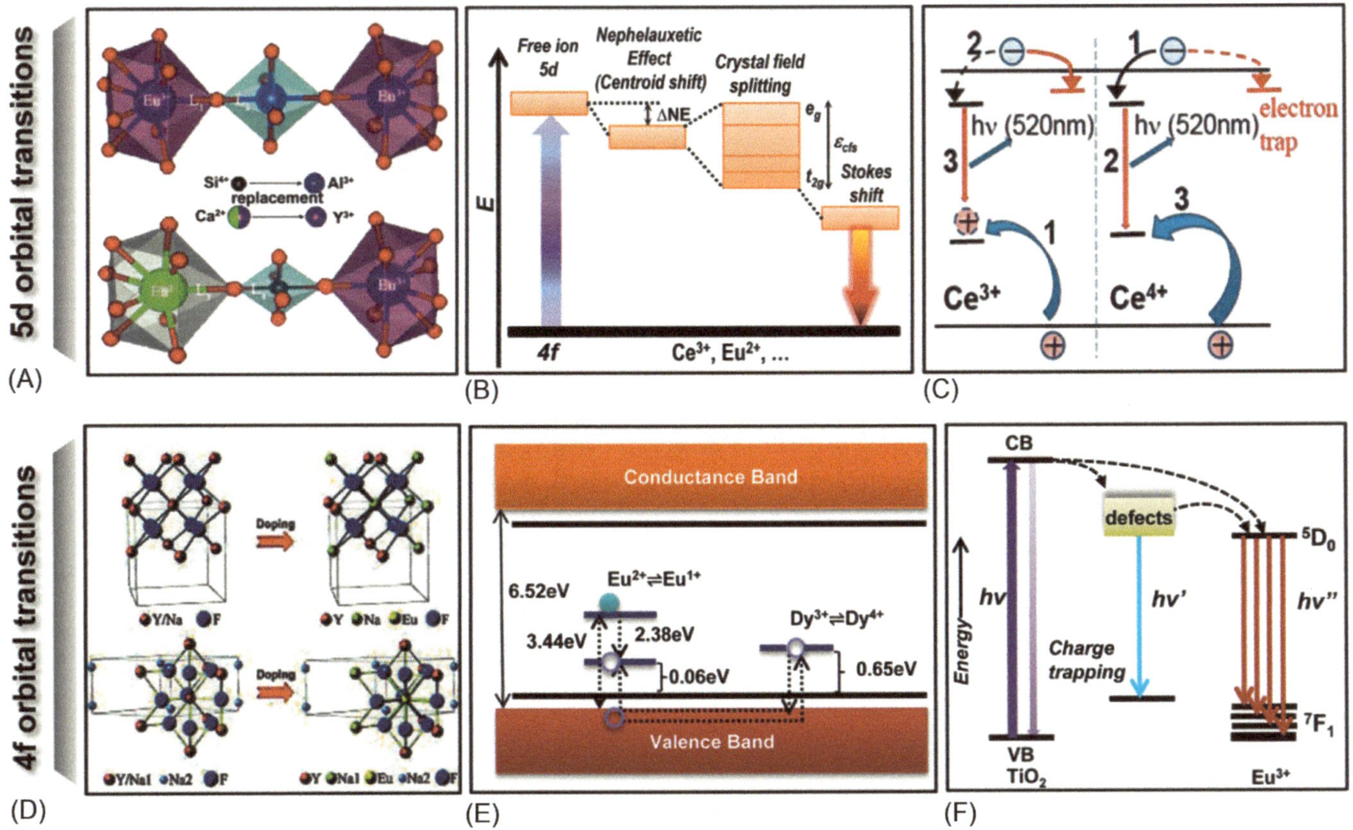

Fig. 6.3 Typical optical performances modulated through defect engineering in persistent phosphors. (A–C) Influences of defect on the reduction of lanthanide ions (A), 5d level shifting and broadening (B), scintillation behavior (C). (D–F) Influences of defect on the symmetry of lanthanide ions and probabilities of 4f transitions (D), persistent luminescence behavior (E), charge trapping and energy-transfer in lanthanide-doped semiconductors (F).
Reproduced with permission from M.D. McCluskey, E.H. Eugene Dopants and Defects in Semiconductors, CRC Press, 2018, Copyright 2018, CRC Press; Reproduced with permission R.J.D. Tilley, Principles and Applications of Chemical Defects, Routledge, 2019, Copyright 2019, Routledge; Reproduced with permission Pelleg, Defects in Materials. Diffusion in Ceramics, Springer, Cham, 2016, Copyright 2016, Springer.

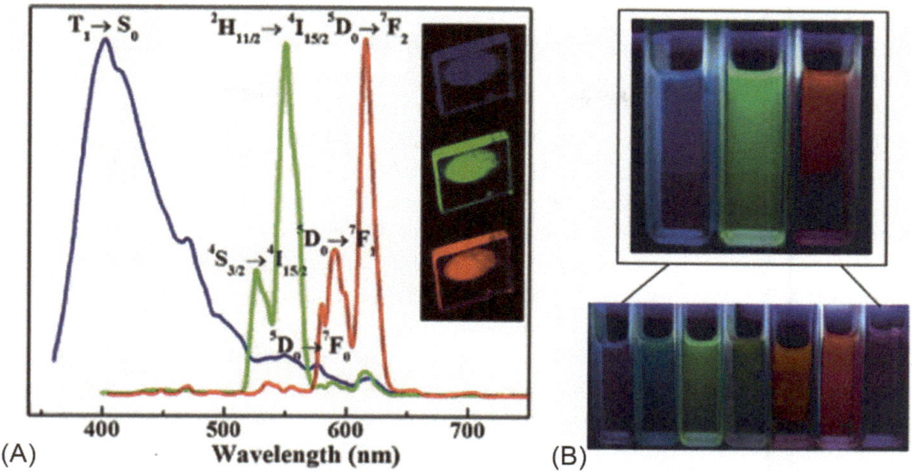

Fig. 6.4 (A) Emission spectra (excited at 340, 379, and 395 nm) of $2Er^{3+}$-$0.5Eu^{3+}$ codoped glass sample; the inset shows the emission of three primary colors excited by 340-, 379-, and 395-nm UV lights, respectively. (B) Full-color images of the three fluorescent dispersions of RE-doped Ca-PO_4-CO_3 particles (up) and their mixtures (down).
(A) From G. Lin, et al., Three primary colors emitting from Er^{3+}–Eu^{3+} co-doped oxygen-deficient glasses, J. Alloys Compd. 509 (22) (2011) 6462–6466. (B) From J. Hui, Q. Yu, Y. Long, Z. Zhang, Y. Yang, P. Wang, B. Xu, X. Wang, Ultrathin Ca-PO_4-CO_3 solid-solution nanowires: a controllable synthesis and full-color emission by rare-earth doping, Chem. Eur. J. 18 (43) (2012) 13702–13711.

respectively [46]. By leveraging the energy transfer between the defects and the lanthanide centers, a Tb^{3+}-doped oxygen-deficient calcium aluminogermanate glasses was also reported, which exhibited bright three-color (sky-blue, green or yellow, and red) persistent luminescence under 300-, 380-, and 395-nm excitation, respectively. Similarly, Rodrigues et al. reported red, pink, blue, green, and nearly white persistent luminescence from the $CdSiO_3$ host doped with Pr^{3+}, Sm^{3+}, Gd^{3+}, Tb^{3+}, and Dy^{3+}, respectively. The energy transfer efficiency from the defect to Ln^{3+} determines the final emission color of the phosphor, which implies the persistent luminescence comes from the lanthanides' 4f-4f transitions, defect, separately, or a combination of these two emissions, depending on the time elapse after cease of excitation [47].

(The following discussions can be removed if not related with persistent luminescence.)

In liquid dispersions containing nanoscale phosphor particles, multicolor emission could be alternatively obtained by simply mixing of solutions containing different emitters. For instance, in the liquid dispersion containing RE-doped Ca-PO_4-CO_3 particles, strong blue luminescence due to abundant defects associated with CO_3^{2-} was observed by naked eyes, and green and red colors were produced when the energy transfers occurred from defects to Tb^{3+} and Eu^{3+}, respectively (Fig. 6.3B). By simply mixing of the three primary colors, a series of new full color–emitting solutions could be obtained.

Tailoring the emission color can be realized by selectively increasing or decreasing the emission intensity of specific peaks, which is benefited from the discreteness of lanthanide emissions. For certain applications, emission energy is expected to be restrictively isolated in a specific wavelength region. For instance, single-band red or NIR upconversion emission is very promising for deep in vivo imaging and cell tracking. Additionally, spectral conversion from NIR to ultraviolet by a composite that is composed of upconversion nanoparticles and semiconductors (TiO_2, ZnO, etc.) facilitates solar energy harvest in photovoltaic or photocatalysis (Fig. 6.5) [48, 51, 52].

The composition design by defect engineering may enable the tailoring of emission color for persistent phosphors. Both in situ chemical processes and physical post-processing could enable tuning of emission color by generating defects that are active in a selected spectral range. In 2013 Chen et al. found that high Ln^{3+} content and/or the temperature for the synthesis of a CaF_2-LnF_3 solid solution facilitate the generation of structural inhomogeneity, that is, Ln^{3+} clusters, and the concomitant interstitial F^- ions. Consequently, emission colors including green, yellow, and red can be easily tuned in the disordered solid solutions. Also, Yan et al. reported that significant difference in the red-to-green emission ratio, which varied from 1.9 to 71 and 1.6 to 116, was observed for Na_xYF_{3+x}:Yb,Er and Na_xGdF_{3+x}:Yb, Er nanocrystals by simply controlling the Na^+ and F^- vacancies in the host lattices [49, 50]. (These are not persistent phosphors.)

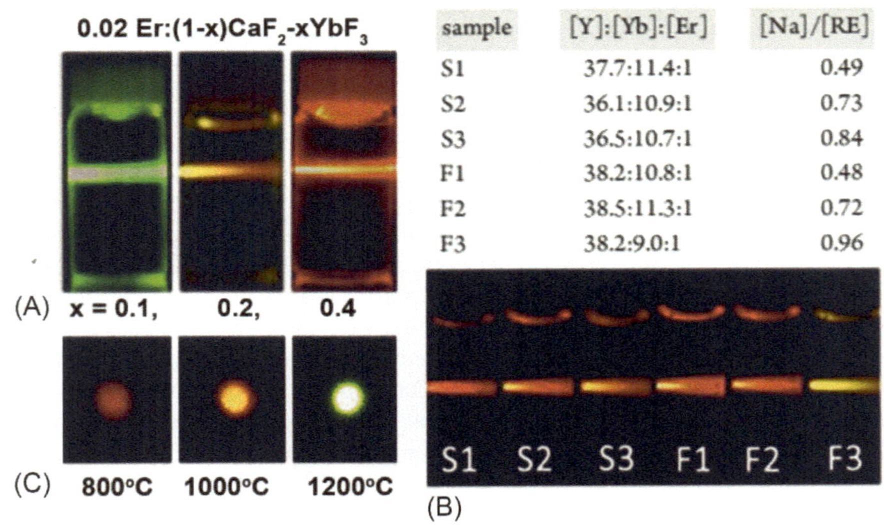

Fig. 6.5 Defects tailored emission colors. Luminescent photos of (A) Er (2 mol%): (1 − x) CaF_2–xYbF_3 (x = 0.1,0.2, and 0.4) nanocrystals dispersed in cyclohexane solutions, (B) Na_xREF_{3+x} nanocrystals with composition differences as is listed, (C) Y_2O_3: Yb, Er nanoparticles after calcination at 800°C, 1000°C, and 1200°C.
From S. Khiari, M. Velázquez, R. Moncorgé, J.L. Doualan, P. Camy, A. Ferrier, M. Diaf, Red-luminescence analysis of Pr^{3+} doped fluoride crystals, J. Alloys Compd. 451 (1-2) (2008), 128–131.

Besides the aforementioned in situ composition design, postprocessing like thermal treatment is an important approach to control the defect concentration in the system. For example, Lu et al. synthesized Y_2O_3: Yb^{3+}, Er^{3+} nanoparticles in the presence of a surfactant (cetyltrimethylammonium bromide, CTAB) and induced surface defect formation in the as-obtained products. By introducing the defects through use of a CTAB and removing the defects by changing the calcination temperature, the green and red emission of Er^{3+} were enhanced selectively, and thus the color output could be tailored from red to green [51, 52]. (These are not persistent phosphors.)

6.2.2 *Energy transfer control*

As discussed earlier, many works indicated that afterglow color can be spectrally designed to meet the requirement of specific applications by persistent energy transfer (PET) processes. Such PET concept between two different emitting centers can be used to design the PPs with desired emission colors that are otherwise difficult to obtain by other mechanisms, especially for the persistent phosphors with long emission wavelength. Typical PET is based on the resonant-type process, which is dictated by the spectral overlap between the emission spectrum of sensitizer and the excitation spectrum of activator. In the scenario of resonant energy transfer, the energy transfer rate (k) from the donor to the acceptor for dipole-dipole interaction is related with their separation (R) by [53].

$$k_{dd}(R) = \frac{1}{\tau_D}\left(\frac{R_0}{R}\right)^6 \tag{6.14}$$

where τ_D is the lifetime of the sensitizer (donor) recorded for samples without the acceptor. From Eq. (6.14) the rate of energy transfer scales with r^{-6} and the critical distance R_0 are described as [54]

$$R_0 = \frac{3c^4\hbar^4\sigma_A}{4\pi n^4}\int\frac{f_D(E)f_A(E)}{E^4}dE \tag{6.15}$$

where c is the speed of light, h is the Planck constant, σ_A is the absorption cross section of the acceptor center, and n is the refractive index of the host. And $f_D(E)$ ' $f_A(E)$ are the normalized emission spectra of the donor and the absorption spectrum of the acceptor. Typical R_0 values are less than 10 nm for lanthanide-activated phosphors. For energy transfer in the much smaller donor-acceptor separation, exchange integration that relies on overlap of their wave functions could become dominate, and this happens only in phosphors with a high doping concentration. The rate of this energy transfer due to exchange interaction is derived by Dexter: [54, 55].

$$k_{ex}(R) = (2\pi/\hbar)K^2\exp(-2R/L)\int f_D(E)f_A(E)dE \tag{6.16}$$

where K is a constant related with the overlap of the wave function and L is the effective Bohr radius of the system. Here the ET rate varied as $\exp.(-2R/L)$ with R, which is used to describe the radial distribution of the wave functions. It is indicated by Eq. (6.16) that the Dexter-type energy transfer occurs dominantly for donor-acceptor separation below 1–2 nm, which is less than 10 times of metal-ligand bond length.

By leveraging energy transfer between optical centers, green, yellow, red, white, and NIR emissions have been achieved in rare earth– and transition metal ion–doped PPs. Most obvious energy transfer in persistent phosphors is found between rare earth ions, as exemplified in the following cases:

(a) $Eu^{2+} \rightarrow Nd^{3+}$: Our group synthesized the $SrAl_2O_4$ phosphors codoped with Eu^{2+}, Dy^{3+}, and Nd^{3+} ions. Strong persistent phosphorescence at 882 nm attributed to Nd^{3+} 4f-4f transition is observed (after irradiation by ultraviolet light at 370 nm), which is due to the persistent energy transfer from Eu^{2+} ions to Nd^{3+} ions [56].

(b) $Ce^{3+} \rightarrow Dy^{3+}$: A white-light long-lasting phosphor $Sr_2Al_2SiO_7$: Ce^{3+}, Dy^{3+} was prepared by a high-temperature solid-state reaction. Energy transfer from Ce^{3+} to Dy^{3+} results in the observation of white light afterglow.

(c) $Eu^{2+} \rightarrow Er^{3+}$: A NIR persistent phosphorescence at 1530 nm due to the persistent energy transfer process from Eu^{2+} ions to Er^{3+} ions has been observed in Eu^{2+}, Dy^{3+}, and Er^{3+} codoped $SrAl_2O_4$ phosphors [57].

(d) Energy transfer between pairs of ions, such as $Eu^{2+} \rightarrow Sm^{3+}$, $Ce^{3+} \rightarrow Tb^{3+}$, and $Ce^{3+} \rightarrow Eu^{2+}$, has also been observed in silicate and aluminate phosphors [58].

In addition to rare earth ions, transition metal ions are also frequently selected as activators. As an independent emission center, the NIR phosphorescence of Cr^{3+} has been observed in $Y_3Al_{5-x}Ga_xO_{12}$ ($x = 2.5$, 3, and 3.5): Ce^{3+}, Cr^{3+}, $Sr_4Al_{14}O_{25}$: Eu^{2+}, Dy^{3+}, Cr^{3+}, $SrAl_2O_4$: Eu^{2+}, Dy^{3+}, Cr^{3+}, via the energy transfer from RE to Cr^{3+}. We also observe energy transfer from RE (Eu^{2+} and Ce^{3+}) to Mn^{2+}, which results in NIR phosphorescence of Mn^{2+} in $Ca_{0.2}Zn_{0.9}Mg_{0.9}Si_2O_6/Eu^{2+}$, Dy^{3+}, Mn^{2+}; $CaAl_2O_4$: Eu^{2+}, Nd^{3+}, Mn^{2+} phosphors; and $CaAl_2O_4$: Ce^{3+}, Mn^{2+} phosphors [59–61].

6.2.3 *Crystalline field control*

Defects enable the modulation of the emission bands with predictable spectral shifting or broadening since 5d orbitals of lanthanide ions are strongly affected by the surrounding environment. The interactions between defects and 5d electrons are influenced by different factors, including the nephelauxetic effect (NE) and crystal fielding splitting (CFS) effect. The NE describes the centroid shift, which is a lowering in energy of the 5d level of activators (Ce^{3+}, Eu^{2+}, ...) due to a decrease in the interelectron repulsion. CFS refers to the energy gap between the highest and lowest 5d level. This effect depends on the bond length from the activators to the coordinating anions, the molecular orbital overlap or degree of covalency between the activators and ligands, and the coordination environment and the symmetry of the activator site. A quantized description is given by the following equation: [62].

$$D_q = \frac{Ze^2r^4}{6R^5} \tag{6.17}$$

where Z is the charge or valence of the anion, e is the charge of an electron, r is the radius of the d wave function, and R is the bond length. Generally, hybridization of 5d orbitals happens with the interaction between molecular orbitals of neighboring anions, in which bonding orbitals at lower energies and antibonding orbitals at higher energies are created. As the degree of covalency increases, both the bonding and antibonding orbitals are displaced to lower energies according to NE effect. This may cause CFS to either increase or decrease. Therefore the emission bands from 5d orbitals are easy to be tuned by making use of the NE and the well-known Stokes shift effect.

6.3 Control of excitation band

Different from the emission band, here in this part, we discuss the methods that enable us to control the excitation band. The excitation of high energy rays and ultraviolet light has been reported many times. However, long-wavelength excitation indeed occurs for many material systems, resulting in the NIR phosphorescence.

6.3.1 High-energy rays

Persistent phosphors for x-ray or cathode-ray tubes have been intensively studied due to their wide applications. One of the important applications is for displaying radar echoes on the screen due to the long period of electron beam rotation over the screen. Zhang and coworkers found that the KY_3F_{10}: Tb^{3+} phosphor exhibits good long persistent phosphorescence after irradiation with x-ray or cathode-ray tubes and can be applied for the radar tubes [61]. In addition, $SrAl_2O_4$: Eu^{2+}, Dy^{3+}; $Sr_4Al_{14}O_{25}$: Eu^{2+}, Dy^{3+}; $Ca_3(PO_4)_2$: Mn^{2+}; $CaMgSi_2O_6$: Mn^{2+}; ZnS: Cu^+, Co^{2+}; Ba_2SiO_4: Eu^{2+}; and $MgGa_2O_4$: Cr^{3+} also can be activated by x-ray or β-ray [63–66].

6.3.2 UV-Vis-NIR light

Up to now, most of persistent phosphors can be activated by UV light, but only a few PPs can be activated by visible or NIR light. Eu^{2+}-doped sulfide is the first reported case that can be excited under visible light irradiation [67]. Afterward, persistent phosphors employing Ce^{3+} ions as the emission centers were reported. For instance, Su et al. reported a $Y_3Al_5O_{12}$: Ce^{3+} persistent phosphor that can be charged under the excitation of UV light at 254 nm [68]. But they do not observe that the phosphorescence also can be excited by visible light. Tanabe et al. developed bright persistent phosphors of Ce^{3+}-doped $Y_3Al_{5-x}Ga_xO_{12}$ ($x = 2.5$, 3, and 3.5) ceramics with green luminescence (Ce^{3+}: 5d → 4f) via blue-light excitation [69]. The persistent luminance intensity recorded 5 min after ceasing blue-light excitation for the Ce^{3+}-Cr^{3+}-doped $Y_3Al_2Ga_3O_{12}$ sample is approximately 3900 times higher than that of a Ce^{3+}-doped $Y_3Al_2Ga_3O_{12}$ sample and better than that from compact $SrAl_2O_4$: Eu^{2+}-Dy^{3+} powders.

In fact, by early 2011, they report the persistent luminescence at around 500 nm in Ce^{3+} singly doped yttrium scandium gallium garnet ceramics with the composition of

$(Y_{0:995}Ce_{0:005})_3Sc_2Ga_3O_{12}$ [4]. The afterglow luminescence was observed for 1 h after the 440-nm excitation in the $(Y_{0:995}Ce_{0:005})_3Sc_2Ga_3O_{12}$ ceramic prepared under vacuum. Meanwhile, Wang and coworkers also develop the similar efficient blue light-activated Ce^{3+}, Pr^{3+}: $Y_3Al_{5-x}Ga_xO_{12}$ (x = 0–5) phosphors, which demonstrate long-persistent phosphorescence lasting for 1 h [70].

In 2012 Pan and coworkers reported sunlight-activated long-persistent phosphorescence in the NIR region from Cr^{3+}-doped zinc gallogermanates [71]. They demonstrated that the NIR persistent phosphorescence could be produced efficiently by charging under ultraviolet illumination (300–400 nm), but less efficiently under visible light (400–630 nm) illumination, despite that the visible-light excitation is very effective for NIR photoluminescence.

Recently, they fabricated $Zn_3Ga_2GeO_8$: 1%Cr^{3+}, 5%Yb^{3+}, 0.5%Er^{3+} phosphors [72]. After being excited by a 980-nm laser, the phosphor emits long-lasting (>24 h) NIR persistent emission peaking at 700 nm. The generation of persistent luminescence in this phosphor is benefited by combining the unique features of upconversion luminescence and long-persistent phosphorescence processes. Our group also developed a lot of persistent phosphors that can be charged under excitation of NIR light. Our research demonstrated that the NIR photostimulated luminescence and phosphorescence of $Zn_3Ga_2Ge_2O_{10}$: Cr^{3+} and $Zn_3Ga_2GeO_8$: Cr^{3+} phosphors can be repeatedly excited under the excitation of NIR light from 780 to 900 nm [73].

6.3.3 Laser and nonlaser excitation sources

Xe lamp is usually used as the light source to charge persistent phosphors. This is a typical incoherent light source that exhibits limited spectral intensity in the UV region. In 2012 Pan and coworkers reported a series of Cr^{3+}-doped zinc gallogermanate NIR all-weather persistent phosphors ($Zn_3Ga_2Ge_2O_{10}$: Cr^{3+}) which can be rapidly, effectively, and repeatedly charged by natural sunlight in almost all kinds of outdoor environment. Maldiney also verified that the persistent luminescence occurred in nanosized $ZnGa_{2(1-x)}/Cr_{2x}O_4$, activated through the animal tissues after illumination under orange/red light-emitting diodes (LEDs) [70].

In fact, there are few reports that focus on the influence of type of excitation sources for generation of persistent luminescence in different materials. Holsa and coworkers reported persistent upconversion luminescence excited by a 980-nm laser diode (LD) in ZrO_2: Yb^{3+}, Er^{3+} nanoparticles [74]. Our group reported photostimulated luminescence and persistent phosphorescence in $Zn_3Ga_2Ge_2O_{10}$:Cr^{3+} and $Zn_3Ga_2SnO_3$: Cr^{3+} phosphors under the excitation by a 980- and 808-nm LD. We found that the NIR PSL could be repeatedly generated by stimulation with a NIR laser. Moreover, we showed that the PSL intensity can be tuned by changing the power density of the excitation laser, which would provide more benefits for the application in bioimaging [15]. Liu and coworkers subsequently reported a new luminescence process called upconverted persistent phosphorescence (UCPL), which enabled the generation of persistent phosphorescence with an emission energy higher than the excitation energy. They created the first UCPL phosphor $Zn_3Ga_2GeO_8$: 1%Cr^{3+}, 5%Yb^{3+}, 0.5%Er^{3+} by incorporating an upconverting ion pair Yb^{3+}/Er^{3+} into a $Zn_3Ga_2GeO_8$: 1%Cr^{3+} NIR phosphorescent phosphor. After being excited by a 980-nm laser, the phosphor emits long-lasting (>24 h) near-infrared persistent emission peaking at 700 nm [75].

6.4 Control of afterglow time

The dynamics of persistent luminescence have been thoroughly investigated in recent decades for different materials. For example, it is reported that the afterglow attenuates exponentially in $SrAl_2O_4$: Eu^{2+}, Dy^{3+} PPs, and their intensity decay behavior follows a first-order, second-order, or general-order kinetic behavior:

$$I = I0/(1+\gamma t) \quad \text{First} - \text{order} \tag{6.18}$$

$$I = I0/(1+\gamma t)^2 \quad \text{Second} - \text{order} \tag{6.19}$$

$$I = I0/(1+\gamma t)^n \quad \text{General} - \text{order} \tag{6.20}$$

Multiple exponential equations have been successfully fitted to experimental decay curves for $SrAl_2O_4$: Eu^{2+}, Dy^{3+} PPs. The calculated decay constants in such equations are used to assess the afterglow characteristics of a persistent phosphor. The decay times of a persistent phosphor may be changeable, and they are dependent on host, temperature, irradiation intensity, sample thickness, and parking density of samples. In addition, the use of different numbers of exponential components in fitting leads to different numerical results for decay constants. The appearance of the luminescence intensity is less correlated to the decay constants than to the time-invariant constants of the equations [76].

Generally speaking the afterglow time is determined by two important parameters of persistent phosphors, that is, the decay rate and luminescence intensity. A slow decay rate and strong luminescence intensity result in a long afterglow time. In general a slow decay rate is resulted when appropriate traps slowly release the trapped charge carrier, while the luminescence intensity is determined by the transition nature of emission center and ratio of radiative to nonradiative recombination.

6.4.1 *Luminescence intensity control*

Atoms and ions, such as optically active ions in laser gain media, usually have a set of energy levels. Transitions of such atoms and ions from high- to lower-lying levels are often associated with the emission of photons (light). The generated photons carry with them the difference of energy between the involved energy levels. However, there are also mechanisms that allow for nonradiative transitions (or nonradiative or radiationless transitions), that is, transitions without involving light emission. The excess energy is then dissipated in some other ways—in most cases, in the form of phonon emission, which are associated with lattice vibrations of a solid. In liquids, similar process can occur but hardly in gases, where the atoms or molecules are not in contact with others for most of the time and therefore hardly have a chance to dissipate excitation energy nonradiatively.

Photoluminescence quantum yield (PLQY) is an important parameter for fluorescent material and determines the luminescence intensity of persistent phosphors. PLQY characterizes the efficiency of photon emission as defined by the ratio of the number of photons emitted to the number of photons absorbed. On the other hand the PLQY can be expressed as the rate of photon emission events over the summed rate constants of all deexcitation events. The deexcitation pathways for a fluorophore are usually more complicated than the photon emission processes. Each deexcitation pathway has an associated quantum yield between 0 and 1, and the sum of all paths' quantum yields is 1. These other deexcitation pathways may be distinguishable through measurement of the heat emitted, differences in emitted photon energy, etc. However, when using the term "quantum yield," we will henceforth refer specifically to the PLQY as that is the most relevant to the topics discussed in this chapter. Quantum yield is determined by the relative contribution of energy transfer efficiency and fluorescence quenching in the deexcitation process, in which both are affected by the types of fluorophores and experimental conditions [77].

PLQY is determined by the rates of radiative transitions and nonradiative transitions. Phonon emission is a very rapid process in solids when the transition energy is comparable with the scale of the phonon energy of the lattices. The radiative transition is then effectively suppressed and cannot be observed. For larger transition energies, only multiphonon transitions are possible, where one transition involves the emission of multiple phonons. The rate of such processes becomes rather small when more than three phonons are emitted.

There are also quenching processes related with impurities or lattice defects, which place additional electronic levels within the bandgap to which excitation energy can be transferred. These impurities or defects may interact with only those ions that are sufficiently close to them, and rapid energy transfer between them could occur.

Nonradiative transitions tend to decrease the upper-state lifetime and the upper-level population of the optical centers, unless the nonradiative transition rates are negligible compared with the radiative ones. In laser crystals the gain efficiency (for continuous pumping) and the maximum achievable laser gain are reduced by nonradiative decay. Note that nonradiative transitions are essential for many solid-state laser gain media: they often facilitate the population of the upper laser level and help to depopulate the lower laser level, which suppresses the reabsorption losses.

Besides emission profile, lifetime is a characteristic dimensionality of emission transitions. Normally, lanthanides and transition metal ions have their 4f-4f and 3d level lifetime from microseconds to milliseconds. Precise control and fine-tuning of the lifetime in such a limited timescale could be realized by defect engineering. On the other hand, defects become an integral part of long persistent phosphors, by extending the lifetime of activators' transitions to a timescale longer than the order of milliseconds. In principle the long decay time of persistent luminescence is due to the storage of the excitation energy by traps, which originates from lattice defects. Subsequent detrapping and recombination yields the delayed luminescence [78].

6.4.2 Miscellaneous

Due to the narrow and weak absorbance by the 4f-4f transition of lanthanides, researchers have constantly searched for efficient method to sensitize the activators by employing energy transfer process. Defect could serve as an energy donor, which has broad and strong absorbance, and it is usually found in semiconductors and inorganic glasses. In a typical system TiO_2: Eu^{3+}, visible color centers are excited, and the excitons are in competition between recombination to produce luminescence and transfer the energy to the surrounding lanthanides [79].

Scintillating luminescence is usually observed in Ce^{3+}- or Pr^{3+}-doped persistent phosphors. Defect engineering has been used to optimize these materials so that the performances of scintillation are significantly improved. Internally, defects participate the subprocess of scintillation transport, in which electrons and holes migrate through the material, repeated trapping at defects may occur, and energy losses are probably due to nonradiative recombination. As a preferred choice, Mg^{2+} is an attractive candidate to be incorporated into the scintillators like Lu_2SiO_5, to produce defects. This effect is usually recognized as the attribution of creation of stable Ce^{4+} centers, which forms a new fast radiative recombination pathway through the processes of (1) the immediate electron capture from the conduction band, (2) radiative deexcitation of the excited Ce^{3+} center, and (3) a hole capture to return to the Ce^{4+} stable ground state [80].

During scintillation the Ce^{4+} centers efficiently compete with electron traps (due to antisite defects and oxygen vacancies) for an immediate capture of electrons from the conduction band ($Ce^{4+} + e_c \rightarrow (Ce^{3+})^*$); In step no. 2 the Ce^{4+} ion transforms to an excited Ce^{3+} center and emits the scintillation photon, that is, contributes to the fastest part of scintillation response ($(Ce^{3+})^* \rightarrow Ce^{3+} + h\vartheta$), simultaneously the Ce^{3+} center, after temporary change to Ce^{4}; captures an electron from the conduction band; and becomes excited. In step no. 3 the return into the initial state (beginning of cycle) is accomplished by the hole capture from the valence band (Ce^{4+} in the right part: $Ce^{3+} + h_v \rightarrow Ce^{4+}$) that result in the emission of scintillation photon (Ce^{3+} in the left part: $(Ce^{3+})^* \rightarrow Ce^{3+} + h\vartheta$) [81].

Point defects can be designed to enhance or quench the luminescence intensity in certain crystal field. Utilization of the defect-enhanced luminescence is one of the most straightforward and efficient approaches to improve the luminescence intensity. Typically, Li^+ ion is an attractive candidate to artificially introduce in persistent phosphors by substitution or formation of interstitial defects due to its small ionic radius. This will break the symmetry of the crystal field around the lanthanides, leading to the enhancement of luminescence intensity. As early as the last century, researchers have explored the usability of Li^+ to enhance the luminescence of lanthanides in phosphors like Li^+-doped ZnS: Tm^{3+}, in which Li^+ greatly improved the energy transfer efficiency from ZnS to Tm^{3+} [82].

Nowadays, more focus has been concentrated on the upconersion persistent phosphors. Enhancement of upconersion intensity has been observed in diverse matrices, and the reported maximum enhancement can be up to two orders of magnitude in Y_2O_3: Yb^{3+}, Er^{3+} phosphor. All of these enhancements are benefited from point defects generation resulted from Li^+ doping in the hosts. Other ions including K^+, Na^+, Zn^{2+}, Bi^{3+},Fe^{3+}, and Sc^{3+} can also provide this enhancement by introducing defects [86, 87].

Apart from the enhancement, many types of defects can be detrimental to the luminescence of lanthanides, in which the defects act as quenching centers. This quenching effect is especially serious in nanosystems, where the large surface-to-volume ratio accommodates easily large amount of surface defects. This issue is of special concern for the upconversion nanoparticles, in which the quenching effect of surface defects on the luminescence of lanthanides has been confirmed and a core-shell architecture is often applied for surface passivation.

References

[1] S. Shionoya, W.M. Yen, H. Yamamoto (Eds.), Phosphor Handbook, CRC Press, 2006.
[2] G. Blasse, B.C. Grabmaier, Luminescent Materials, Springer, Berlin, Heidelberg, 1994.
[3] R.S. Becker, R.S. Becker, Theory and Interpretation of Fluorescence and Phosphorescence, Wiley Interscience, New York, 1969.
[4] Z. Pan, Y.Y. Lu, F. Liu, Sunlight-activated long-persistent luminescence in the near-infrared from Cr^{3+}-doped zinc gallogermanates, Nat. Mater. 11 (1) (2012) 58.
[5] H.H. Wawra, On the role of Grüneisen's elastic parameters in the theory of polycrystal elasticity. A critical assessment, Phys. Status Solidi A 54 (1979) 291–296.
[6] A.C. Lawson, J.C. Lashley, Thermal expansion and transformation behavior of cerium and plutonium alloys: an application of the Aptekar-Ponyatovsky regular solution model, J. Phys.-Condes. Matter 23 (2011) 9.
[7] H.L. Luo, M.F. Merriam, D.C. Hamilton, Superconducting metastable compounds, Science (New York, N.Y.) 145 (1964) 581–583.
[8] A.K. Pathak, T. Mukherjee, D.K. Maity, Theoretical studies on photoelectron and IR spectral properties of Br-2(center dot-)(H_2O)(n) clusters, J. Chem. Phys. 127 (2007) 8.
[9] M.C. Nguyen, K. Hoang, C.Z. Wang, K.M. Ho, Stabilities and defect-mediated lithium-ion conduction in a ground state cubic Li_3N structure, Phys. Chem. Chem. Phys. 18 (2016) 4185–4190.
[10] D.I. Sayago, M. Polcik, R. Lindsay, R.L. Toomes, J.T. Hoeft, M. Kittel, D.P. Woodruff, Structure determination of formic acid reaction products on TiO_2(110), J. Phys. Chem. B 108 (2004) 14316–14323.
[11] J.F. Joly, L.K. Beland, P. Brommer, N. Mousseau, Contribution of vacancies to relaxation in amorphous materials: a kinetic activation-relaxation technique study, Phys. Rev. B 87 (2013) 6.
[12] Antonelli and Bernholc, Pressure effects on self-diffusion in silicon, Phys. Rev. B Condens. Matter 40 (1989) 10643–10646.
[13] Sugino and Oshiyama, Microscopic mechanism of atomic diffusion in Si under pressure, Phys. Rev. B Condens. Matter 46 (1992) 12335–12341.
[14] R. Kabe, C. Adachi, Organic long persistent luminescence, Nature 550 (7676) (2017) 384.
[15] Y. Li, M. Gecevicius, J.R. Qiu, Long persistent phosphors-from fundamentals to applications, Chem. Soc. Rev. 45 (8) (2016) 2090–2136.
[16] Y. Li, Y.Y. Li, R.C. Chen, K. Sharafudeen, S.F. Zhou, M. Gecevicius, H.H. Wang, G.P. Dong, Y.L. Wu, J.R. Qiu, Tailoring of the trap distribution and crystal field in Cr^{3+}-doped non-gallate phosphors with near-infrared long-persistence phosphorescence, NPG Asia Mater. 7 (5) (2015) e180.

[17] A.H. Wako, B.F. Dejene, H.C. Swart, Roles of doping ions in afterglow properties of blue $CaAl_2O_4$: Eu^{2+}, Nd^{3+} phosphors, Physica B 439 (2014) 153–159.
[18] H. Yamamoto, T. Matsuzawa, Mechanism of long phosphorescence of $SrAl_2O_4$: Eu^{2+}, Dy^{3+} and $CaAl_2O_4$: Eu^{2+}, Nd^{3+}, J. Lumin. 72 (1997) 287–289.
[19] Y. Wang, Z. Wang, Characterization of Y_2O_2S: Eu^{3+}, Mg^{2+}, Ti^{4+} long-lasting phosphor synthesized by flux method, J. Rare Earths 24 (1) (2006) 25–28.
[20] A. Abdukayum, J.T. Chen, Q. Zhao, X.P. Yan, Functional near infrared-emitting Cr^{3+}/Pr^{3+} co-doped zinc gallogermanate persistent luminescent nanoparticles with superlong afterglow for in vivo targeted bioimaging, J. Am. Chem. Soc. 135 (38) (2013) 14125–14133.
[21] K.W. Böer, Handbook of the Physics of Thin-Film Solar Cells, Springer Science & Business, 2014.
[22] W. Lu, Y. Fu, Optical properties of fluorescent colloidal quantum dots, in: Spectroscopy of Semiconductors, Springer, Cham, 2018.
[23] T.M. Tritt (Ed.), Thermal Conductivity: Theory, Properties, and Applications, Springer Science & Business Media, 2005.
[24] J. Hinks, C. Geoff, The Technology of Building Defects, Routledge, 2002.
[25] M.D. McCluskey, E.H. Eugene, Dopants and Defects in Semiconductors, CRC Press, 2018.
[26] R.J.D. Tilley, Principles and Applications of Chemical Defects, Routledge, 2019.
[27] Pelleg, Defects in Materials. Diffusion in Ceramics, Springer, Cham, 2016.
[28] E. Finley, M.T. Aria, B. Jakoah, Intrinsic defects drive persistent luminescence in monoclinic $SrAl_2O_4$: Eu^{2+}, J. Phys. Chem. C 122 (2018) 16309–16314.
[29] A.M. Valencia, M.J. Caldas, Single vacancy defect in graphene: insights into its magnetic properties from theoretical modeling, Phys. Rev. B 96 (2017), 125431.
[30] P. Byeon, H.B. Bae, H.S. Chung, S.G. Lee, J.G. Kim, H.J. Lee, J.W. Choi, S.Y. Chung, Li-intercalation oxides: atomic-scale observation of LiFePO4 and LiCoO2 dissolution behavior in aqueous solutions, Adv. Funct. Mater. 28 (45) (2018), 1870320.
[31] D.E.K. Van, P.F. Smet, D. Poelman, Persistent luminescence in Eu^{2+}-doped compounds: a review, Materials 3 (4) (2010) 2536–2566.
[32] V.D.E.K. Van, A.J. Bos, D. Poelman, P.F. Smet, Revealing trap depth distributions in persistent phosphors, Phys. Rev. B 87 (4) (2013), 045126.
[33] G.A. Craig, M. Mark, 3d single-ion magnets, Chem. Soc. Rev. 44 (2015) 2135–2147.
[34] L. Ungur, Introduction to the electronic structure, luminescence, and magnetism of lanthanides, in: Lanthanide-Based Multifunctional Materials, Elsevier, 2018, pp. 1–58.
[35] V.V. Osipov, V.A. Shitov, V.I. Solomonov, K.E. Lukyashin, A.V. Spirina, R.N. Maksimov, Composite Nd:YAG/Cr^{4+}:YAG transparent ceramics for thin disk lasers, Ceram. Int. 41 (2015) 13277–13280.
[36] M. Yamaga, Y. Oda, H. Uno, K. Hasegawa, H. Ito, S. Mizuno, Formation probability of Cr-Nd pair and energy transfer from Cr to Nd in $Y_3Al_5O_{12}$ ceramics codoped with Nd and Cr, J. Appl. Phys. 112 (2012) 12.
[37] C. Wang, Y.H. Zhao, Y. Liu, First-principles calculations of magnetic and optical properties of $Ga_{1-x}Cr_xSb$ ($x=0.25$, 0.50, 0.75), Acta Phys. Sin. 68 (2019) 10.
[38] K.A. Rahman, T. Bak, A. Atanacio, M. Ionescu, J. Nowotny, Toward sustainable energy: photocatalysis of Cr-doped TiO_2: 2. Effect of defect disorder, Ionics 24 (2018) 327–341.
[39] D. Xu, W. Chen, M. Zeng, H. Xue, Y. Chen, X. Sang, Y. Xiao, T. Zhang, R.U. Raymond, K. Xiao, L. Fu, Crystal-field tuning of photoluminescence in two-dimensional materials with embedded lanthanide ions, Angew. Chem. Int. Ed. 57 (2018) 755–759.
[40] L.E. Orgel, Spectra of transition-metal complexes, J. Chem. Phys. 23 (1955) 1004–1014.

[41] K.V. Lamonova, E.S. Zhitlukhina, R.Y. Babkin, S.M. Orel, S.G. Ovchinnikov, Y.G. Pashkevich, Intermediate-spin state of a 3d ion in the octahedral environment and generalization of the Tanabe-Sugano diagrams, J. Phys. Chem. A 115 (2011) 13596–13604.

[42] J.D. Matthew, S. Alexander, P. Santona, G.F. Anthony, L.P. Christy, Y. Kevin, K.Y. Po, M. Uwe, L. Joydeep, PNAS 100 (2003) 389–393.

[43] K. Ogasawara, F. Alluqmani, H. Nagoshi, Multiplet energy level diagrams for Cr^{3+} and Mn^{4+} in oxides with oh site symmetry based on first-principles calculations, ECS J. Solid State Sc. 5 (1) (2016) R3191–R3196.

[44] M.J. Weber, T.E. Varitimos, B.H. Matsinger, Optical intensities of rare-earth ions in yttrium orthoaluminate, Phys. Rev. B 8 (1973) 47.

[45] M. Vollmer, K.P. Möllmann, Infrared Thermal Imaging: Fundamentals, Research and Applications, John Wiley & Sons, 2017.

[46] G. Lin, G.P. Dong, D.Z. Tan, X.F. Liu, Q. Zhang, D.P. Chen, J.R. Qiu, Q.Z. Zhao, Z.Z. Xu, Long lasting phosphorescence in oxygen-deficient zinc-boron-germanosilicate glass-ceramics, J. Alloys Compd. 504 (2010) 177–180.

[47] H. Dong, L.D. Sun, Y.F. Wang, J. Ke, R. Si, J.W. Xiao, G.M. Lyu, S. Shi, C.H. Yan, Efficient tailoring of upconversion selectivity by engineering local structure of lanthanides in Na_xREF_{3+x} nanocrystals, J. Am. Chem. Soc. 137 (2015) 6569–6576.

[48] D. Chen, L. Lei, J. Xu, A. Yang, Y. Wang, Abnormal size-dependent upconversion emissions and multi-color tuning in Er^{3+}-doped CaF_2-YbF_3 disordered solid-solution nanocrystals, Nanotechnology 24 (2013), 085708.

[49] A. Hooda, S.P. Khatkar, A. Khatkar, R.K. Malik, M. Kumar, S. Devi, V.B. Taxak, Reddish-orange light emission via combustion synthesized $Ba_3Y_4O_9$:Sm^{3+} nanocrystalline phosphor upon near ultraviolet excitation, J. Lumin. 217 (2019), 116806.

[50] S. Khiari, M. Velázquez, R. Moncorgé, J.L. Doualan, P. Camy, A. Ferrier, M. Diaf, Red-luminescence analysis of Pr^{3+} doped fluoride crystals, J. Alloys Compd. 451 (1–2) (2008) 128–131.

[51] T. Förster, Zwischenmolekulare energiewanderung und fluoreszenz, Ann. Phys. 2 (1948) 55–75.

[52] D.L. Dexter, A theory of sensitized luminescence in solids, J. Chem. Phys. 21 (1953) 836–850.

[53] M. Inokuti, F. Hirayama, Influence of energy transfer by exchange mechanism on donor luminescence, J. Chem. Phys. 43 (1965) 1978–1989.

[54] T. Wang, Z. Xia, Q. Xiang, S. Qin, Q. Liu, Relationship of 5d-level energies of Ce^{3+} with the structure and composition of nitride hosts, J. Lumin. 166 (2015) 106–110.

[55] Z. Song, Z. Xia, Q. Liu, Insight into the relationship between crystal structure and crystal-field splitting of Ce^{3+} doped garnet compounds, J. Phys. Chem. C 122 (2018) 3567–3574.

[56] Y.M. Yang, Z.Y. Li, J.Y. Zhang, Y. Lu, S.Q. Guo, Q. Zhao, X. Wang, Z.J. Yong, H. Li, J.P. Ma, Y. Kuroiwa, C. Moriyoshi, L.L. Hu, L.Y. Zhang, L.R. Zheng, H.T. Sun, X-ray-activated long persistent phosphors featuring strong UVC afterglow emissions, Light-Sci. Appl. 7 (2018) 11.

[57] J.S. Zhang, Z.D. Hao, X. Zhang, Y.S. Luo, X.G. Ren, X.J. Wang, J.H. Zhang, Color tunable phosphorescence in KY_3F_{10}:Tb^{3+} for x-ray or cathode-ray tubes, J. Appl. Phys. 106 (2009) 6.

[58] S. Gultekin, S. Yildirim, O. Yilmaz, I.C. Keskin, M.I. Kati, E. Celik, Structural and optical properties of $SrAl_2O_4$: Eu^{2+}/Dy^{3+} phosphors synthesized by flame spray pyrolysis technique, J. Lumin. 206 (2019) 59–69.

[59] A.K. Bedyal, V. Kumar, V.K. Singh, S.P. Lochab, F. Singh, O.M. Ntwaeaborwa, H.C. Swart, Thermo-luminescence kinetic parameters of gamma-irradiated $Sr_4Al_{14}O_{25}$:Eu^{2+}, Dy^{3+} phosphor, Radiat. Eff. Defects Solids 168 (2013) 1022–1029.
[60] K. Li, Y. Zhang, X.J. Li, M.M. Shang, H.Z. Lian, J. Lin, Tunable blue-green emission and energy transfer properties in beta-$Ca_3(PO_4)_2$: Eu^{2+}, Tb^{3+} phosphors with high quantum efficiencies for UV-LEDs, Dalton Trans. 44 (2015) 4683–4692.
[61] H. Li, X.Q. Chen, P.C. Zhang, B. Bin Tang, S.Q. Yu, J.W. Lu, S.R. Zhang, Influence of Mn^{2+} introduction on microwave dielectric properties of $CaMgSi_2O_6$ ceramic, Ceram. Int. 45 (2019) 24425–24430.
[62] N. Chandrasekhar, K. Gartia, Gaussian approximation of thermoluminescence (TL) peaks: a common misconception for analysis of TL of persistent luminescent materials, J. Alloys Compd. 745 (2018) 773–778.
[63] J. Sun, H. Lin, D.W. Zhang, R.J. Hong, C.X. Tao, Z.X. Han, Green emitting spinel/ Ba_2SiO_4:Eu^{2+}/spinel sandwich structure robust ceramic phosphor prepared by spark plasma sintering, Ceram. Int. 45 (2019) 23643–23650.
[64] X.L. Duan, J. Liu, X.Q. Wang, H.D. Jiang, Cation distribution and optical properties of Cr-doped $MgGa_2O_4$ nanocrystals, Opt. Mater. 37 (2014) 854–861.
[65] R.L. Nyenge, H.C. Swart, O.M. Ntwaeaborwa, Luminescent properties, intensity degradation and X-ray photoelectron spectroscopy analysis of CaS:Eu^{2+} powder, Opt. Mater. 40 (2015) 68–75.
[66] S. Zhang, C.Y. Li, R. Pang, L.H. Jiang, L.L. Shi, Q. Su, Long-lasting phosphorescence study on $Y_3Al_5O_{12}$ doped with different concentrations of Ce^{3+}, J. Rare Earths 29 (2011) 426–430.
[67] J. Ueda, K. Kuroishi, S. Tanabe, Bright persistent ceramic phosphors of Ce^{3+}-Cr^{3+}-codoped garnet able to store by blue light, Appl. Phys. Lett. 104 (2014) 4.
[68] J. Ueda, K. Aishima, S. Nishiura, S. Tanabe, Afterglow luminescence in Ce^{3+}-doped $Y_3Sc_2Ga_3O_{12}$ ceramics, Appl. Phys. Express 4 (2011) 3.
[69] J. Ueda, A. Meijerink, P. Dorenbos, A.J.J. Bos, S. Tanabe, Thermal ionization and thermally activated crossover quenching processes for 5d-4f luminescence in $Y_3Al_{5-x}Ga_xO_{12}$: Pr^{3+}, Phys. Rev. B 95 (2017) 8.
[70] F. Liu, Y.J. Liang, Z.W. Pan, Detection of up-converted persistent luminescence in the near infrared emitted by the $Zn_3Ga_2GeO_8$: Cr^{3+}, Yb^{3+}, Er^{3+} phosphor, Phys. Rev. Lett. 113 (2014) 5.
[71] Y.L. Wu, Y. Li, X.X. Qin, R.C. Chen, D.K. Wu, S.J. Liu, J.R. Qiu, Near-infrared long-persistent phosphor of $Zn_3Ga_2Ge_2O_{10}$: Cr^{3+} sintered in different atmosphere, Spectroc. Acta Pt. A-Molec. Biomolec. Spectr. 151 (2015) 385–389.
[72] E. Teston, S. Richard, T. Maldiney, N. Lievre, G.Y. Wang, L. Motte, C. Richard, Y. Lalatonne, Non-aqueous sol-gel synthesis of ultra small persistent luminescence nanoparticles for near-infrared in vivo imaging, Chem. Eur. J. 21 (2015) 7350–7354.
[73] L. Pihlgren, T. Laihinen, L.C.V. Rodrigues, S. Carlson, K.O. Eskola, A. Kotlov, M. Lastusaari, T. Soukka, H.F. Brito, J. Holsa, On the mechanism of persistent up-conversion luminescence in the ZrO_2:Yb^{3+}, Er^{3+} nanomaterials, Opt. Mater. 36 (2014) 1698–1704.
[74] S.S. Zhao, B. Yu, Y.Y. Lin, Z.B. Wang, X.P. Gao, W.S. Liu, Research progress of red long afterglow materials, Chin. Sci. Bull.-Chin. 64 (2019) 3717–3729.
[75] B. Milicevic, V. Dordevic, K. Vukovic, G. Drazic, M.D. Dramicanin, Effects of Li^+ co-doping on properties of Eu^{3+} activated TiO_2 anatase nanoparticles, Opt. Mater. 72 (2017) 316–322.

[76] T.D. Gustafson, E.D. Milliken, L.G. Jacobsohn, E.G. Yukihara, Progress and challenges towards the development of a new optically stimulated luminescence (OSL) material based on MgB_4O_7:Ce,Li, J. Lumin. 212 (2019) 242–249.
[77] H. Zimmermann, R. Boyn, Excitation und deexcitation of Tm^{3+} centres in ZnS involving "external" excited states, Phys. Status Solidi B 135 (1986) 379–387.
[78] S.H. Yang, Y.H. Ling, J.H. Yan, Luminescence enhancement and potential application of sky-blue sulfide phosphor doped with promoter, J. Solid State Electrochem. 18 (2014) 89–95.
[79] H. Wu, Z.D. Hao, L.L. Zhang, X. Zhang, G.H. Pan, Y.S. Luo, H.J. Wu, H.F. Zhao, H. Zhang, J.H. Zhang, Enhancing IR to NIR upconversion emission in Er^{3+}-sensitized phosphors by adding Yb^{3+} as a highly efficient NIR-emitting center for photovoltaic applications, CrystEngComm 22 (2020) 229–236.
[80] B.X. Xu, J.C. Liu, K.S. Zou, The photo-switch effect and the energy-level population change of Li^{+} doping in Yb^{3+}/Er^{3+} co-doped Y_2O_3 upconversion films, Appl. Phys. A-Mater. Sci. Process. 125 (2019) 12.
[81] E.F. Huerta, S. Carmona-Tellez, J.G. Cabanas-Moreno, C. Falcony, Visible and near infrared upconversion photoluminescence from Y_2O_3:Er^{3+}, Yb^{3+}, Li^{+} phosphors under 1532 nm excitation light, J. Alloys Compd. 732 (2018) 422–428.
[82] B.X. Xu, G.Z. Dong, J.C. Liu, K.S. Zou, D.P. Wang, The luminescence regulation effect of Na^{+} on the Yb^{3+}/Er^{3+} co-doped Y_2O_3 up-conversion films, J. Lumin. 203 (2018) 16–25.

Applications

7

Persistent phosphors have found diverse applications in industrial, military, medicine, firefighting, daily life, etc. [1] For instance, luminescent paints, luminescent plastics, glasses, ceramics, and luminous productions have been used widely in garnishing, traffic, transportation, military facility, firefighting emergency, and many other areas (Fig. 7.1). Especially, persistent phosphors have been widely used as weak light sources in the dark. Many persistent phosphorescence minerals are used in the manufacturing of different artworks, such as "luminous pearl" (Fig. 7.2A), "luminous wine cup" (Fig. 7.2B), and "phosphorescent bird" (Fig. 7.2C). Commercial applications of persistent phosphors include glowing toys, paint, and clock dials. The applications of persistent luminescent materials as luminous paint began in the early 20th century based on metal sulfide ZnS/Cu^{2+} [2, 3]. Modern persistent phosphors entered into the commercial market since alkaline earth aluminate phosphor ($SrAl_2O_4/Eu^{2+}$, Dy^{3+}) was developed in 1995. [4] $Sr_2MgSi_2O_7/Eu^{2+}$, Dy^{3+} (blue); $SrAl_2O_4/Eu^{2+}$, Dy^{3+} (green); and Y_2O_2S/Eu^{3+}, Mg^{2+}, Ti^{4+} (red) are the three commercially available persistent phosphors used for various applications. [5] Fig. 7.3A shows the photoluminescence spectra of these phosphors. Mixing of these phosphors in designed ratios results in afterglow in different colors, and white-light afterglow is also attainable under an appropriate mixing ratio. Fig. 7.3B presents photographs for these three phosphors taken under ultraviolet irradiation, in dark after excitation and in daylight.

By dispersion of persistent phosphor particles in several matrices (e.g., polymers), they can be used for the manufacturing of different items, such as luminous paint, luminous ink, luminous plastic, luminous thread, luminous pottery, and porcelain. The persistent phosphors themselves can be used directly in the field of architectural decoration, safety signs, and lighting sources. Furthermore, in the bioimaging field, the persistent phosphorescent nanomaterials have been used as sensitive luminescent probes for real-time bioimaging that attracted growing attention since 2007. [6] In this chapter, we introduce the applications of persistent phosphors in safety signs, luminous fiber, artwork, and their use as optical probes in bioimaging, molecular sensing, etc.

7.1 Color display

The oldest report on persistent luminescence was about luminous pearls, a mysterious material emitting green phosphorescence in the dark. [7] However, the luminous pearl was made up of natural ores rather than the persistent phosphors we mentioned earlier. Until 1866 Theodore Sidot was the first to prepare a yellow-green persistent phosphor, ZnS/Cu. Thereafter, persistent phosphors with different compositions have been synthesized and used in diverse fields, such as color display, safety indicator, road traffic signs, luminous decoration, and alternating current light-emitting diode (AC-LED).

Persistent Phosphors. https://doi.org/10.1016/B978-0-12-818637-4.00007-0

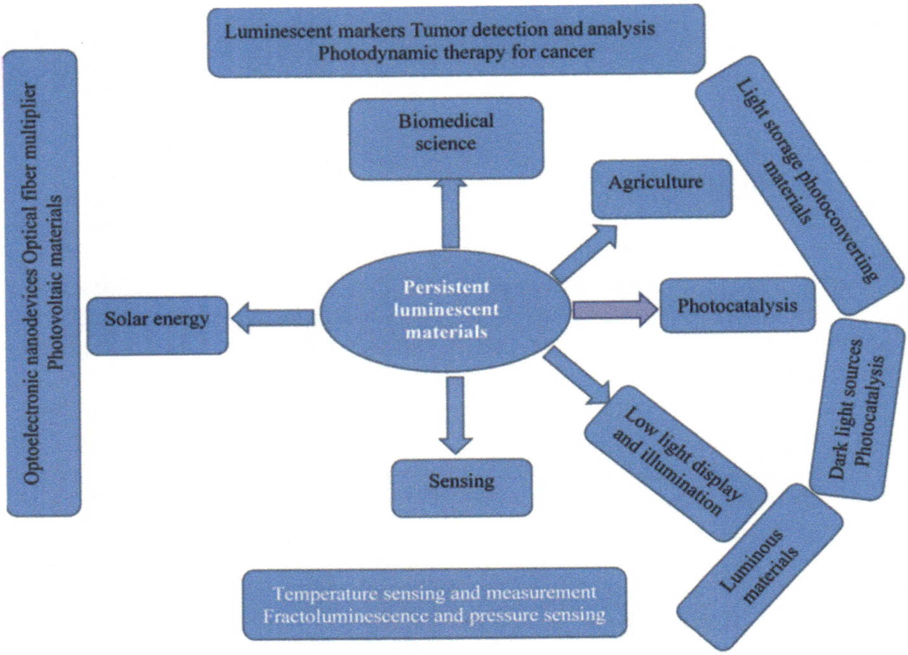

Fig. 7.1 Schematic diagram for the applications of persistent phosphors.

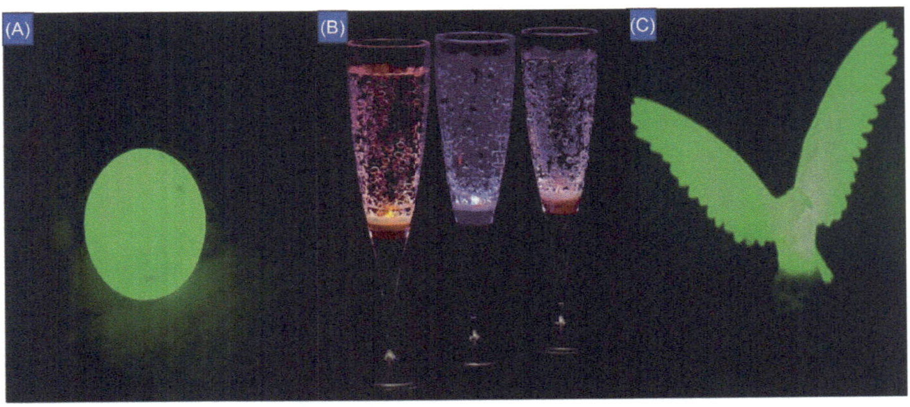

Fig. 7.2 Artworks made from natural ores with persistent phosphorescence. (A) Legendary luminous pearl (https://sg.carousell.com/p/ye-ming-zhu-legendary-glowing-stone-luminous-38281598). (B) Luminous wine cup (https://fr.dhgate.com/product/6-8-18cm-liquid-active-acrylic-led-light/251457223.html). (C) Phosphorescent bird (https://en.wikipedia.org/wiki/Phosphorescence).

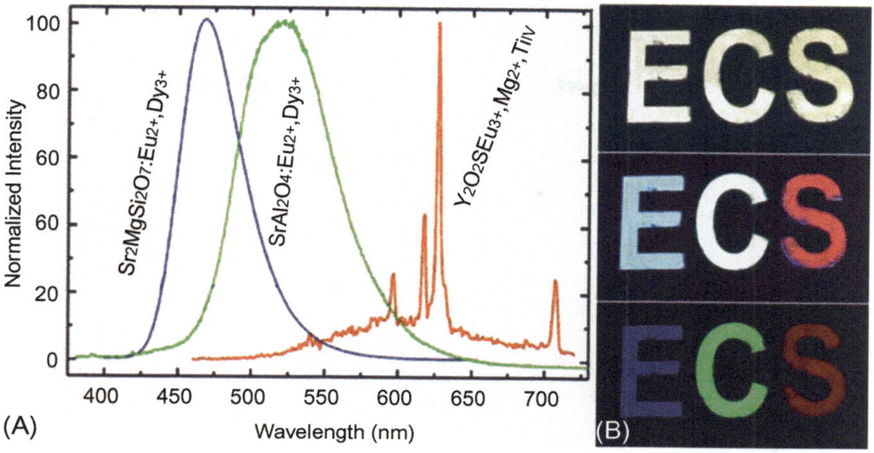

Fig. 7.3 (A) Luminescence spectra of the blue-emitting $Sr_2MgSi_2O_7/Eu^{2+}$, Dy^{3+}; green-emitting $SrAl_2O_4/Eu^{2+}$, Dy^{3+}; and red-emitting Y_2O_2S/Eu^{3+}, Mg^{2+}, Ti^{4+} phosphors after ultraviolet excitation. (B) The persistent luminescence behavior of the blue-emitting $Sr_2MgSi_2O_7/Eu^{2+}$, Dy^{3+}. (E) Green-emitting $SrAl_2O_4/Eu^{2+}$, Dy^{3+}. (C) Red-emitting Y_2O_2S/Eu^{3+}, Mg^{2+}, Ti^{4+} (S) phosphors in daylight, under ultraviolet excitation, and in the dark.
Reproduced with permission from J. Hölsä, Persistent luminescence beats the afterglow: 400 years of persistent luminescence, Electrochem. Soc. Interface 18 (2009), 42–45, Copyright 2009, Electrochemical Society.

The alkaline aluminate persistent phosphors have been widely used since they are chemically stable and their optical properties are superior to that of contemporary persistent phosphors.

7.1.1 *Safety indicator*

Emergency lighting with a certain brightness can guide people to move orderly in an emergency to reduce chaos and save energy. Many researchers have focused on improving the stability and extending the afterglow time of persistent phosphors, to improve their performance for display applications. [8–12] Ueda et al. prepared a transparent ceramic phosphor of YAGG/Ce^{3+}-Yb^{3+} by solid-state reaction. The result showed that Yb^{3+} ions serve as the electron trap at an energy depth of 1.01 eV, which leads to persistent phosphor for over 138.8 "h" after blue light excitation (as shown in Fig. 7.4A). [9] Pan et al. reported a new $LiGa_5O_8$:Cr^{3+} phosphor exhibiting a super-long persistent phosphorescence of over 1000 h (as shown in Fig. 7.4B). [12] Besides, Lu and his coworkers synthesized $SrAl_2O_4$:Eu @ SiO_2 nanoparticles with super-hydrophobic surfaces, which not only prevent exposure to water but also provide a good self-cleaning property. [11] This approach can be extended to improve the aqueous stability of persistent phosphors that are important for display applications, particularly in high humidity environments.

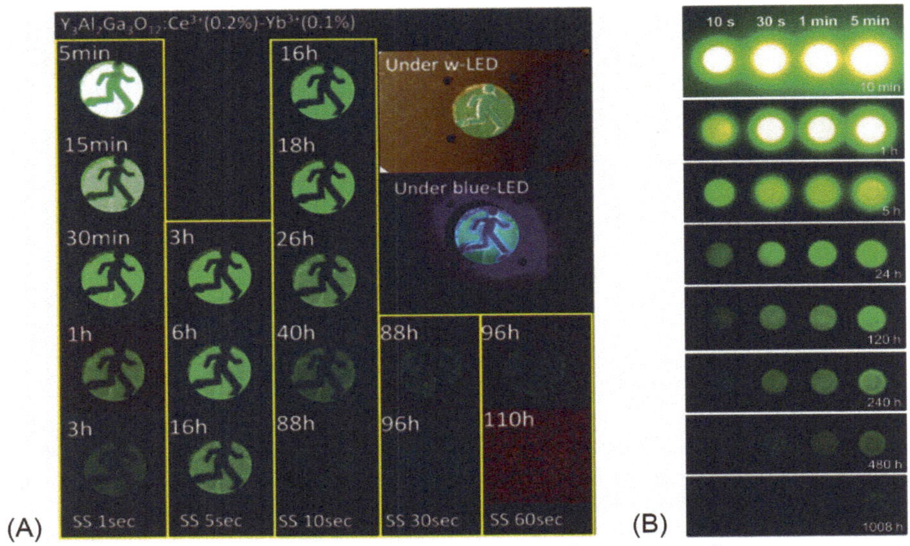

Fig. 7.4 (A) Photograph of as-made YAGG (x = 3.0): Ce^{3+}(0.2%) – Yb^{3+}(0.1%) transparent ceramics under white LED and blue LED illumination and at different times after switching off the blue LED excitation. (B) Photograph of $LiGa_5O_8$: Cr^{3+} phosphor discs taken at different persistent luminescence times (10 min to 1080 h) after irradiated by a 254-nm lamp for 10 s to 5 min.

A: Reproduced with permission from J. Ueda, S. Miyano, S. Tanabe, Formation of deep electron traps by Yb^{3+} codoping leads to super-long persistent luminescence in Ce^{3+}-doped yttrium aluminum gallium garnet phosphors, ACS Appl. Mater. Interfaces 10 (2018), 20652–20660, Copyright 2018, American Chemical Society; B: Reproduced with permission from F. Liu, W. Yan, Y.-J. Chuang, Z. Zhen, J. Xie, Z. Pan, Photostimulated near-infrared persistent luminescence as a new optical read-out from Cr^{3+}-doped $LiGa_5O_8$, Sci. Rep. 3 (2013), 1554, Copyright 2013, Nature Publishing Group.

7.1.2 Road traffic signs

Persistent phosphors are widely used in the fabrication of road traffic signs including, in particular, warning signs, directional signs, and tourist signs. These signs store solar energy during the day and are effectively used as emitting sources at night. It can be regarded as intelligent energy-saving light-emitting materials without manual operation or electrical driving. Road traffic sign traditionally consists of a graphic symbol and/or character, providing legal information and admonition to drivers and pedestrians. It is efficacious to command and direct traffic, which plays a significant role in modern highway transportation management. [13, 14] There has been a luminous road in the Netherlands in which luminous paints are used for making the signs as shown in Fig. 7.5. The persistent phosphors contained in paint absorb the daylight and act as lighting sources at night. By using these road traffic signs, traffic capacity could be upgraded, and the number of traffic accidents could be reduced.

Fig. 7.5 Luminous road (https://www.asme.org/engineering-topics/articles/energy/glowing-the-roads).
Image: Studio Roosegaarde.

7.1.3 Luminous decoration

As persistent phosphors emit phosphorescence in the dark, they have been widely utilized in decoration for toys, jewelry, ornaments, and apparel. Polymer composite–containing persistent phosphors are usually prepared by, for example, casting, into different shapes for decorative purposes. [15–18] For instance, Kumar et al. reported a sunlight-activated composite with green phosphorescence made up of $Sr_{0.7}Al_2O_4$: Eu^{2+}/Dy^{3+} phosphor doped in commercial polyvinyl chloride resin, which can be used as both spraying material and transparent, bendable, and waterproof phosphorescent layers (as shown in Fig. 7.6). [17] Recently, Qiu et al. developed a new strategy to combine the persistent phosphors with resin matrix by the stereolithography (SLA) technology. The strong fluorescence, persistent phosphorescence, and multicolor luminescence make these composite materials attractive for applications in luminous decoration. [15]

7.1.4 Luminous fibers

Luminous fibers that are commonly called noctilucent fibers are typically prepared by dispersion of rare earth–activated persistent phosphor particles into polyamide fibers. [5, 19] The luminous fibers have many extraordinary features including high luminance, persistent phosphorescence, and no radioactivity, [6, 20] which facilitate an extensive range of applications in several fields, involving transportation, night work products, safety protective clothing, etc. By using spinning raw materials as a substrate, the fiber is made of rare earth–doped persistent phosphor (e.g., aluminate and silicate). Noctilucent fibers can continuously glow for more than 10 h in the dark after being charged by visible light illumination for 10 min. In luminous fibers,

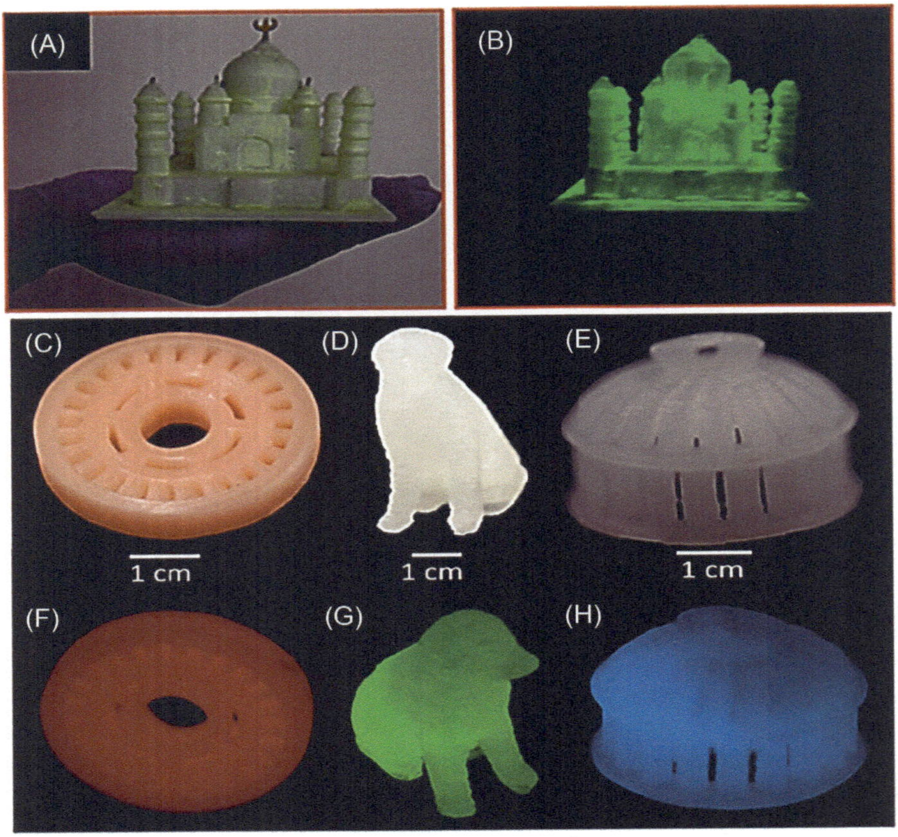

Fig. 7.6 (A) An optical image of the $Sr_{0.7}Al_2O_4$: $Eu_{0.1}^{2+}/Dy_{0.2}^{3+}$ phosphor-coated Taj Mahal monument under ambient light illumination; (B) the optical image of the afterglow emission from the same Taj Mahal monument taken in dark, after the white fluorescent tube light was switched off after 5 min. (C–H) Optical images of the complex-shaped parts printed by SLA photographed in daylight illumination and in the darkroom right after 5-min excitation by a xenon lamp through a digital camera: (C, F) a gear wheel printed with resin slurry containing CaS: Eu^{2+} phosphors; (D, G) a hollow dog printed with resin slurry containing $SrAl_2O_4$: Eu^{2+}, Nd^{3+} phosphors; and (E, H) a hollow birdcage hanging decoration printed with resin slurry containing $CaAl_2O_4$: Eu^{2+}, Nd^{3+} phosphors.
A and B: Reproduced with permission from A. Kumar, G. Kedawat, P. Kumar, J. Dwivedi, B.K. Gupta, Sunlight-activated Eu^{2+}/Dy^{3+} doped $SrAl_2O_4$ water resistant phosphorescent layer for optical displays and defence applications, New J. Chem. 39 (2015), 3380–3387, Copyright 2015, Royal Society of Chemistry; C–H: Reprinted with permission from R. Ni, B. Qian, C. Liu, X. Liu, J. Qiu, Three-dimensional printing of hybrid organic/inorganic composites with long persistence luminescence, Opt. Mater. Express 8 (2018), 2823–2831. © The Optical Society.

$SrAl_2O_4$: Eu^{2+}, Dy^{3+} is usually employed, which produces strong greenish afterglow emission. [21, 22] By controlling the particle size the composition of the polymeric host and/or the inorganic luminous materials, the concentration of the luminous dopant, and the charging (excitation) time, the afterglow properties of the luminous fibers can be finely tuned, and their performance can be optimized. Moreover, luminous fibers can be employed to design child safety clothing that helps to enhance traffic safety during crossing the street in the evening/night. In addition, luminous fiber–manufactured signs have been used in cinemas to help people exit and escape from accidents. The colored luminous fibers are presented in Fig. 7.7. For these commercial applications the afterglow characteristics that are robust against change in environmental conditions should be possessed by fibers. For instance, the afterglow characteristics should not degrade notably after sunlight irradiation, repeated washing and/or ironing, contact with acidic or basic cleaning chemicals, or simply a long period of storage. For these reasons an ideal luminous fiber should be thermally stable and possesses long-term stability against ultraviolet irradiation and chemicals including acids and bases. Furthermore, its performances should remain unaffected after multiple cycles of water washing such that the afterglow properties still meet the requirements for practice use. Ge's group investigated the structure, mechanical, and thermal properties of luminescent fibers and studied the characteristics when used for fiber spinning. [23] Xu et al. examined the mechanical and thermal properties of luminescent fibers, which they considered might be used in industrial fields as normal chemical fibers. [24] Zhao et al. studied the effects of the illumination excitation on the afterglow emission process of luminescent fibers. [25] Zhang et al. investigated the working conditions for the luminescent fiber to be used for anticounterfeiting based on the spectrum fingerprint of the fibers. [26] Guo et al. [27] investigated the effects of several parameters on the stability of luminescence properties of the luminous fibers. These factors include long-term storage under ambient conditions, artificial sunlight irradiation, multiple cycles of water washing, high temperature, and treatment with acid or base solutions. They concluded that the luminescent fibers exhibit good thermal

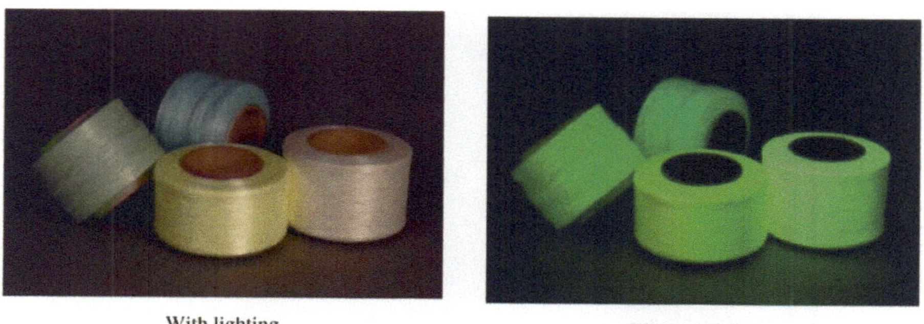

Fig. 7.7 Luminous fibers under the light and without light illumination.
Reproduced with permission from Y. Yan, C. Liu, X. Ding, Study on the equivalent luminance of luminous fibers in photopic vision and scotopic vision, Text. Res. J. 88 (2018), 1157–1163, Copyright 2018, SAGE Publishing.

characteristics under the high-temperature working conditions. The best performance was observed at near 80°C, below which the number of trapped electrons is small and weak afterglow is observed. At higher temperatures, however, thermal annihilation of the trapped states leads to shortened afterglow duration with reduced brightness.

7.1.5 Luminous watch

Persistent phosphors have been long used to decorate the pointers and dial of waters to enable the visibility in the dark (Fig. 7.8). Zinc sulfide was the first persistent phosphor used for luminous watch. Nevertheless, this material has a short period of afterglow. To enhance the afterglow emission duration, radioactive components are selected. The first element to be used was radium, which was afterward replaced by a meagerly detrimental but less radioactive element, tritium. On the other hand, radioactive elements can result in radiation damage to the luminescent materials. Hence the afterglow emission of such systems degrades steadily during the lifetime, and this process is almost irreversible. More significantly, radioactive elements have a serious impact on the human body. Therefore the use of tritium to generate persistent luminescence was eliminated in watches in the early 1990s. The use of modern persistent phosphors in watches was started by a Japanese company Nemoto, by which rare earth–doped alkaline earth aluminate persistent phosphors have been used and they are much better than conventional zinc sulfide persistent phosphors in terms of chemical stability and brightness afterglow duration.

7.2 Light-emitting diode

White light-emitting diode (w-LED) is now gradually taking the market share of a traditional incandescent light bulb and fluorescent lamps due to their superiority in energy conservation, stability, and environmental friendliness. Unlike the ordinary

Fig. 7.8 Luminous watch (https://www.reisnichols.com/blog/2016/09/07/luminous-watches).

DC-driven w-LEDs, high-power alternating current-driven light-emitting diode (AC-LED) has been extensively used in architectural interior and exterior space lightings, due to their relatively lower cost (due to the reduced extra electronic components), higher energy efficiency, smaller volume, and longer lifetime. [28–32] However, the strobe frequency of the alternating current light-emitting diode in alternating current cycles (5–20 ms in every cycle) results in several problems. The persistent phosphors can be used for overcoming the drawbacks of the stroboscopic light source by the persistent phosphorescence. Lin et al. prepared a $Gd_3Al_2Ga_3O_{12}:Ce^{3+}$ persistent phosphor that was embedded into a phosphor-in-glass (PiG) converter to building a white AC-LED. The persistent phosphorescence of the PiG composite would partly compensate the alternating current intervals, and the percent flicker could be reduced to 69% (Fig. 7.9A–C). [32] Hereafter, their research team fabricated a $Lu_{1.98}CaMg_2Si_{1.5}Ge_{1.5}O_{12}:0.02Ce^{3+}$-based phosphor-in-silicone composite and successfully reduced the percent flicker to 64.1%. [30]

In addition, Yu and his coworkers reported a motion-driven phosphorescence material by mixing an electroluminescent ZnS: Cu with fluorescent dye/persistent phosphorescence ($Sr_2MgSi_2O_7:Eu^{2+}/Dy^{3+}$) in the polydimethylsiloxane-flexible matrix, which can be used for display-related applications. [29]

7.3 Optical information storage

Energy storage phosphors are materials in which irradiation induces ionization of ions followed by the capture of the liberated electrons by vacancies, forming color centers, or oxidizing cations. By irradiation in the absorption bands of the centers to induce photostimulation or temperature rise, the trapped electrons can be released, and the recombination energy with holes transferred to a luminescent ion.

Persistent phosphors can be used to store radiation energy from visible light, ultraviolet, and x-ray by immobilizing the generated charge carriers in traps and then liberate the preserved energy upon thermal, optical, or mechanical stimulation. Therefore this process in principle can be exploited for optical information storage. This process can be mainly divided into two steps: "write-in" and "read-out." For the application in optical information storage, the persistent phosphors are exposed to the light beam to form a potential image in the form of captured electrons by the "write-in" step. The quantity of captured charge carriers (electrons or holes) is equal to the number of radiation energy absorbed, and the potential image is optically readable within a definite period of time after the exposure. The "read-out" process is the method to regenerate stored images that can be visible to human eyes or detected by instruments.

Table 7.1 lists the persistent phosphors used for optical information storage. For optical information storage the persistent phosphors are first stimulated by high-energy ultraviolet light (called write-in) to encode the information into the material. The read-out process is usually carried out optically by irradiation at an appropriate wavelength or heat treatment at a designated temperature, such that the data or image encoded in the persistent phosphors will become visible. For example, Hu et al.

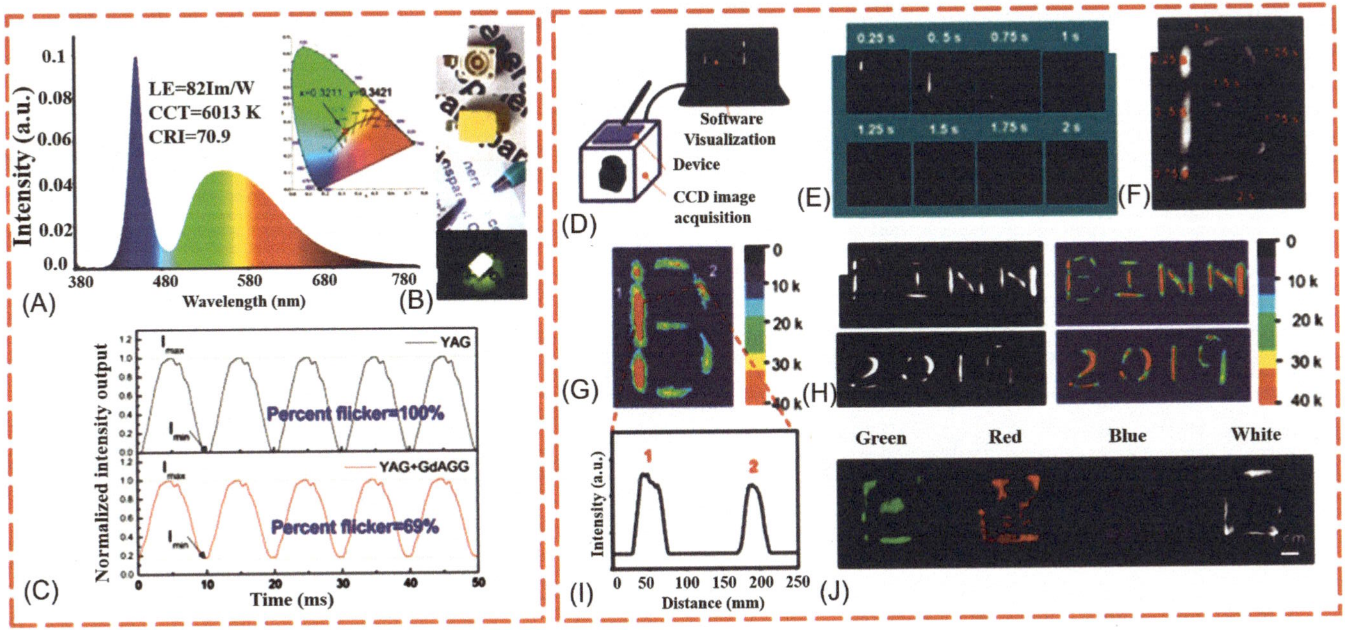

Fig. 7.9 (A) Photoluminescence spectrum and colorimetry parameters (including LE, CCT, CRI, and chromaticity coordination) of the remote-type AC-LED encapsulated by PiG at an operating current of 350 mA. (B) From top to bottom: photographs of the used blue chip with a groove-shaped sample holder, the assembled LED device covered with PiG on the top, the white luminescence when power is on, and the yellowish-green afterglow luminescence when power is off. (C) Luminescence intensity variations of the labeled LED devices driven in AC periodic cycles. Demonstration of the applications of color-tunable TIEL. (D) Schematic of the acquisition system for motion tracking. (E) Live images captured by the camcorder in consecutive frames at an interval of 0.25 s. (F) Superimposed image showing the complete trajectory of the letter "B." (G) Luminescence intensity mapping after being processed by Mat Lab. (H) A continuous trajectory showing the word "BINN" and "2019." Left section, live luminescence image; right section, corresponding mapping of the luminescence intensity. (I) Luminescence intensity distribution along the dashed line defined in (G). (J) Color-tunable TIEL induced by a large-area object.

A–C: Reproduced with permission from H. Lin, B. Wang, J. Xu, R. Zhang, H. Chen, Y. Yu, Y. Wang, Phosphor-in-glass for high-powered remote-type white AC-LED. ACS Appl. Mater. Interfaces 6 (2014), 21264–21269, Copyright 2014, American Chemical Society; D–J: Reproduced with permission from Q. Cheng, Y. Wang, L. Su, H. Wang, G. Zhu, W. Yu, Wide-spectrum manipulation of triboelectrification-induced electroluminescence by long afterglow phosphors in elastomeric zinc sulfide composites, J. Mater. Chem. C 7 (2019), 4567–4572, Copyright 2019,

Table 7.1 Optical information storage by using different persistent phosphors.

Phosphor	Write-in	Time interval	Read-out	Ref.
$SrAl_2O_4$:Eu^{2+}/Dy^{3+}	350 nm for 2 min	30 s	760 nm for 70 s	[33]
$SrSi_2O_2N_2$:Ln^{2+}/Ln^{3+} (Ln^{2+} =Yb, Eu, Ln^{3+} = Dy, Ho, Er)	254 nm for 20 s	15 days	100°C or 980-nm laser	[34]
Ba_2SiO_4:Eu^{2+}/Ho^{3+}	254 nm for 3 min	15 days	100°C or 980-nm laser	[35]
$Y_3Al_{5-x}Ga_xO_{12}$: Ce^{3+}/V^{3+}	450 nm	30 days	250 °C or 808-nm laser	[36]
$NaBaScSi_2O_7$:Eu^{2+}	Ultraviolet light for 20 min	30 h	980-nm laser	[37]
$Mg_4Ga_4Ge_3O_{16}$:Cr^{3+}	230 nm for 5 min	3 months	Visible or 980-nm light	[38]
$Ca_3Ga_4O_9$:Bi^{3+}	285 nm		100°C, 300°C, or 980 nm	[39]
$BaSi_2O_5$:Eu^{2+}/Nd^{3+}	350 nm for 1 min	48 h	200°C	[40]

reported a persistent phosphors Ba_2SiO_4:Eu^{2+}/Ho^{3+} and its application for optical information storage. They showed that the image recorded by the persistent phosphorescent film could be read out by irradiation with a 980-nm laser even after 15 days (Fig. 7.10). [35] Zhuang et al. successfully synthesized two groups of novel persistent phosphors: orange red-emitting $SrSi_2O_2N_2$:Yb,Ln_B and green-emitting $SrSi_2O_2N_2$: Eu,Ln_B (Ln_B = Dy, Ho, and Er). These materials showed excellent chemical stability and contained a large number of traps with tunable depths in the range of 0.8–1.2 eV. The trap depth increased in the order of Er, Ho, and Dy, which is well consistent with the energy levels indicated in the host-referred binding energy (HRBE) scheme. Furthermore, they demonstrated the use of these deep-trap persistent phosphors for information storage. These $SrSi_2O_2N_2$: Yb, Dy and $SrSi_2O_2N_2$: Eu, Dy phosphors with the large trap depth showed great promise as optical information storage media. [34] Moreover, Viswanath et al. observed persistent phosphorescence by the bluish emission of Ce^{3+} for nearly 4 h by in the $Na_3ScSi_3O_9$ (NSSO) host. The effect of Tb^{3+}, Dy^{3+}, and Mn^{2+}codoping has been studied and codoping with Ln^{3+} ions has been used to control the electron trap depth. From this study, it is demonstrated that NSSO:Ce^{3+}, Dy^{3+}as a persistent phosphor that continuously gives emission for 17 h after stoppage of the excitation can be an ideal medium for information storage. [41]

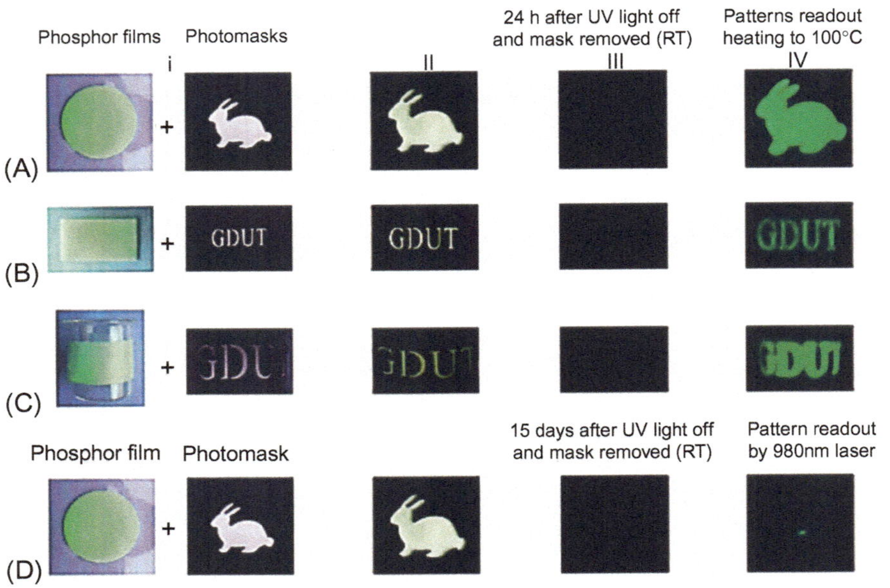

Fig. 7.10 (A) The write-in process using ultraviolet (254 nm) irradiation and read-out (heating at 100°C) of a rabbit pattern on the phosphor film. (B) The write-in process using ultraviolet (254 nm) irradiation and read-out (heating at 100°C) of the letters "GDUT" on the phosphor film. (C) The write-in using ultraviolet (254 nm) irradiation and read-out (heating at 100°C) of the letters "GDUT" on the flexible phosphor film. (D) The write-in using ultraviolet (254 nm) irradiation and read-out (980-nm laser) of the rabbit pattern on the flexible phosphor film. (i) Phosphor films and photomasks (rabbit pattern or letters "GDUT"). (ii) Phosphor films are covered by photomasks when using ultraviolet 254-nm irradiation to write the patterns. (iii) The phosphor films were kept at room temperature in the dark for 24 h or 15 days, after removing the ultraviolet light and photomasks. (iv) The phosphor films were heated to 100°C or irradiated by a 980-nm laser to read out the stored patterns.
Reproduced with permission from C. Wang, Y. Jin, Y. Lv, G. Ju, D. Liu, L. Chen, Z. Li, Y. Hu, Trap distribution tailoring guided design of super-long-persistent phosphor Ba_2SiO_4: Eu^{2+}, Ho^{3+} and photostimulable luminescence for optical information storage, J. Mater. Chem. C 6 (2018) 6058–6067, Copyright 2018, Royal Society of Chemistry.

7.4 Sensors

At present, optical pyrometers based on a noncontact measurement have been widely used in different fields. During the measurement, usually, a probe beam is used, which may result in the change in sample temperature due to optically induced thermal effect. In addition, real-time excitation may result in the perturbation of scattered light due to a Stokes or anti-Stokes shift, which imparts the measurement precision. In this case the temperature-dependent phosphorescence of persistent phosphors can be exploited for optical temperature measurement, because the

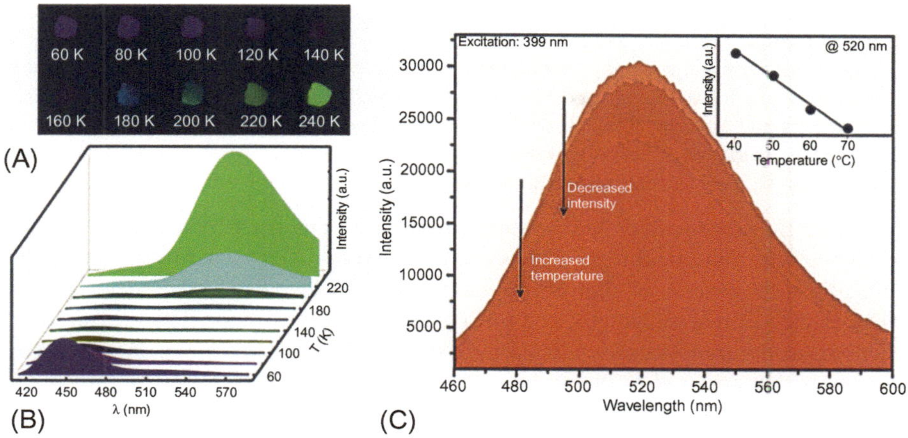

Fig. 7.11 (A) Digital camera images and (B) afterglow spectra of $SrAl_2O_4$: Eu^{2+}, Dy^{3+}, Tb^{3+} phosphor recorded at different temperatures. (C) Luminescence spectra of a resin composite containing $SrAl_2O_4$: Eu^{2+}, Dy^{3+} phosphor recorded at different temperatures. The excitation wavelength is 399 nm. The inset picture shows the dependence of emission intensity on sample temperature at 520 nm.

Reprinted with permission from R. Ni, B. Qian, C. Liu, X. Liu, J. Qiu, Three-dimensional printing of hybrid organic/inorganic composites with long persistence luminescence, Opt. Mater. Express 8 (2018), 2823–2831. © The Optical Society.

influence of excitation light on the temperature of the test sample can be completely eliminated. [15, 42, 43]

Zhao and coworkers synthesized a $SrAl_2O_4$: Eu^{2+}, Dy^{3+}, Tb^{3+} persistent phosphor, which could be used as a sensor in the low-temperature range from 60 K to 240 K (Fig. 7.11A). The phosphorescence of $SrAl_2O_4$: Eu^{2+}, Dy^{3+}, Tb^{3+} changes from blue to green with the increase of temperature due to the different thermal quenching behavior of blue/green emission band, and the normalized phosphorescence spectrum recorded at different temperatures shows larger changes in phosphorescence intensity ratio between blue and green (Fig. 7.11B). [42] Ni et al. reported a simple method to fabricate a 3-D organic resin structure with persistent phosphorescence by additive manufacturing. The as-fabricated 3-D objects containing $SrAl_2O_4$: Eu^{2+}, Dy^{3+} phosphors can be used for optical temperature sensing from 40 to 70 °C (Fig. 7.11C). [15]

For temperature sensing the temperature dependence of the afterglow intensity and lifetime can be both exploited. The benefits of using persistent phosphors for temperature sensing include the following:

- Passive contactless sensor. Through monitoring the emission spectra, the temperature of the examined object can be determined directly.
- Wide application range. Compared with other temperature sensors, the persistent phosphor-based temperature sensors can be used in a wide range of fields where excitation light cannot

be provided. They can also be managed remotely as long as the emission spectra can be collected or routed through optical fibers to detectors.
- High sensitivity and accuracy. According to the previous experiments carried out on these materials, it is found that the persistent phosphor-based temperature sensors show excellent sensitivity and accuracy in a definite temperature range. Moreover, they show strong luminescence intensity, persistent phosphorescence, and large temperature coefficients and hence can be used as ideal probe materials in optical temperature sensing system. [44–46]

To create fluorescent fiber temperature sensors, the persistent phosphor is usually combined with an optical fiber. The sensing system normally consists of three major parts: the probe material that emits the fluorescent signals, a photodetector that receives the fluorescent signals, and the signal analysis processor. The measurement is based on the temperature-induced intensity or lifetime changes of the persistent phosphors that are used as the probe material.

Another important application of persistent phosphors is pressure sensing based on their mechanoluminescence. The light emission generated under mechanical stimuli like stress, friction, scratching, breakage, and ultrasonic oscillation is known as mechanoluminescence. For many years, it is confirmed that ordinary fluorescent materials exhibit a low-intensity mechanoluminescence, which cannot be used for sensing applications. The development of persistent luminescent materials with strong visible mechanoluminescence and the enhancement of detection efficiency of weak light have made mechanoluminescence attractive for potential applications in mechanical engineering, materials science, earthquake forecasting, life science, medical diagnostics, and other areas. A growing number of reports on the use of mechanoluminescence for stress sensors have been published in recent years. [47–49] The employment of the mechanoluminescent membrane for measurement of the stress distribution on the material surfaces has been developed by Li et al. [50] In 2011 Wang et al. demonstrated the use of mechanoluminescence from $SrAl_2O_4$: Eu to track the structure damages to evaluate the fracture and stress intensity. [51] Zhang et al. provided a novel design for the high integration of colorful (red–orange–yellow–green), bitemporal (fluorescent and delayed), and four-modal (thermo –/mechano-motivated and upconverted/downshifted) emissions in a single material. This capability is acquired by multiplexing the red-delayed emission of Pr^{3+} dopants and green fluorescent emission of Er^{3+} dopants in $NaNbO_3$ piezoelectric microparticles. They concluded that the luminescence of the composite film enables excellent multidimensional identifiability (including color, lifetime, and excitation mode) by using simple and common tools. Moreover the abundance of available candidate functionalization hosts and luminescent ions may benefit the creation of advanced materials and devices and extend their applications to anticounterfeiting, information encryption, and photonic displays. [52] Kim et al. studied the sudden transmission proceeding through the cracks of concrete by utilizing mechanoluminescent effects. [53] They provided evidence of the formation and transmission of cracks during the fracture process and also observed some interesting phenomena among the crack transmission with the shedding speed during the bending and fracture process concretes. These results have made mechanoluminescent sensors attractive for application in diverse fields. Jiang et al. provided a new design for the integration of trap-controlled color change and multimodal luminescence through oxygen vacancy engineering and doping in the piezoelectric lattice. More importantly, they have developed a novel strategy of combining thermal, photonic, and mechanical stimulation to achieve the modulation and differential display of bright-field color change and dark-field persistent luminescence. [54] The mechanoluminescence has usually been interpreted by two different mechanisms related with electricity or heat. [55] In the electricity mechanism, there are free electrons that are typically generated by the mechanical force, and these electrons create an electric field via the compression of the crystals. These free electrons can be liberated when the regional electric field forms at the crystal distortion region and trigger the light emission. In the heating mechanism the crystal disfiguration arises due to mechanical force that liberates free electrons, a portion of which (or holes) can be captured by traps in the crystal and results in light emission under by thermal stimulation.

7.5 Bioapplications

7.5.1 In vivo imaging

Fluorescence-based bioimaging is a very powerful imaging technique widely used in biomedical field because it is fast and flexible to obtain real-time data. Therefore bioimaging has captured growing attention of scientists from different communities. With the development of fluorescence-based bioimaging in the last decades, an increasing number of the fluorescent imaging agent, like fluorescent dyes, semiconductor quantum dot fluorescent nanoparticles, upconversion nanoparticles, and persistent phosphors have been reported and successfully used in bioimaging. Among the aforementioned materials, persistent phosphors have an obvious advantage since their afterglow time is long enough to permit time-lapse imaging that avoids damage of biological tissues by excitation light and improves the signal-to-noise ratio.

A large number of persistent phosphors have been used for bioimaging in recent years (Table 7.2), and most of them exhibit one emission band in the first biologically transparent window (from 650 to 950 nm). Some of them not only have a persistent phosphorescence but also are rechargeable in vivo. Xue and coworkers reported a novel persistent phosphor $Zn_3Ga_2GeO_8$: $Cr^{3+}/Yb^{3+}/Er^{3+}$ synthesized by sol-gel method, and this phosphor was rechargeable in vivo (Fig. 7.12A). As shown in Fig. 7.12B, the fluorescence signal of a mouse was still detectable after 10 h of excitation. More importantly the as-synthesized persistent phosphors can be effectively reexcited by a 980-nm laser for more than four times in vivo (Fig. 7.12C). [56] Xu et al. synthesized sub-10-nm persistent phosphorescent nanocrystals, which can be recharged by a white LED in vitro and in vivo. [62] Xu et al. reported a red persistent phosphor of Y_3Al_5 %xGa_xO_{12}: Cr^{3+} garnet (YAGG: Cr^{3+}, x from 0 to 5); they found that the persistent luminescence of the YAGG: Cr^{3+} (x = 3) phosphor was nearly five times higher than that of the widely used $ZnGa_2O_4$: Cr^{3+} red persistent phosphor. These novel red persistent phosphors have great potential for an improved in vivo bioimaging application [71]. Moreover, Lv et al. developed novel red-to-NIR-emitting

Table 7.2 Persistent phosphors used for bioimaging.

Phosphor	Excitation wavelength	Emission wavelength	Afterglow time	Reexcitation	Tissue thickness	Ref.
$Zn_3Ga_2GeO_8$:Yb/Er/Cr	980 nm	700 nm	10 h	980 nm	–	[56]
$Sr_3Sn_2O_7:Nd^{3+}$	254 nm	1079 nm	–	–	2 mm	[57]
$SrAl_2O_4:Eu^{2+},Dy^{3+}$	365 nm	520 nm	30 min	w-LED	–	[58]
$MAlO_3$: Mn^{4+}, Ge^{4+} (M = La, Gd)	325 nm	719 nm	12 h	–	3 mm	[59]
$Zn_3Ga_2Sn1O_8:0.5\ Cr^{3+}$	–	700 nm	2 h	980 nm	–	[60]
β^-NaYF_4:Yb,Er@$NaYF_4$	980 nm	–	20 min	980 nm	–	[61]
$ZnGa_2O_4:Cr^{3+}$	w-LED	696 nm	30 min	w-LED	1 cm	[62]
FA- $Zn_3Ga_2GeO_8:Cr^{3+},Yb^{3+},Er^{3+}$	254 nm	700 nm	90 min	650 nm	–	[63]
$Li_5Zn_8Ga_5Ge_9O_{36}$: Cr^{3+}, Ti^{4+}	Ultraviolet light	700–800 nm	700 min	–	2 mm	[64]
$LaAlO_3:Er^{3+},Cr^{3+},Sm^{3+}$	ultraviolet light	700 nm	60 min	–	–	[65]
SiO_2@$ZnGa_2O_4:Cr^{3+}$@SiO_2	430 nm	700 nm	2 h	Soft x-rays	1 cm	[66]
$SrAl_{12}O_{19}:0.5\%Cr^{3+}$, $0.5\%Ti^{4+}$	254 nm	650–950 nm	4 h	–	–	[67]
$ZnGa_2O_4:Cr^{3+}/Sn^{4+}$	254 nm	700 nm	15 min	w-LED	–	[68]
$Zn_3Ga_2Ge_2O_{10}:Cr^{3+}/Eu^{3+}$	254 nm	697 nm	1 h	–	Cell size	[69]
$BaZrGe_3O_9$: Pr^{3+}	254 nm	615 nm	20 min	980 nm	5 mm	[70]

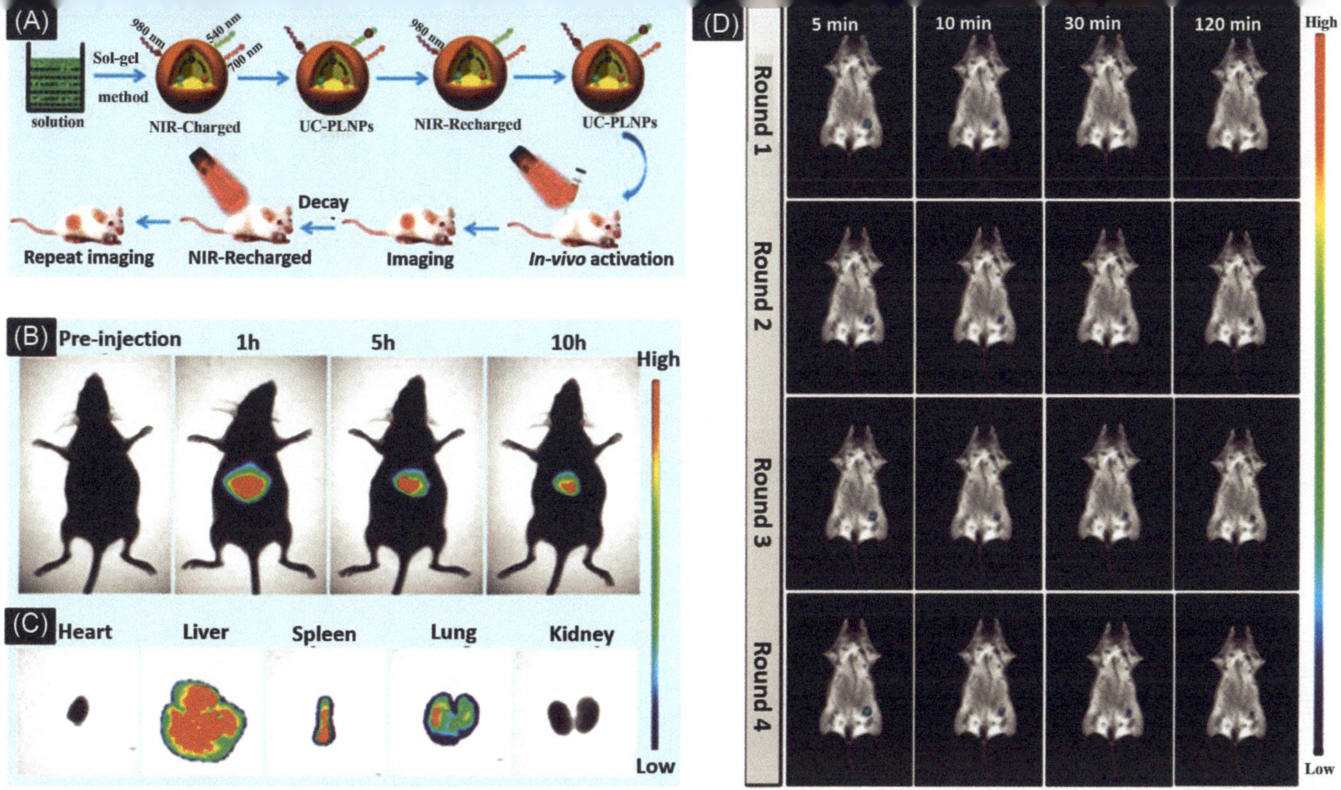

Fig. 7.12 (A) Schematic illustration of the sol-gel method and NIR-to-NIR rechargeable in vivo bioimaging based on 980-nm laser-activated NIR-emitting upconversion-persistent phosphorescent nanoparticles. (B) In vivo upconversion bioimaging of a mouse intravenously injected with poly(ethylene glycol)-modified upconversion persistent phosphorescent NPs at different time intervals under 980-nm laser (150 mW cm^{-2}) excitation, (C) ex vivo imaging (heart, liver, spleen, lung, and kidney) after 10 h of intravenous injection under the 980-nm laser (150 mW cm^{-2}) excitation. (D) In vivo rechargeable bioimaging based on the nanosized $Zn_3Ga_2GeO_8$: Yb/Er/Cr upconversion-persistent phosphorescence nanoparticles for four cycles at different decay times (5 min, 10 min, 30 min, and 120 min) after in situ excitation by a 980-nm laser. In situ excitation was performed using a 980-nm laser as the light source with a power density of 150 mW cm^{-2} for 120 s. Reproduced with permission from Z. Xue, X. Li, Y. Li, M. Jiang, G. Ren, H. Liu, S. Zeng, J. Hao, A 980 nm laser-activated upconverted persistent probe for NIR-to-NIR rechargeable in vivo bioimaging, Nanoscale 9 (2017), 7276–7283, Copyright 2017, Royal Society of Chemistry.

Yb-doped ($Ba_{1-x}Sr_x$) $AlSi_5O_2N_7$ persistent phosphors, which were considered to be a promising candidate for in vivo bioimaging applications. [72] Zhuang et al. devolved a Zn $(Ga_{1-x}Al_x)_2O_4$: Cr, Bi red persistent phosphors in which the trap depths for the persistent phosphor can be tuned using the Al concentration, and they concluded that at 20°C, the trapped electrons in the Al-containing phosphors can be efficiently released with the assistance of additional NIR photostimulation. This NIR photostimulated red persistent luminescence (NIR-to-red mode) as observed in the Al-containing $ZnGa_2O_4$: Cr, Bi phosphors could enable the development of new in vivo imaging systems [73].

7.5.2 Molecular sensing

Since persistent phosphors possess low background autofluorescence and high detection sensitivity compared with conventional fluorescent probes, they can be used as optical probes for molecular sensing. Recently, nanoscale persistent phosphors also have evoked widespread attention in biosensing. Since the lifetime of autofluorescence is very short (10–50 ns), the autofluorescence can be efficiently eliminated by using time-gated detection technique, while the afterglow signal is left for imaging. Therefore persistent phosphors offer a particularly high signal-to-noise ratio and sensitivity for biosensing. Up to now, persistent phosphors have been applied in the biosensing of molecules including glutathione, avidin protein, and lysozyme. [74–76]

7.5.2.1 Homogeneous fluorescence resonance energy transfer assay

Yan's group was the first to demonstrate a fluorescence resonance energy transfer assay based on persistent phosphorescent nanoparticles. [77] They fabricated biocompatible polyethyleneimine functionalized $Ca_{1.86}Mg_{0.14}ZnSi_2O_7$: Eu^{2+}, Dy^{3+} persistent phosphorescence nanoparticle-based probes to detect alpha-fetoprotein (AFP). In their experiment the combination with polyethyleneimine-modified persistent phosphorescence nanoparticles and alpha-fetoprotein–antibody-gold nanoparticle conjugates (Ab-AuNPs) enables the construction of a fluorescence resonance energy transfer-based reserve assay. Alpha-fetoprotein was used to adjust the efficiency of fluorescence resonance energy transfer via inhibition of the electrostatic interactions between antibody-gold nanoparticle conjugates and polyethyleneimine-persistent phosphorescence nanoparticles (Fig. 7.13). The determination of fluorescence resonance energy transfer efficiency is carried out by calculating the spectral overlap between the emission band of polyethyleneimine-persistent phosphorescence nanoparticles and the absorption band of antibody-gold nanoparticle conjugates (Fig. 7.14A). The luminescence quenching of polyethyleneimine-persistent phosphorescence nanoparticles takes place owing to the formation of polyethyleneimine-persistent phosphorescence nanoparticle/antibody-gold nanoparticle conjugates combined with fluorescence resonance energy transfer reserve characteristics (Fig. 7.14B), which is due to the electrostatic interaction between polyethyleneimine-persistent phosphorescence

Fig. 7.13 Schematic diagram of the fluorescence resonance energy transfer inhibition assay for alpha-fetoprotein based on luminescence quenching effect of polyethyleneimine-persistent phosphorescence nanoparticles by antibody-gold nanoparticle conjugates.
Reproduced with permission from B.-Y. Wu, H.-F. Wang, J.-T. Chen, X.-P. Yan, Fluorescence resonance energy transfer inhibition assay for α-fetoprotein excreted during cancer cell growth using functionalized persistent luminescence nanoparticles, J. Am. Chem. Soc. 133 (2010), 686–688, Copyright 2010, American Chemical Society.

nanoparticles and AuNPs. The persistent luminescence of polyethyleneimine-persistent phosphorescence nanoparticles can be improved by alpha-fetoprotein, which is a finalizer for the conjugation structure of the antibody-gold nanoparticle conjugates and polyethyleneimine-persistent phosphorescence nanoparticles. The strong connection of alpha-fetoprotein to the antibody results in stripping of antibody-gold nanoparticle conjugates from polyethyleneimine-persistent phosphorescence nanoparticles (Figs. 7.13 and 7.14C). A good linear relationship is observed by plotting the enhanced luminescence intensity (ΔP) of the persistent phosphorescence nanoparticles against the alpha-fetoprotein concentration (Fig. 7.14D). Due to the high affinity of antibody-gold nanoparticle conjugates to alpha-fetoprotein, the probe displays high selectivity to alpha-fetoprotein (Fig. 7.17E), indicating that this probe is suitable for in vitro detection of alpha-fetoprotein in human serum samples.

7.5.2.2 Homogeneous reaction assay

A bioprobe based on the chemical reaction between cobalt oxyhydroxide (CoOOH) and $Sr_2MgSi_2O_7$:0.01Eu^{2+}, 0.02Dy^{3+} persistent phosphorescence nanoparticles was developed by Li and coworkers for the identification of ascorbic acid (AA) in cell extracts. [78] By depositing cobalt oxyhydroxide nanoflakes onto the persistent phosphorescence nanoparticle surfaces, efficient quenching of the persistent phosphorescence through fluorescence resonance energy transfer was observed. Moreover the luminescence of persistent phosphorescence nanoparticles was restored due to the reduction of cobalt oxyhydroxide to Co^{2+} (Fig. 7.15). In addition, ascorbic acid was oxidized to dehydroascorbic acid during the redox reaction. The nanoprobe showed high selectivity and prompt response to the presence of ascorbic acid with a concentration range of 2.5–100 μM and a detection limit of 2.20 μM.

Wang et al. reported a Zn_2GeO_4: Mn (ZGO: Mn) persistent phosphor nanorods by direct hydrothermal synthesis. [76] Fig. 7.16A explained the serum lysozyme biosensing works. Firstly the ZGO: Mn NRs were modified with lysozyme-binding aptamer (marked as ZGO: MnLBA), and black hole quencher-labeled DNA (designated as BHQ-DNA) was loaded as an emission quencher for ZGO: Mn-LBA through

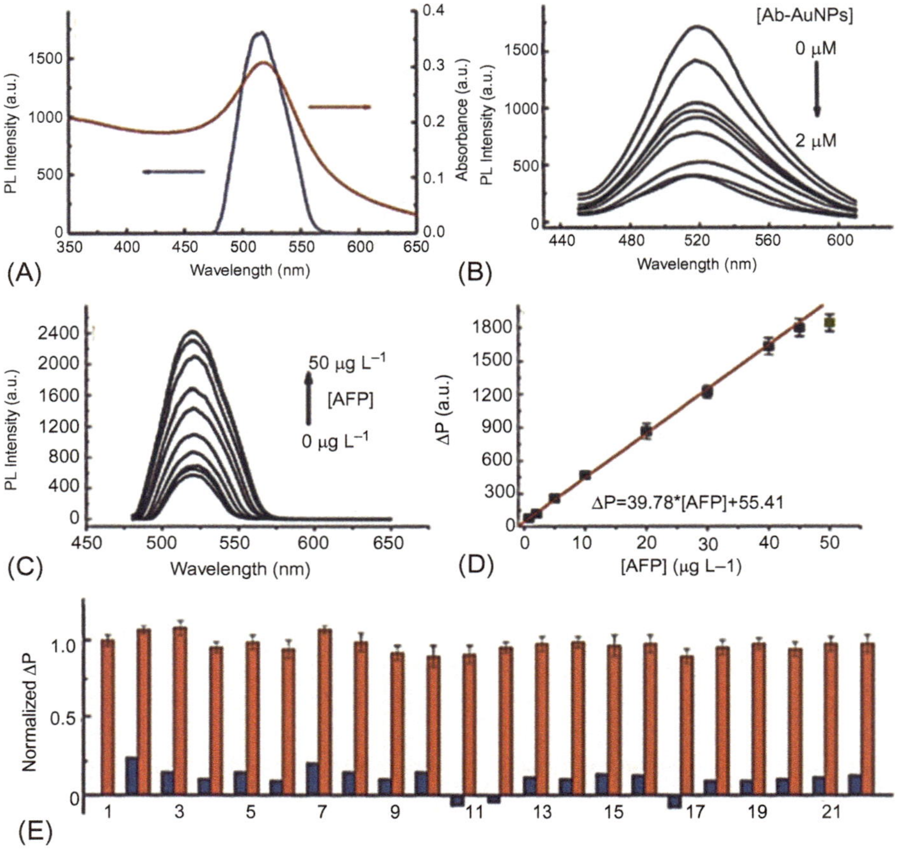

Fig. 7.14 (A) Emission spectrum (blue curve) of polyethyleneimine-persistent phosphorescence nanoparticles and absorption spectrum (red curve) of antibody-gold nanoparticle conjugates. (B) The persistent luminescence quenching effect of polyethyleneimine-persistent phosphorescence nanoparticles by the antibody-gold nanoparticle conjugates. (C) Emission recovery of polyethyleneimine-persistent phosphorescence nanoparticle/antibody-gold nanoparticle conjugates by the addition of alpha-fetoprotein. (D) The linear relationship between enhanced emission intensity (ΔP) and alpha-fetoprotein concentration. (E) Emission responses of diverse analytes to polyethyleneimine-persistent phosphorescence nanoparticle/antibody-gold nanoparticle conjugates in PBS buffer (pH 7.4). Reproduced with permission from B.-Y. Wu, H.-F. Wang, J.-T. Chen, X.-P. Yan, Fluorescence resonance energy transfer inhibition assay for α-fetoprotein excreted during cancer cell growth using functionalized persistent luminescence nanoparticles, J. Am. Chem. Soc. 133 (2010), 686–688, Copyright 2010, American Chemical Society.

Fig. 7.15 Schematic diagram of ascorbic acid detection using cobalt oxyhydroxide–modified persistent phosphorescence nanoparticles.
Reproduced with permission from N. Li, Y. Li, Y. Han, W. Pan, T. Zhang, B. Tang, A highly selective and instantaneous nanoprobe for detection and imaging of ascorbic acid in living cells and in vivo, Anal. Chem. 86 (2014), 3924–3930, Copyright 2014, American Chemical Society.

DNA hybridization. The aptamer folds into a unique 3-D structure, which binds with lysozyme, and the BHQ-DNA is detached from the ZGO: Mn-LBA. Consequently the recovery of the phosphorescence of ZGO: Mn-LBA can be used for sensing. As shown in Fig. 7.16B–H, the ZGO: Mn phosphor nanoparticles are used for bioimaging in the analysis of clinical medicine molecules.

7.5.3 Contrast agent

Multimodal imaging has received much attention in biomedical applications since it supplies more accurate, complete, and decisive information for diagnosis. [79] Each imaging modality has its advantages and disadvantages in terms of sensitivity and temporal penetration depth. The NIR imaging shows a high sensitivity but instead has a limited resolution. On the contrary, magnetic resonance imaging (MRI) offers superior resolution and unlimited depth penetration yet remains at inherently low sensitivity. [80] The combination of NIR-emitting persistent phosphorescence nanoparticles and magnetic resonance imaging contrast agent into one single nanoplatform would offer attractive synergistic advantages in biomedical imaging with high sensitivity, fine spatial resolution, high signal-to-noise ratio (SNR), and free of ionizing radiation. Maldiney et al. developed a method to prepare persistent phosphorescence nanoparticle-based bimodal imaging agents. In their strategy, rather than using the surface of the persistent phosphorescence nanoparticles to link diethylenetriaminepentaacetic acid–Gd complexes, the authors incorporated various amounts of Gd^{3+} ions at the beginning of the synthesis. In doing so a negative contrast was obtained, with an r2 value close to 60 mM-1 s-1 (Fig. 7.17A). The developed nanomaterials were intravenously injected into active mice, enabling both optical imaging (Fig. 7.17B) and negative magnetic resonance imaging contrast of the liver (Fig. 7.17C and D). [81] In 2016 Zhang and coworkers fabricated $Zn_{0.97}Ga_2O_{3.97}$: Cr^{3+}, Bi^{3+}(ZGOCB) persistent phosphorescence nanoparticles with enhanced red emission

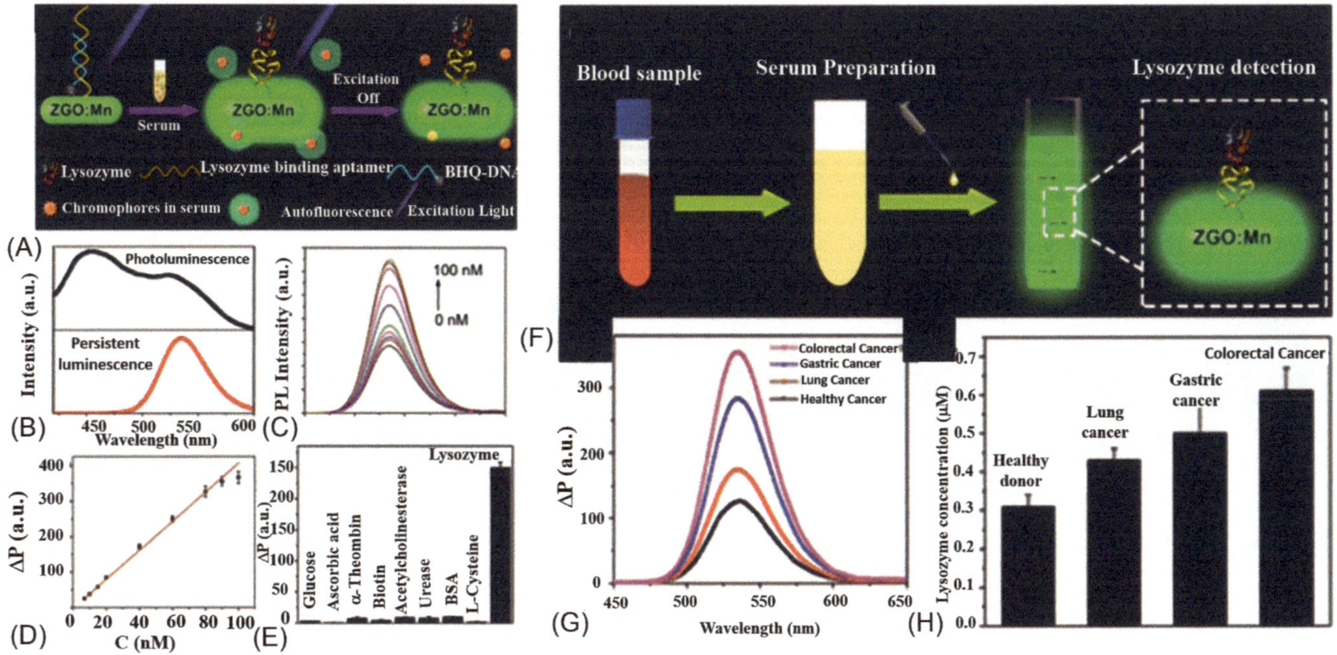

Fig. 7.16 (A) Schematic illustration of autofluorescence-free serum lysozyme biosensing. (B) Photoluminescence and persistent luminescence spectra of human serum containing ZGO: Mn-LBA. (C) Recovery of persistent luminescence in the presence of lysozyme with different concentrations. (D) A plot of enhanced persistent luminescence intensity (ΔP) versus lysozyme concentration. (E) Responses of the biosensing probes to different kinds of biomolecules. (F) Schematic illustration of the preparation of serum samples from cancer patients. (G) The response of the biosensing probes to different kinds of serum samples. (H) Concentrations of lysozyme in different kinds of serum determined with the biosensing probes.

Reproduced with permission from J. Wang, Q. Ma, W. Zheng, H. Liu, C. Yin, F. Wang, X. Chen, Q. Yuan, W. Tan, One-dimensional luminous nanorods featuring tunable persistent luminescence for autofluorescence-free biosensing, ACS Nano 11 (2017) 8185–8191, Copyright 2017, American Chemical Society.

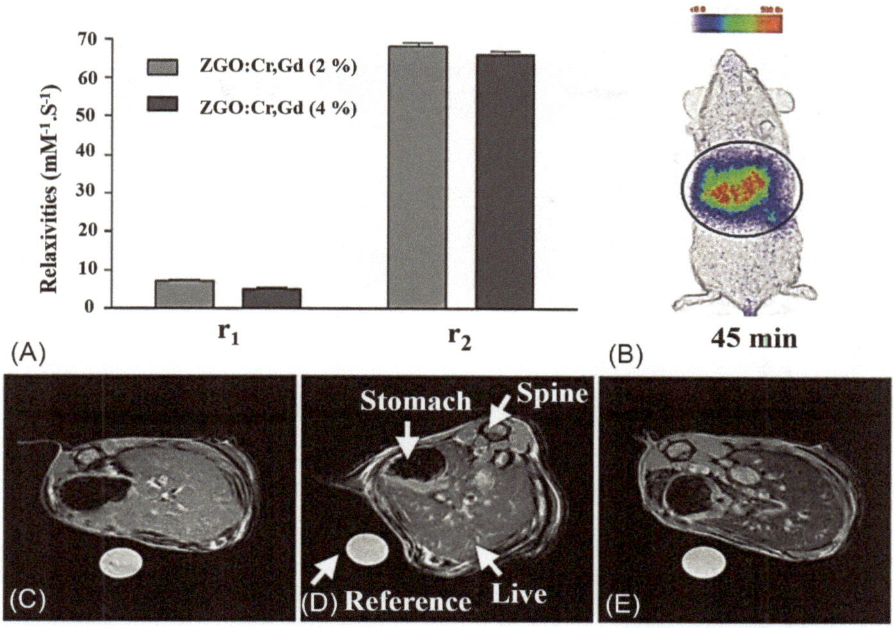

Fig. 7.17 ZGO: Cr^{3+}, Gd^{3+} persistent phosphorescence nanoparticles as optical and negative magnetic resonance imaging contrast agent. (A) Relaxivities r1 and r2 of ZGO nanoparticles suspended in NaOH at 7T. (B) Persistent luminescence images recorded 45 min after the injection of ultraviolet-excited ZGO:Cr,Gd (4%) dispersed in 5% glucose solution. *Black circle* corresponds to a global region of interest comprising liver and spleen. (C–E) Magnetic resonance imaging of mice liver recorded at 7T. (C) Reference mouse before the injection of ZGO nanoparticles. (D) Reference mouse, 1 h after the injection of ZGO:Cr,Gd (2%) nanoparticles. (E) Reference mouse, 1 h after the injection of ZGO:Cr,Gd (4%) nanoparticles). Reproduced with permission from T. Maldiney, B.T. Doan, D. Alloyeau, M. Bessodes, D. Scherman, C. Richard, Gadolinium-doped persistent nanophosphors as versatile tool for multimodal in vivo imaging, Adv. Funct. Mater. 25 (2015), 331–338, Copyright 2015, Wiley Online Library.

by a biphasic synthesis route. [82] Simultaneously, mesoporous SiO_2 was also utilized as a drug carrier and as a morphology controlling template in the synthesis of the core/shell structured nanoprobe Gd_2O_3@mSiO_2/ZGOCB. Due to the presence of Gd_2O_3, Gd_2O_3@mSiO_2/ZGOCB also possesses magnetic properties, which allows for the use as a magnetic resonance imaging agent (longitudinal contrast agent). Furthermore, through suitable surface functionalization with folic acid (FA) as a tumor-targeting group, Gd_2O_3@mSiO_2/ZGOCB shows a high signal-to-noise ratio when used for in vitro and in vivo imaging. Employing doxorubicin (Dox) as a drug model, the authors further demonstrated that this nanoprobe could be used to load an anticancer agent and consequently initiate its progressive release in situ to cure tumors (Fig. 7.18).

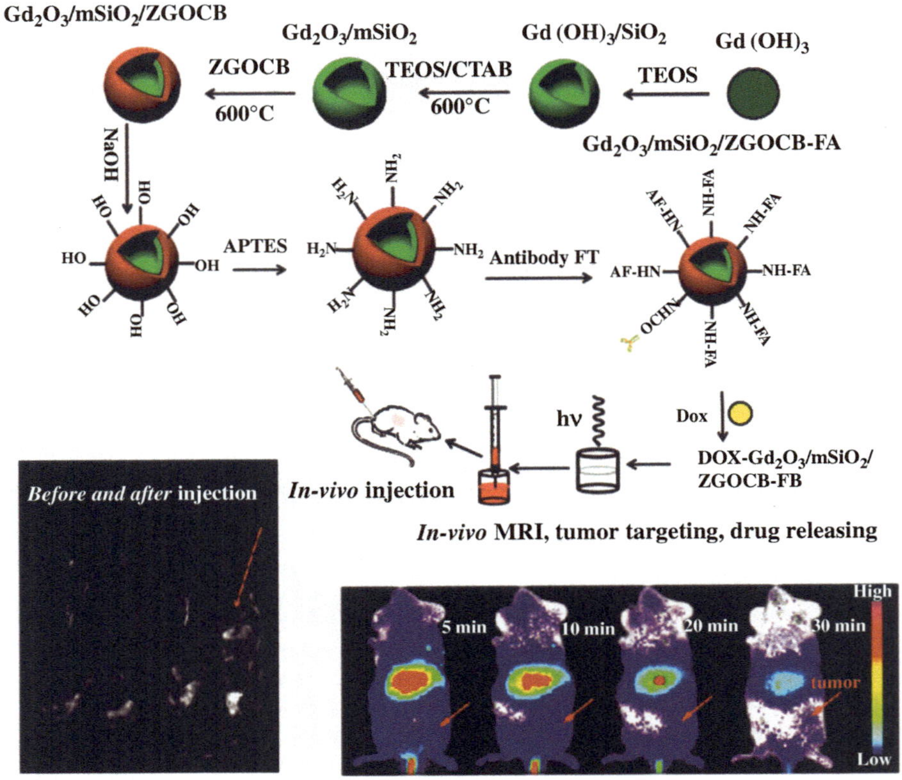

Fig. 7.18 (A) Schematic illustration of the synthesis, surface functionalization, and application of the core/shell structure nanoprobe Gd_2O_3@$mSiO_2$/ZGOCB. (B) T1-weighted MR images of a mouse before (left) and after (right) intravenous injection of Gd_2O_3@$mSiO_2$/ZGOCB-NH_2 (0.15 mg). The *red arrow* points to the liver. (C) In vivo luminescence images of H22 tumor-bearing mice (a) and normal mice (b) after intravenous injection of Dox-Gd_2O_3@$mSiO_2$/ZGOCB-FA. (D) Ex vivo image of liver, spleen, kidneys, heart, lung, and tumor corresponding biodistribution for each organ 6 h after systemic injection.
Reproduced with permission from W. Dai, Y. Lei, S. Ye, E. Song, Z. Chen, Q. Zhang, Mesoporous nanoparticles Gd_2O_3@ $mSiO_2/ZnGa_2O_4$: Cr^{3+}, Bi^{3+} as multifunctional probes for bioimaging, J. Mater. Chem. B 4 (2016) 1842–1852, Copyright 2016, Royal Society of Chemistry.

Yan and coworkers reported the fabrication of bimodal imaging agents by surface functionalization with diethylenetriaminepentaacetic acid (DTPA)-Gd complexes, which can be detected optically (Fig. 7.19, left) and used as a positive T1 magnetic resonance imaging contrast agent (Fig. 7.19, right). [83]

Yan and coworkers reported the fabrication of core-shell structured $Zn_{2.94}Ga_{1.96}Ge_2O_{10}$: Cr^{3+}, Pr^{3+}@TaO_x@SiO_2 nanoparticles for optical and x-ray computed tomography (CT) imaging. [84] Persistent phosphorescence nanoparticles based on $Zn_{2.94}Ga_{1.96}Ge_2O_{10}$: Cr^{3+}, Pr^{3+} were used as the core to provide near-infrared

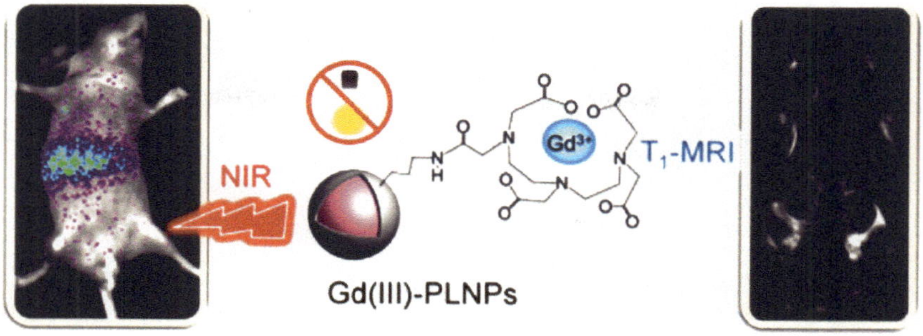

Fig. 7.19 Bimodal imaging agent based on $Zn_3Ga_2GeO_8$ nanoparticles coated with diethylenetriaminepentaacetic acid–Gd complexes for in vivo optical (left) and magnetic resonance imaging (right).
Reproduced with permission from A. Abdukayum, C.-X. Yang, Q. Zhao, J.-T. Chen, L.-X. Dong, X.-P. Yan, Gadolinium complexes functionalized persistent luminescent nanoparticles as a multimodal probe for near-infrared luminescence and magnetic resonance imaging in vivo, Anal. Chem. 86 (2014) 4096–4101, Copyright 2014, American Chemical Society.

luminescence, and a TaOx layer was grown onto the core to serve as the contrast agent for CT (Fig. 7.20A). Tantalum oxide (TaOx) nanoparticles have been demonstrated to be promising candidates as CT contrast agents. [85] However, CT has inherently low sensitivity due to its poor soft-tissue contrast. Thus it is desirable to introduce a complementary imaging method for CT to gain an accurate diagnosis.

7.5.4 Photodynamic therapy

Persistent phosphors can be also used for photodynamic therapy because it can serve as drug carriers [86, 87] and as optical markers for in vivo tracking. [88, 89] Shi and coworkers reported the synthesis of $ZnGa_2O_4$: Cr^{3+}@HMS nanospheres by emulsifier-free emulsion polymerization and their application in ibuprofen delivery and in vivo tracking with high sensitivity. [87] Chen et al. developed a reusable mesoporous nanocomposite by combining $SiO_2/ZnGa_2O_4$: Cr^{3+} and indocyanine green for photothermal therapy and in vivo imaging (Fig. 7.21). [89] Therefore the nanoparticles could be tracked in vivo by photoluminescence imaging, and the nanoparticles could also be excited by ultraviolet light for many times. Subsequently the ICG@mZGC nanoparticles could arrive at tumors tissue by enhanced permeation retention effect (EPR effect) and enter into tumors cells finally. In tumors cells, ICG could be released from the ICG@mZGC nanoparticles. In step 3 the site of the tumor was irradiated by an 808-nm NIR laser, and then ICG could generate heat to kill tumor cells.

7.5.5 Lateral flow assay

The lateral flow assay (LFA) is a paper-based platform for the disclosure and quantifying analytes in complex mixtures in which the sample is rested on a test device and the results are revealed within 5–30 min. In recently years, lower growth costs and

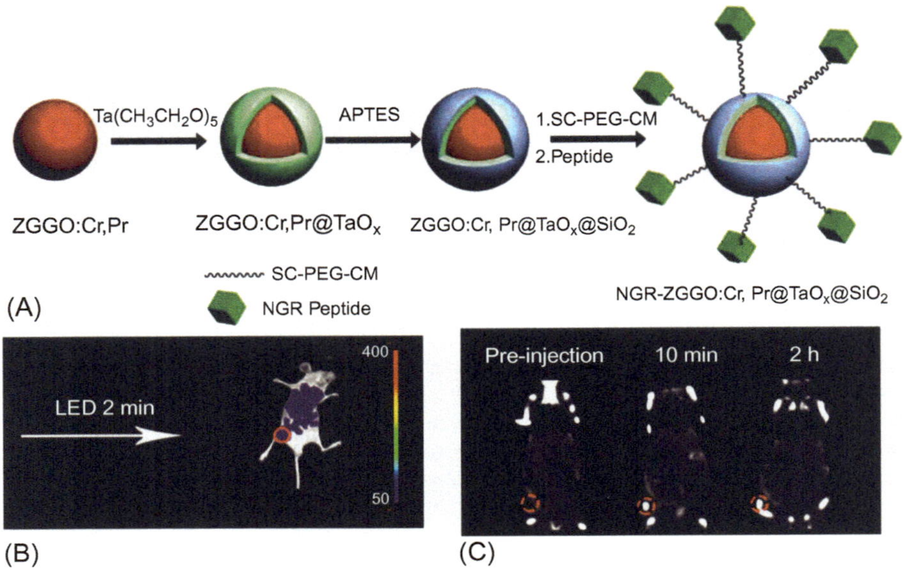

Fig. 7.20 $Zn_3Ga_2GeO_8$: Cr, Pr@TaOx@SiO_2 nanoparticles for optical imaging and CT. (A) Schematic protocol for the preparation of core-shell-structured $Zn_3Ga_2GeO_8$: Cr, Pr@TaOx@SiO_2. (B) In vivo imaging of HepG2 tumor-bearing nude mice after tail vein injection of poly(ethylene glycol)-$Zn_3Ga_2GeO_8$: Cr, Pr@TaOx@SiO_2 nanoparticles. (C) In vivo CT imaging of HepG2 tumor-bearing mice with poly(ethylene glycol)–$Zn_3Ga_2GeO_8$: Cr, Pr@TaOx@SiO_2 nanoparticles via orthotopic injection: preinjection, 10 min, and 2 h postinjection.
Reproduced with permission from Y.-C. Lu, C.-X. Yang, X.-P. Yan, Radiopaque tantalum oxide coated persistent luminescent nanoparticles as multimodal probes for in vivo near-infrared luminescence and computed tomography bioimaging, Nanoscale 7 (2015) 17929–17937, Copyright 2015, Royal Society of Chemistry.

simpler fabrication processes for lateral flow assays have been realized, which will enable the scaling up of its approaches to multiple fields in which swift analyses are required. Lateral flow assay-based tests are widely used in hospitals, physician's offices, and clinical laboratories for the qualitative and quantitative diagnosis of distinct antigens [90] and antibodies [91], as well as products of gene amplification [92, 93]. Photoluminescent labels, such as conventional gold nanoparticles, fluorescent nanoparticles, and quantum dots, are typically used in lateral flow assays. [94, 95] The principle behind the lateral flow assay is simple: a liquid sample (or its extract) containing the analyte of interest moves without the assistance of external forces (capillary action) through various zones of polymeric strips, on which molecules that interact with the analyte are attached. A typical lateral flow test strip (Fig. 7.22A) consists of overlapping membranes that are mounted on a backing card for better stability and handling. The operation of a lateral flow immunoassay is presented in Fig. 7.22B.

Paterson et al. demonstrated lateral flow assays based on strontium aluminate ($SrAl_2O_4$: Eu^{2+}, Dy^{3+}) persistent phosphorescence nanoparticles. [96] These particles

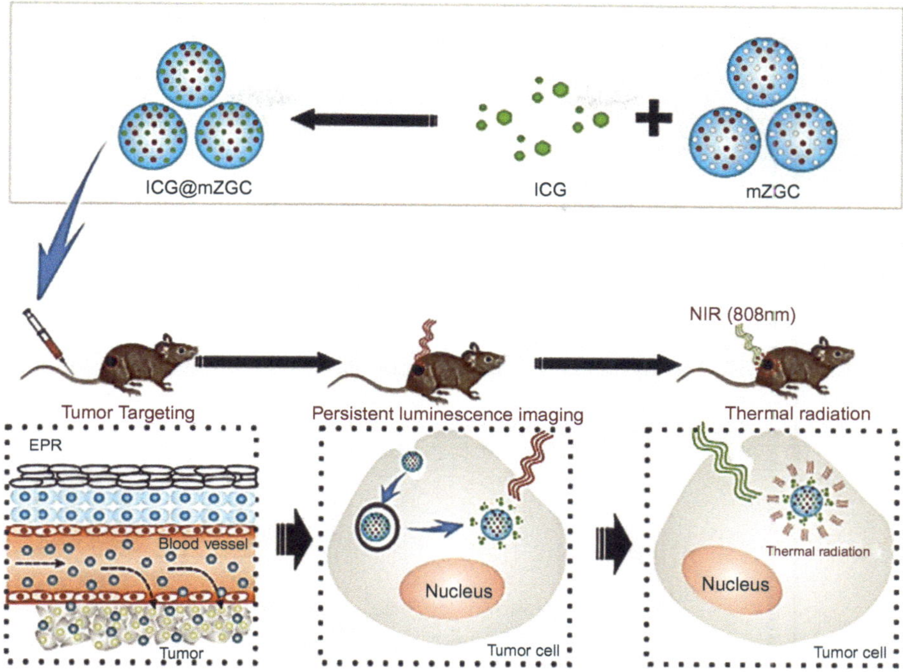

Fig. 7.21 Schematic illustration of the applications of the persistent luminescent nanoparticles for imaging-guided PTT in vivo. In this schematic illustration the fabrication of the ICG and persistent phosphors coloaded mesoporous silica nanoparticle consisted of two steps: (1) Step 1, synthesis of mesoporous mZGC and the loading of ICG to form ICG@mZGC nanoparticles, and Step 2, the ICG@mZGC nanoparticles were firstly excited by an ultraviolet lamp for 5 min, and then intravenous injection in healthy mice was performed.
Reproduced with permission from H. Chen, B. Zheng, C. Liang, L. Zhao, Y. Zhang, H. Pan, W. Ji, X. Gong, H. Wang, J. Chang, Near-infrared persistent luminescence phosphors $ZnGa_2O_4$:Cr^{3+} as an accurately tracker to photothermal therapy in vivo for visual treatment, Mater. Sci. Eng.: C 79 (2017) 372–381, Copyright 2017, Elsevier.

are highly photostable, and the silica shell preserves the luminescence properties by preventing hydrolytic degradation. They demonstrated that the particles can be utilized in lateral flow assays for rapid identification benefited by the high sensitivity of the persistent phosphors without the need for continuous excitation. In their experiment, NeutrAvidin persistent phosphorescence nanoparticles were applied in lateral flow assays (LFAs) with biotinylated lysozyme as a model analyte in the buffer and monoclonal antilysozyme HyHEL-5 antibodies at the trial bar. Bovine serum albumin (BSA) was biotinylated and adsorbed on the nitrocellulose to aid as a guideline to capture NeutrAvidin persistent phosphorescence nanoparticles that streamed past the test line as shown in Fig. 7.23A. By comparing with lateral flow assays run with NeutrAvidin persistent phosphorescence nanoparticles with streptavidin gold nanoparticles, it was found that the perception limit for lateral flow assays run with

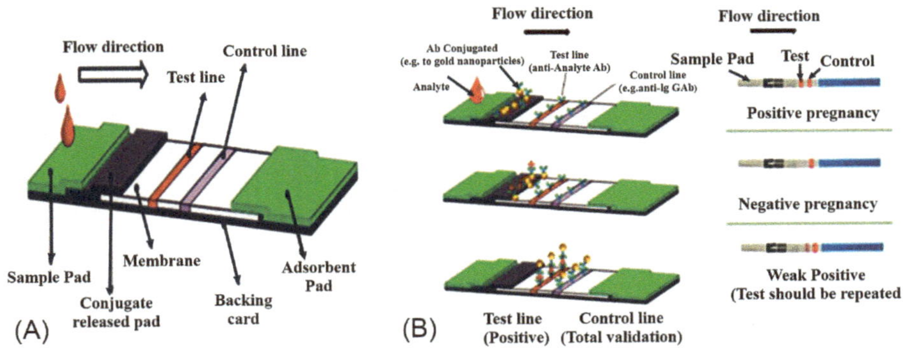

Fig. 7.22 (A) Typical configuration of a lateral flow immunoassay test strip and (B) operation of lateral flow immunoassays.
Reproduced with permission from K.M. Koczula, A. Gallotta, Lateral flow assays, Essays Biochem. 60 (2016) 111–120, Copyright 2016, Portland Press.

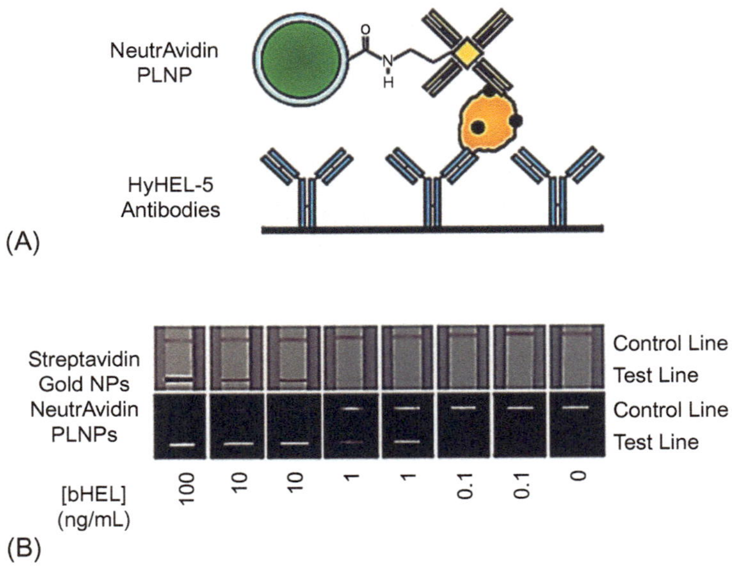

Fig. 7.23 (A) Schematic of test line showing NeutrAvidin phosphor bound to bHEL captured by anti-HEL antibodies. (B) Lateral flow assay strips showing biotinylated hen egg lysozyme serial dilutions with duplicates and detection with streptavidin gold nanoparticles (top) and NeutrAvidin persistent phosphorescence nanoparticles (bottom).
Reproduced with permission from A.S. Paterson, B. Raja, G. Garvey, A. Kolhatkar, A. E. Hagström, K. Kourentzi, T.R. Lee, R.C. Willson, Persistent luminescence strontium aluminate nanoparticles as reporters in lateral flow assays, Anal. Chem. 86 (2014) 9481–9488, Copyright 2014, American Chemical Society.

persistent phosphorescence nanoparticles is more sensitive than the streptavidin gold nanoparticles (Fig. 7.23B).

7.5.6 Sterilization

As a biological reagent the sterilization ability of persistent phosphors was very important, which has been investigated by several researchers recently. [97, 98] Sun et al. developed a Cs_2NaYF_6: Pr^{3+} phosphor with ultraviolet C afterglow emission (which was referred to as an ultraviolet C persistent phosphors) by judiciously selecting defect-bearing fluoride elpasolite (i.e., Cs_2NaYF_6) as a host and Pr^{3+} ions as emitters. This phosphor shows bright ultraviolet C emissions that can last for over 2 h after x-ray irradiation. The authors showed that the ultraviolet C persistent luminescence of this phosphor is strong enough to be used for sterilization. When used for sterilization of a common gram-negative and monoflagellated bacterium *Pseudomonas aeruginosa* PAO1, sterilization rate as high as 60.4% within 16 min was obtained, [98] as sown in Fig. 7.24. Importantly the phosphors can be charged repeatedly by x-ray or post stimulated by NIR light in the first biological window. The ultraviolet C persistent phosphors are also promising for different applications beyond sterilization, such as in vivo killing of pathogens and cancer cells.

7.6 Solar cells

Solar cells convert solar radiation directly into electric energy based on semiconducting materials. They have been widely used in diverse areas, such as aerospace, photovoltaic plants, and intelligent transportation system. By light illumination, electron-hole pairs are generated, which are then separated under bias to generate an electric voltage. The energy conversion efficiency has been improved steadily by optimizing the fabrication techniques and materials. Since all solar cells can only produce electricity under daylight illumination, the combination with a persistent phosphorescent layer may allow the generation of electricity without light illumination in the dark.

Table 7.3 lists the photon to the electricity conversion efficiency of different persistent phosphors in recent years. By combining the persistent phosphors and dye-sensitized TiO_2 cell, it has been shown that the power conversion efficiency of the solar cell is enhanced not only under daylight illumination but also in the dark. Chakrapani et al. integrated inorganic phosphors with a dye-sensitized solar cell and demonstrated improvement in light utilization and photoelectric conversion efficiency. From their experimental results the power conversion efficiency of composite photovoltaic cells reached 3.29% with a 54% increase. Besides the as-fabricated solar cell severs as a “night solar cell” with a 52-μW cm^{-2} power density at night. [100] As shown in Fig. 7.25, Tang and his coworkers reported a novel strategy by coating persistent phosphors onto TiO_2 to construct an all-weather solar cell. [105]

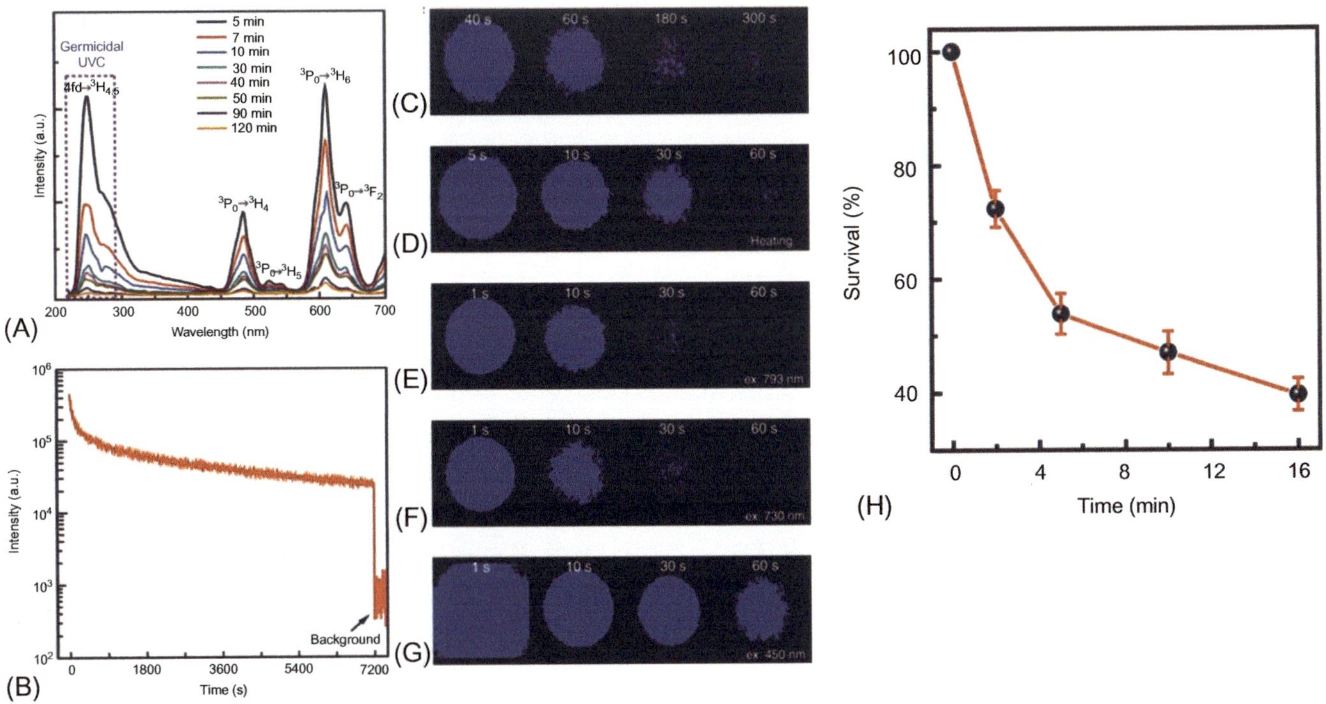

Fig. 7.24 (A) Afterglow spectra recorded at different times after ceasing x-ray irradiation. (B) Afterglow decay curve detected at 250 nm as a function of time. The data were taken 5 min after stopping the x-ray irradiation. (C) Ultraviolet C images of phosphors taken at different afterglow times. (D) Ultraviolet C images of the 24 h-decayed phosphors heated to 200°C on a hot plate. (E–G) Ultraviolet C images of the 24 h-decayed phosphors under laser irradiation with different wavelengths of (E) 793 nm, (F) 730 nm, and (G) 450 nm. The excitation power density is 1.77 W/cm^2 for the 793-, 730-, and 450-nm excitation wavelengths. (H) The dependence of *P. aeruginosa* PAO1 survival ratios on the x-ray irradiation time of the given phosphor sheets. One hundred percent viability can be maintained when keeping *P. aeruginosa* PAO1 under ambient conditions. The error bars represent the standard error and are obtained based on three independent experiments.

Reproduced with permission from Y.-M. Yang, Z.-Y. Li, J.-Y. Zhang, Y. Lu, S.-Q. Guo, Q. Zhao, X. Wang, Z.-J. Yong, H. Li, J.-P. Ma, X-ray-activated long persistent phosphors featuring strong UVC afterglow emissions. Light: Sci. Appl. 7 (2018) 88, Copyright 2018, Nature Publishing Group.

Table 7.3 The power conversion efficiencies of solar cells incorporating different persistent phosphors.

Phosphor	Original solar cells	Power conversion efficiency (before)	Power conversion efficiency (after)	Increase in efficiency	Power conversion efficiency (in the dark)	Ref.
$SrAl_2O_4$:Eu,Dy	CdSe/TiO_2	0.82%	1.22%	48%	0.04%	[99]
$SrAl_2O_4$:Eu,Dy	N719/TiO_2	2.27%	3.49%	54%		[100]
$SrAl_2O_4:Eu^{2+}, Dy^{3+}$	FTO/TiO_2	5.04%	6.85%	36%	–	[101]
$SrAl_2O_4:Eu^{2+}, Dy^{3+}$	N719/TiO_2	6.620%	7.938%	19.90%	0.0076%	[102]
$SrAl_2O_4:Eu^{2+}, Dy^{3+}$	SiO_2	10.91%	11.12%	1.92%	1.16%	[103]
$Ca_3Ga_2Ge_3O_{12}: Pr^{3+}, Yb^{3+}$	Dye-sensitized TiO_2	5.902%	8.658%	47%	–	[104]
Persistent phosphorescence	N719/m-TiO_2	8.08%	10.08%	25%	26.69%	[105]
$SrAl_2O_4:Eu^{2+}, Dy^{3+}$	FTO/TiO_2	16.7%	17.8%	6%	–	[106]
$CaAl_2O_4:Eu^{2+}, Nd^{3+}$	FTO/TiO_2	5.96%	6.61%	11%	–	[107]
$Sr_4Al_{14}O_{25}$: Eu,Dy	CdS/TiO_2	1.02%	1.40%	38%	0.09%	[108]

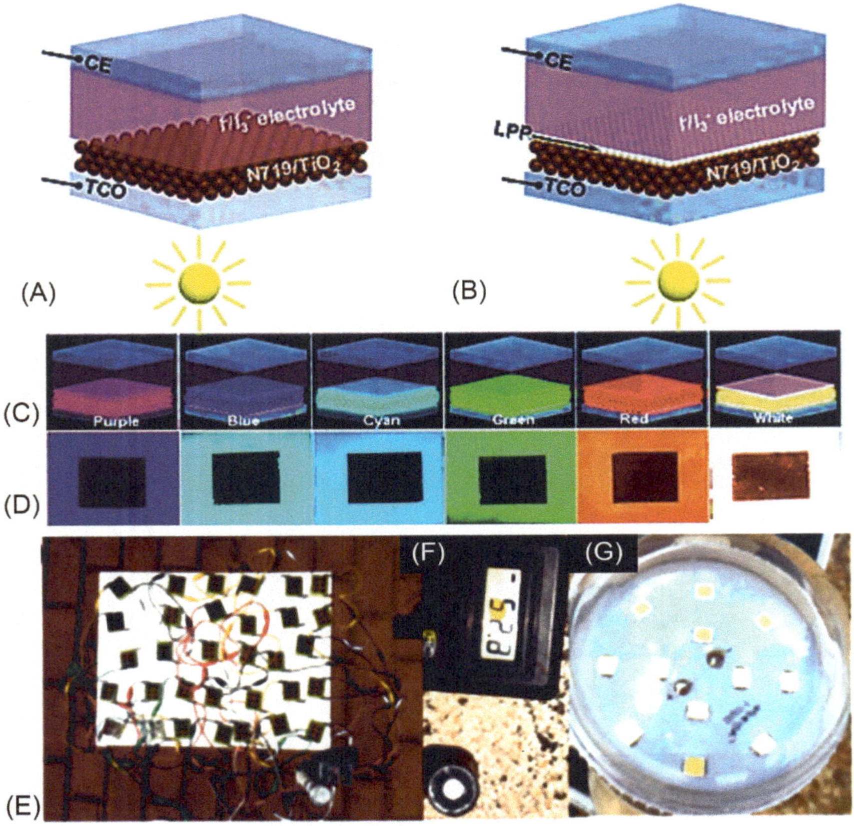

Fig. 7.25 (A) Sandwiched structure of state-of-the-art DSSC covered with a layer of persistent phosphor particles. (B) All-weather solar cell by incorporating a persistent phosphorescent layer with the TiO_2 photoanode. (C) Schematic diagrams of the design for the all-weather solar cells at nights. (D) The optical images of six m-TiO_2/persistent phosphor photoanodes taken at persistent luminescence time of 5 s after irradiating by simulated sunlight (AM1.5, 100 mW cm^{-2}) for 1 min. The dark regions are dye-sensitized m-TiO_2 cells cover by a layer of persistent phosphor, while the fluorescent regions are only covered by persistent phosphorescent layers. (E) The solar modules fabricated by 32 all-weather solar cells. (F) Recorded light intensity at outdoors. (G) The LED driven by solar modules under sun illumination at outdoors.

Reproduced with permission from Q. Tang, J. Wang, B. He, P. Yang, Can dye-sensitized solar cells generate electricity in the dark? Nano Energy 33 (2017) 266–271, Copyright 2017, Elsevier.

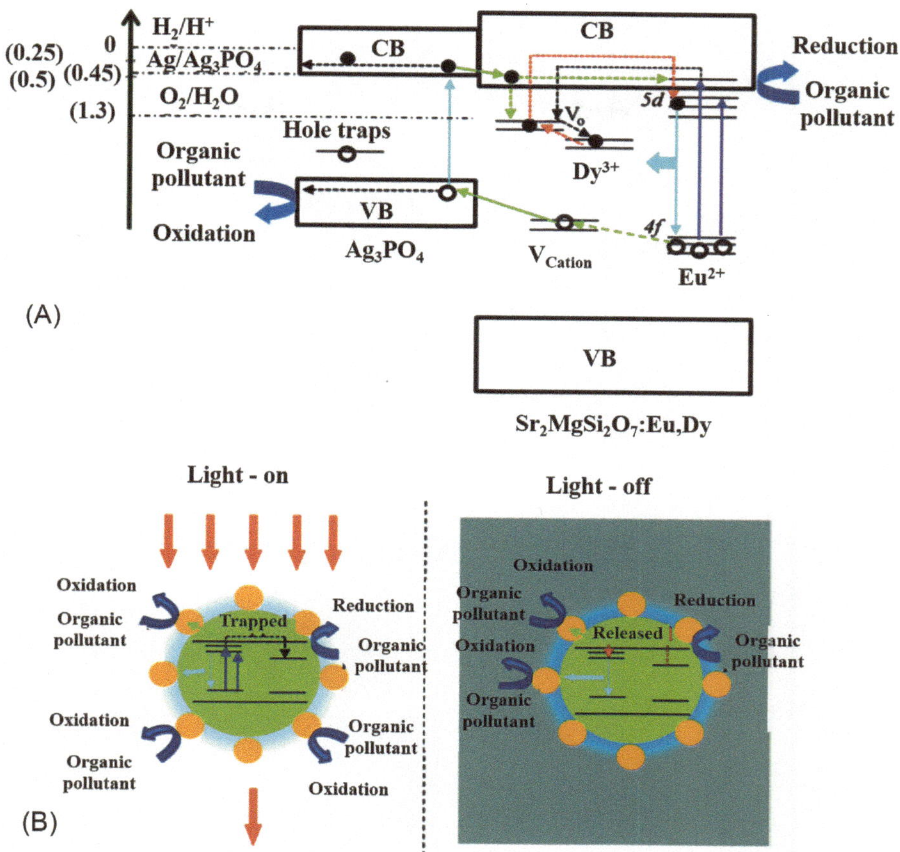

Fig. 7.26 Schematic illustrations of the (A) charge transfer mechanism and (B) photocatalytic processes under light illumination and after light-off.
Reproduced with permission from H. Wu, Z.-M. Wang, K. Koike, N. Negishi, Y. Jin, Hybridization of silver orthophosphate with a melilite-type phosphor for enhanced energy-harvesting photocatalysis, Catal. Sci. Technol. 7 (2017) 3736–3746, Copyright 2017, Royal Society of Chemistry.

7.7 Photocatalysis

Nowadays, there has been a growing concern about global environmental pollution, which has attracted extensive attention from the science community. Water pollution is one of the most serious problems because clean water is indispensable for human existence and development. Photocatalysis is an environmentally friendly technology for water purification, and particularly, solar energy is inexhaustible clean energy in the outdoor environment. However, unpredictable weather condition has significant impacts on the solar energy system, which limits its application as a constantly dependable energy source. This problem might be circumvented through the use of

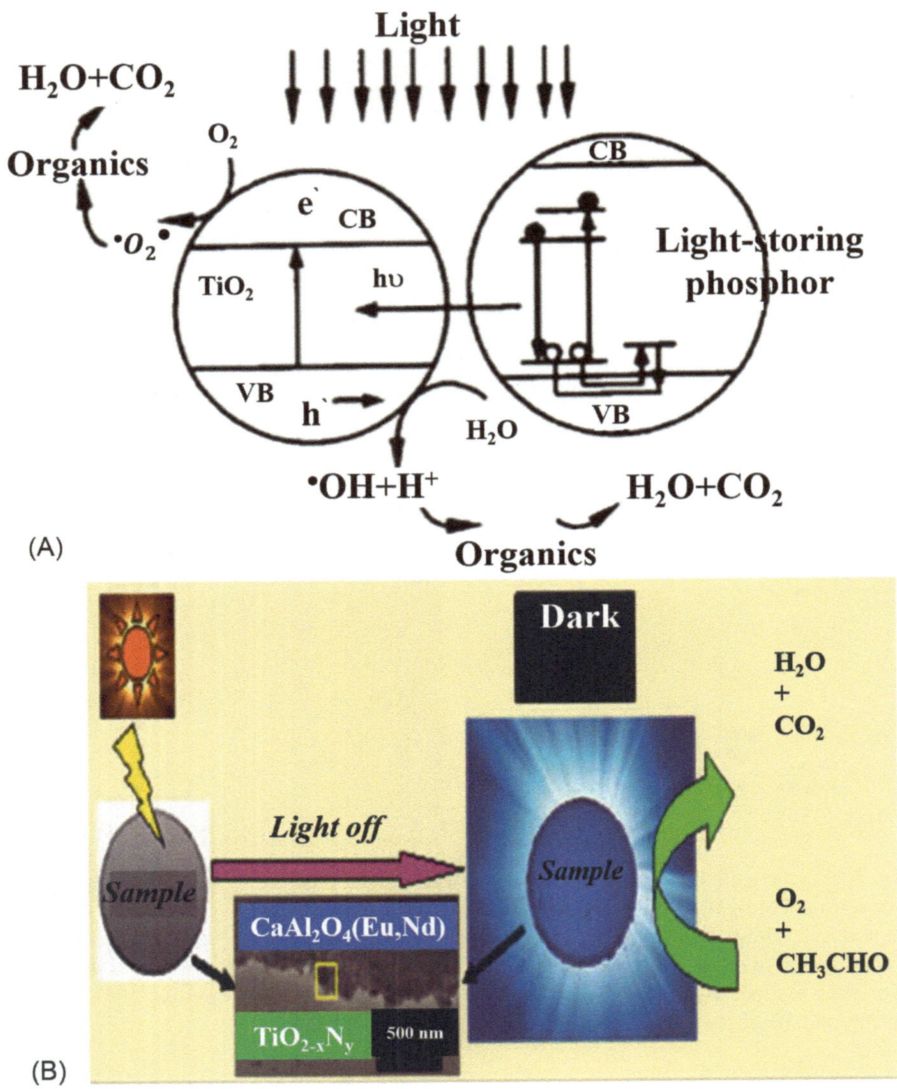

Fig. 7.27 (A) Mechanism of the light-storing process in photocatalysis, (B) persistent luminescence-assisted $TiO_{2-x}N_y$-based photocatalyst for gaseous acetaldehyde degradation. A: Reprinted from J. Zhang, F. Pan, W. Hao, Q. Ge, T. Wang, Light-storing photocatalyst. Appl. Phys. Lett. 85 (2004) 5778–5780, with permission of AIP Publishing LLC; B: Reproduced with permission from H. Li, S. Yin, Y. Wang, T. Sato, Persistent fluorescence-assisted TiO_{2-x} N_y-based photocatalyst for gaseous acetaldehyde degradation, Environ. Sci. Technol. 46 (2012), 7741–7745, Copyright 2012, American Chemical Society.

persistent phosphor as a light storage medium, which may enable photocatalysis to be performed without irradiation by external light sources. The use of persistent phosphors for photocatalysis may also offer the following advantages:

(1) The energy match between the emission band of persistent phosphors and the absorption band of a photocatalyst may allow for the photocatalysis under illumination by afterglow emission of the persistent phosphors in darkness.
(2) Many persistent phosphors with different spectral properties are available, such that photons with different energy can be stored and finally being used to drive photocatalysis.
(3) Persistent phosphors can enhance interfacial charge transfer efficiency that might be beneficial for enhancing photocatalytic activity.

An example of charge transfer reaction and photocatalytic processes are illustrated in Fig. 7.26. In the Dy^{3+}-doped $Sr_2MgSi_2O_7$: Eu^{2+} persistent phosphor, the presence of cation vacancies facilitates charge transfer process in the composite sample. This

Table 7.4 Photocatalytic degradation of chemicals assisted by different persistent phosphors developed in recent years.

Phosphor	Photocatalyst	Dye molecules	Decomposition rate	Ref.
$ZnGa_2O_4$		Rhodamine B	90% for 90 min	[116]
$Zn_{0.8}Ga_2O_{3.8}$		Rhodamine B	98.5% for 90 min	[117]
$SrAl_4O_2$:Eu^{2+},Dy^{3+},Bi	TiO_2	Methylene blue	100% for 210 min	[118]
$ZnGa_2O_4$:Cr^{3+}		Rhodamine B	60% for 110 min	[119]
Ga_2O_3:Cr^{3+}/In^{3+}		Rhodamine B	90% for 70 min	[120]
$Sr_2MgSi_2O_7$: (Eu, Dy)	g-C_3N_4	Rhodamine B	58.3% for 60 min	[121]
$Ca_2Al_2SiO_7$:Ce^{3+}	Ag/AgCl	Methyl orange	79% for 35 min	[122]
Zn_2GeO_4: Mn^{2+}, Eu^{3+}		Rhodamine B	73% for 90 min	[123]
$CdSiO_3$:Gd^{3+}, Bi^{3+}	TiO_2	Methylene blue	96.6% for 30 min	[124]
$Sr_2MgSi_2O_7$:Eu^{2+}, Dy^{3+}	Ag_3PO_4	Methyl orange	91% for 35 min	[109]
$MgGa_2O_4$		Rhodamine B	86% for 60 min	[125]
$Sr_4Al_{14}O_{25}$: Eu, Dy	ZnO:Co/Ag	Methyl orange	91.4% for 210 min	[126]
$MgAl_2O_4$:Pr^{3+}, Dy^{3+}	Cr–TiO_2	Methyl orange	70% for 120 min	[127]
$Sr_2MgSi_2O_7$:Eu^{2+}, Dy^{3+}		Methanol	5.81%	[128]
$ZnGa_2O_4$: Cr/Bi		Rhodamine B	99.2% for 100 min	[129]
$Sr_{2.90}Eu_{0.03}Dy_{0.07}Al_4SiO_{11}$	N-TiO_2	Crystal violet (CV) dye	90% for 30 min	[130]

composite with a 3-D trap-hole structure is beneficial for charge separation, which could increase the efficiency of visible-light photocatalysis greatly. [109] TiO_2 is a prototype photocatalytic semiconductor that has been widely used for water splitting under solar light illumination. A composite consists of a persistent phosphor, and TiO_2 has been considered for photocatalysis in the dark. For instance, recent studies have confirmed that the persistent phosphor-TiO_2 photocatalyst can notably enhance the photocatalysis properties of TiO_2 and realize the degradation of organic pollutes, formaldehyde, ammonia, and other chemicals in the dark [110–115]. The mechanisms of photocatalytic degradation of organic chemicals assisted by persistent phosphors are given in Fig. 7.27A and B.

Ko et al. proposed the use of persistent phosphors as the light source to drive photocatalytic degradation of ammonia in the dark by TiO_2 [115]. Zhang et al. [113] proposed a concept of energy storage in photocatalytic materials made up of a persistent phosphor and a semiconductor. The photocatalysis process by a semiconductor like TiO_2 generally involves charge carrier generation and separation, which is similar to the generation of trapped carriers that occurs after excitation of persistent phosphors. The process of photocatalysis by a persistent phosphor-modified TiO_2 photocatalyst consists of three steps: (1) optical charging of the composite by illumination with ultraviolet or visible light, (2) trapping of carries by the persistent phosphors, and (3) afterglow emission of persistent phosphor that activate the photocatalysis of TiO_2 in the dark (Table 7.4).

References

[1] E.N. Harvey, A History of Luminescence From the Earliest Times Until 1900, vol. 44, American Philosophical Society, 1957.

[2] R. Xie, N. Hirosaki, M. Mitomo, W. Yen, S. Shionoya, H. Yamamoto, Phosphor Handbook, CRC Press, Boca Raton, 2007.

[3] W. Hoogenstraaten, H. Klasens, Some properties of zinc sulfide activated with copper and cobalt, J. Electrochem. Soc. 100 (1953) 366.

[4] T. Matsuzawa, Y. Aoki, N. Takeuchi, Y. Murayama, A new long phosphorescent phosphor with high brightness, $SrAl_2O_4$: Eu^{2+}, Dy^{3+}, J. Electrochem. Soc. 143 (1996) 2670–2673.

[5] J. Hölsä, Persistent luminescence beats the afterglow: 400 years of persistent luminescence, Electrochem. Soc. Interface 18 (2009) 42–45.

[6] Q.L.M. de Chermont, C. Chanéac, J. Seguin, F. Pellé, S. Maîtrejean, J.-P. Jolivet, D. Gourier, M. Bessodes, D. Scherman, Nanoprobes with near-infrared persistent luminescence for in vivo imaging, Proc. Natl. Acad. Sci. 104 (2007) 9266–9271.

[7] Y. Zhuang, J. Ueda, S. Tanabe, Multi-color persistent luminescence in transparent glass ceramics containing spinel nano-crystals with Mn^{2+} ions, Appl. Phys. Lett. 105 (2014) 191904.

[8] H. Guo, Y. Wang, G. Li, P. Feng, D. Liu, Q. Ye, Activating persistent luminescence and thermal stability of $K_2Ba_7Si_{16}O_{40}$: Eu^{2+} via traps modulation, J. Alloys Compd. 801 (2019) 295–301.

[9] J. Ueda, S. Miyano, S. Tanabe, Formation of deep electron traps by Yb^{3+} codoping leads to super-long persistent luminescence in Ce^{3+}-doped yttrium aluminum gallium garnet phosphors, ACS Appl. Mater. Interfaces 10 (2018) 20652–20660.

[10] S. Yoon, J. Bierwagen, M. Trottmann, B. Walfort, N. Gartmann, A. Weidenkaff, H. Hagemann, S. Pokrant, The influence of boric acid on improved persistent luminescence and thermal oxidation resistance of $SrAl_2O_4$: Eu^{2+}, J. Lumin. 167 (2015) 126–131.
[11] X. Shi, R. Dou, T. Ma, W. Liu, X. Lu, K.J. Shea, Y. Song, L. Jiang, Bioinspired lotus-like self-illuminous coating, ACS Appl. Mater. Interfaces 7 (2015) 18424–18428.
[12] F. Liu, W. Yan, Y.-J. Chuang, Z. Zhen, J. Xie, Z. Pan, Photostimulated near-infrared persistent luminescence as a new optical read-out from Cr^{3+}-doped $LiGa_5O_8$, Sci. Rep. 3 (2013) 1554.
[13] Y. Zhuang, L. Wang, Y. Lv, T.L. Zhou, R.J. Xie, Optical data storage and multicolor emission readout on flexible films using deep-trap persistent luminescence materials, Adv. Funct. Mater. 28 (2018) 1705769.
[14] J. Botterman, P.F. Smet, Persistent phosphor $SrAl_2O_4$: Eu, Dy in outdoor conditions: saved by the trap distribution, Opt. Express 23 (2015) A868–A881.
[15] R. Ni, B. Qian, C. Liu, X. Liu, J. Qiu, Three-dimensional printing of hybrid organic/inorganic composites with long persistence luminescence, Opt. Mater. Express 8 (2018) 2823–2831.
[16] S. Tian, J. Wen, H. Fan, Y. Chen, J. Yan, P. Zhang, Sunlight-activated long persistent luminescent polyurethane incorporated with amino-functionalized $SrAl_2O_4$: Eu^{2+}, Dy^{3+} phosphor, Polym. Int. 65 (2016) 1238–1244.
[17] A. Kumar, G. Kedawat, P. Kumar, J. Dwivedi, B.K. Gupta, Sunlight-activated Eu^{2+}/Dy^{3+} doped $SrAl_2O_4$ water resistant phosphorescent layer for optical displays and defence applications, New J. Chem. 39 (2015) 3380–3387.
[18] D. Kulesza, E. Zych, Managing the properties of Lu_2O_3: Tb, Hf storage phosphor by means of fabrication conditions, J. Phys. Chem. C 117 (2013) 26921–26928.
[19] Y. Yan, C. Liu, X. Ding, Study on the equivalent luminance of luminous fibers in photopic vision and scotopic vision, Text. Res. J. 88 (2018) 1157–1163.
[20] M.C. Gather, A. Köhnen, K. Meerholz, White organic light-emitting diodes, Adv. Mater. 23 (2011) 233–248.
[21] Z. Pang, Y. Zhu, M. Ge, Study on the structural and luminous properties of two kinds self-luminous fibers, J. Text. Inst. 111 (3) (2019) 345–351.
[22] J. Yang, L. Xiaoyun, Z. Yanan, G. Mingqiao, Optical performance study of $Sr_2ZnSi_2O_7$: Eu^{2+}, Dy^{3+}, $SrAl_2O_4$: Eu^{2+}, Dy^{3+} and Y_2O_2S: Eu^{3+}, Mg^{2+}, Ti^{4+} ternary luminous fiber, J. Rare Earths 34 (2016) 1206–1212.
[23] M. Ge, X. Guo, Y. Yan, Preparation and study on the structure and properties of rare-earth luminescent fiber, Text. Res. J. 82 (2012) 677–684.
[24] Y.N. Xu, Study on the Performance of the Luminous Fiber, MS Thesis, Jiangnan University, Wuxi, China, 2008.
[25] J. Zhao, X. Guo, Y. Xu, M. Ge, Luminescence properties of rare-earth strontium aluminate luminescent fiber, J. Text. Res. 29 (2008) 1–5.
[26] X. Guo, M. Ge, The afterglow characteristics and trap level distribution of chromatic rare-earth luminous fiber, Text. Res. J. 83 (2013) 1263–1272.
[27] X. Guo, K. Zhang, H. Zhang, M. Ge, Working conditions on the afterglow characteristics of rare-earth luminous fibers, Fibers Polym. 19 (2018) 531–537.
[28] A. Mao, Z. Zhao, J. Wang, C. Yang, J. Ren, Y. Wang, Crystal structure and photoluminescence of $Gd_3Ga_2(Al_{3-y}Si_y)(O_{12-y}N_y)$: Ce^{3+} phosphors for AC-warm LEDS, Chem. Eng. J. 368 (2019) 924–932.
[29] Q. Cheng, Y. Wang, L. Su, H. Wang, G. Zhu, W. Yu, Wide-spectrum manipulation of triboelectrification-induced electroluminescence by long afterglow phosphors in elastomeric zinc sulfide composites, J. Mater. Chem. C 7 (2019) 4567–4572.

[30] H. Lin, B. Wang, Q. Huang, F. Huang, J. Xu, H. Chen, Z. Lin, J. Wang, T. Hu, Y. Wang, $Lu_2CaMg_2(Si_{1-x}Ge_x)_3O_{12}$:$Ce^{3+}$ solid-solution phosphors: bandgap engineering for blue-light activated afterglow applicable to AC-LED, J. Mater. Chem. C 4 (2016) 10329–10338.
[31] H. Lin, J. Xu, Q. Huang, B. Wang, H. Chen, Z. Lin, Y. Wang, Bandgap tailoring via Si doping in inverse-garnet $Mg_3Y_2Ge_3O_{12}$: Ce^{3+} persistent phosphor potentially applicable in AC-LED, ACS Appl. Mater. Interfaces 7 (2015) 21835–21843.
[32] H. Lin, B. Wang, J. Xu, R. Zhang, H. Chen, Y. Yu, Y. Wang, Phosphor-in-glass for high-powered remote-type white AC-LED, ACS Appl. Mater. Interfaces 6 (2014) 21264–21269.
[33] H. Liu, B. Feng, L. Luo, C. Han, P.A. Tanner, Near infrared photostimulated persistent luminescence and information storage of $SrAl_2O_4$: Eu^{2+}, Dy^{3+} phosphor, Opt. Mater. Express 6 (2016) 3375–3385.
[34] Y. Zhuang, Y. Lv, L. Wang, W. Chen, T.-L. Zhou, T. Takeda, N. Hirosaki, R.-J. Xie, Trap depth engineering of $SrSi_2O_2N_2$: Ln^{2+}, Ln^{3+} (Ln^{2+}= Yb, Eu; Ln^{3+}= Dy, Ho, Er) persistent luminescence materials for information storage applications, ACS Appl. Mater. Interfaces 10 (2018) 1854–1864.
[35] C. Wang, Y. Jin, Y. Lv, G. Ju, D. Liu, L. Chen, Z. Li, Y. Hu, Trap distribution tailoring guided design of super-long-persistent phosphor Ba_2SiO_4: Eu^{2+}, Ho^{3+} and photostimulable luminescence for optical information storage, J. Mater. Chem. C 6 (2018) 6058–6067.
[36] W. Li, Y. Zhuang, P. Zheng, T.-L. Zhou, J. Xu, J. Ueda, S. Tanabe, L. Wang, R.-J. Xie, Tailoring trap depth and emission wavelength in $Y_3Al_{5-x}Ga_xO_{12}$: Ce^{3+}, V^{3+} phosphor-in-glass films for optical information storage, ACS Appl. Mater. Interfaces 10 (2018) 27150–27159.
[37] Z. Liu, L. Zhao, W. Chen, X. Fan, X. Yang, S. Tian, X. Yu, J. Qiu, X. Xu, Multiple anti-counterfeiting realized in $NaBaScSi_2O_7$ with a single activator of Eu^{2+}, J. Mater. Chem. C 6 (2018) 11137–11143.
[38] Y. Zhan, Y. Jin, H. Wu, L. Yuan, G. Ju, Y. Lv, Y. Hu, Cr^{3+}-doped $Mg_4Ga_4Ge_3O_{16}$ near-infrared phosphor membrane for optical information storage and recording, J. Alloys Compd. 777 (2019) 991–1000.
[39] Z. Long, Y. Wen, J. Zhou, J. Qiu, H. Wu, X. Xu, X. Yu, D. Zhou, J. Yu, Q. Wang, No-interference reading for optical information storage and ultra-multiple anti-counterfeiting applications by designing targeted recombination in charge carrier trapping phosphors, Adv. Opt. Mater. 7 (2019) 1900006.
[40] S. Lin, H. Lin, Q. Huang, Y. Cheng, J. Xu, J. Wang, X. Xiang, C. Wang, L. Zhang, Y. Wang, A photostimulated $BaSi_2O_5$: Eu^{2+}, Nd^{3+} phosphor-in-glass for erasable-rewritable optical storage medium, Laser Photonics Rev. 13 (2019) 1900006.
[41] N. Viswanath, G.K. Grandhi, H.J. Kim, Y. Zhuang, R.-J. Xie, W.B. Im, A new persistent blue-emitting phosphor: tailoring the trap density for enhancing the persistent time, Appl. Mater. Today 18 (2020) 100518.
[42] L. Zhao, J. Mao, B. Jiang, X. Wei, Y. Chen, M. Yin, Temperature-dependent persistent luminescence of $SrAl_2O_4$: Eu^{2+}, Dy^{3+}, Tb^{3+}: a strategy of optical thermometry avoiding real-time excitation, Opt. Lett. 43 (2018) 3882–3884.
[43] L. Fan, Y. Li, X. Lin, J. Peng, G. Ju, S. Zhang, L. Chen, F. He, Y. Hu, Tailoring light emission properties and optoelectronic and optothermal responses from rare earth-doped bismuth oxide for multifunctional light shielding, temperature sensing, and photo-detection, RSC Adv. 7 (2017) 44908–44914.
[44] T. Takayama, T. Katsumata, S. Komuro, T. Morikawa, Growth and characteristics of a new long afterglow phosphorescent yttrium tantalate crystal, J. Cryst. Growth 275 (2005) e2013–e2017.

[45] T. Katsumata, Y. Kohno, H. Kubo, S. Komuro, T. Morikawa, Low temperature fluorescence thermometer application of long afterglow phosphorescent Sr $Al_{12}O_{19}$: Eu^{2+}, Dy^{3+} crystals, Rev. Sci. Instrum. 76 (2005), 084901.
[46] B. Lei, Y. Liu, G. Tang, Z. Ye, C. Shi, Spectra and long-lasting properties of Sm^{3+}-doped yttrium oxysulfide phosphor, Mater. Chem. Phys. 87 (2004) 227–232.
[47] P. Jha, B. Chandra, Survey of the literature on mechanoluminescence from 1605 to 2013, Luminescence 29 (2014) 977–993.
[48] Y. Jia, M. Yei, W. Jia, Stress-induced mechanoluminescence in $SrAl_2O_4$: Eu^{2+}, Dy^{3+}, Opt. Mater. 28 (2006) 974–979.
[49] M. Akiyama, K. Nishikubo, K. Nonaka, Intense visible light emission from stress-activated $SrMgAl_6O_{11}$: Eu, Appl. Phys. Lett. 83 (2003) 650–652.
[50] C. Li, C.-N. Xu, L. Zhang, H. Yamada, Y. Imai, Dynamic visualization of stress distribution on metal by mechanoluminescence images, J. Vis. 11 (2008) 329–335.
[51] W.X. Wang, Y. Imai, C. Xu, T. Matsubara, Y. Takao, A new smart damage sensor using mechanoluminescence material, in: Materials Science Forum, 2011, pp. 1081–1084.
[52] J.C. Zhang, C. Pan, Y.F. Zhu, L.Z. Zhao, H.W. He, X. Liu, J. Qiu, Achieving thermo-mechano-opto-responsive bitemporal colorful luminescence via multiplexing of dual lanthanides in piezoelectric particles and its multidimensional anticounterfeiting, Adv. Mater. 30 (2018) 1804644.
[53] W.-J. Kim, J.-M. Lee, J.-S. Kim, C.J. Lee, Measuring high speed crack propagation in concrete fracture test using mechanoluminescent material, Smart Struct. Syst. 10 (2012) 547–555.
[54] T. Jiang, Y.F. Zhu, J.C. Zhang, J. Zhu, M. Zhang, J. Qiu, Multistimuli-responsive display materials to encrypt differentiated information in bright and dark fields, Adv. Funct. Mater. 29 (2019) 1906068.
[55] Y.Y.X.Y. Rong, H.D. Dao, C. Bin, T.Z. Xun, Origins and mechanisms of the triboluminescence, Chem. Res. Appl. (in Chinese) (2001) 13.
[56] Z. Xue, X. Li, Y. Li, M. Jiang, G. Ren, H. Liu, S. Zeng, J. Hao, A 980 nm laser-activated upconverted persistent probe for NIR-to-NIR rechargeable in vivo bioimaging, Nanoscale 9 (2017) 7276–7283.
[57] C.X.S. Kmimuar, H. Yamada, G. Marriott, K. Hyodo, T. Ohno, J. Ceram. Soc. Jpn. 7 (2017) 591–595.
[58] M. Sun, Z.-J. Li, C.-L. Liu, H.-X. Fu, J.-S. Shen, H.-W. Zhang, Persistent luminescent nanoparticles for super-long time in vivo and in situ imaging with repeatable excitation, J. Lumin. 145 (2014) 838–842.
[59] Y. Li, Y.-Y. Li, K. Sharafudeen, G.-P. Dong, S.-F. Zhou, Z.-J. Ma, M.-Y. Peng, J.-R. Qiu, A strategy for developing near infrared long-persistent phosphors: taking $MAlO_3$: Mn^{4+}, Ge^{4+}(M= La, Gd) as an example, J. Mater. Chem. C 2 (2014) 2019–2027.
[60] Y. Li, S. Zhou, Y. Li, K. Sharafudeen, Z. Ma, G. Dong, M. Peng, J. Qiu, Long persistent and photo-stimulated luminescence in Cr^{3+}-doped Zn–Ga–Sn–O phosphors for deep and reproducible tissue imaging, J. Mater. Chem. C 2 (2014) 2657–2663.
[61] Z. Li, L. Huang, Y. Zhang, Y. Zhao, H. Yang, G. Han, Near-infrared light activated persistent luminescence nanoparticles via upconversion, Nano Res. 10 (2017) 1840–1846.
[62] Z. Li, Y. Zhang, X. Wu, L. Huang, D. Li, W. Fan, G. Han, Direct aqueous-phase synthesis of sub-10 nm "luminous pearls" with enhanced in vivo renewable near-infrared persistent luminescence, J. Am. Chem. Soc. 137 (2015) 5304–5307.
[63] Y.-J. Li, X.-P. Yan, Synthesis of functionalized triple-doped zinc gallogermanate nanoparticles with superlong near-infrared persistent luminescence for long-term orally administrated bioimaging, Nanoscale 8 (2016) 14965–14970.

[64] X. Zhou, G. Ju, T. Dai, Y. Li, H. Wu, Y. Jin, Y. Hu, $Li_5Zn_8Ga_5Ge_9O_{36}$: Cr^{3+}, Ti^{4+}: a Long persistent phosphor excited in a wide spectral Region from UV to red light for reproducible imaging through biological tissue, Chem. Asian J. 14 (2019) 1506–1514.
[65] J. Xu, D. Murata, Y. Katayama, J. Ueda, S. Tanabe, Cr^{3+}/Er^{3+} co-doped $LaAlO_3$ perovskite phosphor: a near-infrared persistent luminescence probe covering the first and third biological windows, J. Mater. Chem. B 5 (2017) 6385–6393.
[66] X.-H. Lin, L. Song, S. Chen, X.-F. Chen, J.-J. Wei, J. Li, G. Huang, H.-H. Yang, Kiwifruit-like persistent luminescent nanoparticles with high-performance and in situ activable near-infrared persistent luminescence for long-term in vivo bioimaging, ACS Appl. Mater. Interfaces 9 (2017) 41181–41187.
[67] X. Lin, R. Zhang, X. Tian, Y. Li, B. Du, J. Nie, Z. Li, L. Chen, J. Ren, J. Qiu, Coordination geometry-dependent multi-band emission and atypically deep-trap-dominated NIR persistent luminescence from chromium-doped aluminates, Adv. Opt. Mater. 6 (2018) 1701161.
[68] R. Zou, J. Huang, J. Shi, L. Huang, X. Zhang, K.-L. Wong, H. Zhang, D. Jin, J. Wang, Q. Su, Silica shell-assisted synthetic route for mono-disperse persistent nanophosphors with enhanced in vivo recharged near-infrared persistent luminescence, Nano Res. 10 (2017) 2070–2082.
[69] Q. Wang, S. Zhang, Z. Li, Q. Zhu, Near infrared-emitting Cr^{3+}/Eu^{3+} co-doped zinc Gallogermanate persistence luminescent nanoparticles for cell imaging, Nanoscale Res. Lett. 13 (2018) 64.
[70] S. Tian, B. Liu, L. Zhao, Z. Liu, X. Fan, L. Yang, Q. Min, H. Zhang, X. Yu, J. Qiu, Red photo-stimulated luminescence from deep traps of $BaZrGe_3O_9$: Pr^{3+} for optical imaging application, J. Alloys Compd. 800 (2019) 224–230.
[71] J. Xu, J. Ueda, Y. Zhuang, B. Viana, S. Tanabe, $Y_3Al_{5-x}Ga_xO_{12}$: Cr^{3+}: a novel red persistent phosphor with high brightness, Appl. Phys. Express 8 (2015), 042602.
[72] Y. Lv, L. Wang, Y. Zhuang, T.-L. Zhou, R.-J. Xie, Discovery of the Yb^{2+}–Yb^{3+} couple as red-to-NIR persistent luminescence emitters in Yb-activated $(Ba_{1-x}Sr_x)$ $AlSi_5O_2N_7$ phosphors, J. Mater. Chem. C 5 (2017) 7095–7101.
[73] Y. Zhuang, J. Ueda, S. Tanabe, Tunable trap depth in Zn $(Ga_{1-x}Al_x)_2O_4$: Cr, bi red persistent phosphors: considerations of high-temperature persistent luminescence and photostimulated persistent luminescence, J. Mater. Chem. C 1 (2013) 7849–7855.
[74] S. Hu, Z. Li, Q. Luo, Q. Ma, N. Chen, L. Fu, J. Wang, R. Yang, Q. Yuan, Facile synthesis of luminous nanoparticles with tunable size and long-lived luminescence for lifetime-based biosensing, Cryst. Growth Des. 19 (2019) 2322–2328.
[75] Z. Zhou, W. Zheng, J. Kong, Y. Liu, P. Huang, S. Zhou, Z. Chen, J. Shi, X. Chen, Rechargeable and LED-activated $ZnGa_2O_4$: Cr^{3+} near-infrared persistent luminescence nanoprobes for background-free biodetection, Nanoscale 9 (2017) 6846–6853.
[76] J. Wang, Q. Ma, W. Zheng, H. Liu, C. Yin, F. Wang, X. Chen, Q. Yuan, W. Tan, One-dimensional luminous nanorods featuring tunable persistent luminescence for autofluorescence-free biosensing, ACS Nano 11 (2017) 8185–8191.
[77] B.-Y. Wu, H.-F. Wang, J.-T. Chen, X.-P. Yan, Fluorescence resonance energy transfer inhibition assay for α-fetoprotein excreted during cancer cell growth using functionalized persistent luminescence nanoparticles, J. Am. Chem. Soc. 133 (2010) 686–688.
[78] N. Li, Y. Li, Y. Han, W. Pan, T. Zhang, B. Tang, A highly selective and instantaneous nanoprobe for detection and imaging of ascorbic acid in living cells and in vivo, Anal. Chem. 86 (2014) 3924–3930.
[79] D.-E. Lee, H. Koo, I.-C. Sun, J.H. Ryu, K. Kim, I.C. Kwon, Multifunctional nanoparticles for multimodal imaging and theragnosis, Chem. Soc. Rev. 41 (2012) 2656–2672.

[80] L.H. Reddy, J.L. Arias, J. Nicolas, P. Couvreur, Magnetic nanoparticles: design and characterization, toxicity and biocompatibility, pharmaceutical and biomedical applications, Chem. Rev. 112 (2012) 5818–5878.

[81] T. Maldiney, B.T. Doan, D. Alloyeau, M. Bessodes, D. Scherman, C. Richard, Gadolinium-doped persistent nanophosphors as versatile tool for multimodal in vivo imaging, Adv. Funct. Mater. 25 (2015) 331–338.

[82] W. Dai, Y. Lei, S. Ye, E. Song, Z. Chen, Q. Zhang, Mesoporous nanoparticles Gd_2O_3@ $mSiO_2/ZnGa_2O_4$: Cr^{3+}, Bi^{3+} as multifunctional probes for bioimaging, J. Mater. Chem. B 4 (2016) 1842–1852.

[83] A. Abdukayum, C.-X. Yang, Q. Zhao, J.-T. Chen, L.-X. Dong, X.-P. Yan, Gadolinium complexes functionalized persistent luminescent nanoparticles as a multimodal probe for near-infrared luminescence and magnetic resonance imaging in vivo, Anal. Chem. 86 (2014) 4096–4101.

[84] Y.-C. Lu, C.-X. Yang, X.-P. Yan, Radiopaque tantalum oxide coated persistent luminescent nanoparticles as multimodal probes for in vivo near-infrared luminescence and computed tomography bioimaging, Nanoscale 7 (2015) 17929–17937.

[85] N. Lee, H.R. Cho, M.H. Oh, S.H. Lee, K. Kim, B.H. Kim, K. Shin, T.-Y. Ahn, J.W. Choi, Y.-W. Kim, Multifunctional Fe_3O_4/TaO_x Core/Shell nanoparticles for simultaneous magnetic resonance imaging and X-ray computed tomography, J. Am. Chem. Soc. 134 (2012) 10309–10312.

[86] J.-M. Liu, D.-D. Zhang, G.-Z. Fang, S. Wang, Erythrocyte membrane bioinspired near-infrared persistent luminescence nanocarriers for in vivo long-circulating bioimaging and drug delivery, Biomaterials 165 (2018) 39–47.

[87] J. Shi, M. Sun, X. Sun, H. Zhang, Near-infrared persistent luminescence hollow mesoporous nanospheres for drug delivery and in vivo renewable imaging, J. Mater. Chem. B 4 (2016) 7845–7851.

[88] P. Ge, Y. Liang, Y. Cheng, K. Sun, R. Liu, A new persistent luminescent composite for tracing toxic air particulate matter, Environ. Chem. Lett. 16 (2018) 1487–1492.

[89] H. Chen, B. Zheng, C. Liang, L. Zhao, Y. Zhang, H. Pan, W. Ji, X. Gong, H. Wang, J. Chang, Near-infrared persistent luminescence phosphors $ZnGa_2O_4$:Cr^{3+} as an accurately tracker to photothermal therapy in vivo for visual treatment, Mater. Sci. Eng. C 79 (2017) 372–381.

[90] M.L. Boisen, V.H.F. Consortium, D. Oottamasathien, V.H.F. Consortium, A.B. Jones, V. H.F. Consortium, M.M. Millett, V.H.F. Consortium, D.S. Nelson, V.H.F. Consortium, Development of prototype filovirus recombinant antigen immunoassays, J. Infect. Dis. 212 (2015) S359–S367.

[91] K. Nielsen, W. Yu, L. Kelly, R. Bermudez, T. Renteria, A. Dajer, E. Gutierrez, J. Williams, J. Algire, S. Torioni de Eschaide, Development of a lateral flow assay for rapid detection of bovine antibody to Anaplasma marginale, J. Immunoass. Immunochem. 29 (2007) 10–18.

[92] B.A. Rohrman, V. Leautaud, E. Molyneux, R.R. Richards-Kortum, A lateral flow assay for quantitative detection of amplified HIV-1 RNA, PLoS One 7 (2012), e45611.

[93] X. Xiao, Z.-C. Wu, K.-C. Chou, A multi-label classifier for predicting the subcellular localization of gram-negative bacterial proteins with both single and multiple sites, PLoS One 6 (2011), e20592.

[94] L. Lee, E. Nordman, M. Johnson, M. Oldham, A low-cost, high-performance system for fluorescence lateral flow assays, Biosensors 3 (2013) 360–373.

[95] K.M. Koczula, A. Gallotta, Lateral flow assays, Essays Biochem. 60 (2016) 111–120.

[96] A.S. Paterson, B. Raja, G. Garvey, A. Kolhatkar, A.E. Hagström, K. Kourentzi, T.R. Lee, R.C. Willson, Persistent luminescence strontium aluminate nanoparticles as reporters in lateral flow assays, Anal. Chem. 86 (2014) 9481–9488.

[97] T.A. Khattab, M.M. Fouda, M.S. Abdelrahman, S.I. Othman, M. Bin-Jumah, M.A. Alqaraawi, H. Al Fassam, A.A. Allam, Development of illuminant glow-in-the-dark cotton fabric coated by luminescent composite with antimicrobial activity and ultra-violet protection, J. Fluoresc. 29 (2019) 703–710.
[98] Y.-M. Yang, Z.-Y. Li, J.-Y. Zhang, Y. Lu, S.-Q. Guo, Q. Zhao, X. Wang, Z.-J. Yong, H. Li, J.-P. Ma, X-ray-activated long persistent phosphors featuring strong UVC after-glow emissions, Light: Sci. Appl. 7 (2018) 88.
[99] H. Sun, L. Pan, X. Piao, Z. Sun, Enhanced performance of cadmium selenide quantum dot-sensitized solar cells by incorporating long afterglow europium, dysprosium co-doped strontium aluminate phosphors, J. Colloid Interface Sci. 416 (2014) 81–85.
[100] A. Puntambekar, V. Chakrapani, Excitation energy transfer from long-persistent phosphors for enhancing power conversion of dye-sensitized solar cells, Phys. Rev. B 93 (2016) 245301.
[101] W. He, A.S. Timur, H.-K. Kim, Y.-H. Hwang, Energy harvesting of dye-sensitized solar cells assisted with Ti-mesh and phosphor materials, in: IOP Conference Series: Materials Science and Engineering, 2014, p. 012025.
[102] J. Zhang, J. Lin, J. Wu, S. Zhang, P. Zhou, X. Chen, R. Xu, Preparation of long persistent phosphor $SrAl_2O_4$: Eu^{2+}, Dy^{3+} and its application in dye-sensitized solar cells, J. Mater. Sci. Mater. Electron. 27 (2016) 1350–1356.
[103] C. Wang, T. Xuan, J. Liu, H. Li, Z. Sun, Long afterglow $SrAl_2O_4$: Eu^{2+}, Dy^{3+} phosphors as luminescent Down-shifting layer for crystalline silicon solar cells, Int. J. Appl. Ceram. Technol. 12 (2015) 722–727.
[104] W. Dai, Y. Lei, J. Zhou, M. Xu, L. Chu, L. Li, P. Zhao, Z. Zhang, Near-infrared quantum-cutting and long-persistent phosphor $Ca_3Ga_2Ge_3O_{12}$: Pr^{3+}, Yb^{3+} for application in in vivo bioimaging and dye-sensitized solar cells, J. Alloys Compd. 726 (2017) 230–239.
[105] Q. Tang, J. Wang, B. He, P. Yang, Can dye-sensitized solar cells generate electricity in the dark? Nano Energy 33 (2017) 266–271.
[106] C. Chen, H. Li, J. Jin, X. Chen, Y. Cheng, Y. Zheng, D. Liu, L. Xu, H. Song, Q. Dai, Long-lasting nanophosphors applied to UV-resistant and energy storage perovskite solar cells, Adv. Energy Mater. 7 (2017) 1700758.
[107] X. Luo, J.Y. Ahn, S.H. Kim, Aerosol synthesis and luminescent properties of $CaAl_2O_4$: Eu^{2+}, Nd^{3+} down-conversion phosphor particles for enhanced light harvesting of dye-sensitized solar cells, Sol. Energy 178 (2019) 173–180.
[108] H. Sun, L. Pan, G. Zhu, X. Piao, L. Zhang, Z. Sun, Long afterglow $Sr_4Al_{14}O_{25}$: Eu, Dy phosphors as both scattering and down converting layer for CdS quantum dot-sensitized solar cells, Dalton Trans. 43 (2014) 14936–14941.
[109] H. Wu, Z.-M. Wang, K. Koike, N. Negishi, Y. Jin, Hybridization of silver orthophosphate with a melilite-type phosphor for enhanced energy-harvesting photocatalysis, Cat. Sci. Technol. 7 (2017) 3736–3746.
[110] H. Li, S. Yin, Y. Wang, T. Sato, Persistent fluorescence-assisted $TiO_{2-x}N_y$-based photocatalyst for gaseous acetaldehyde degradation, Environ. Sci. Technol. 46 (2012) 7741–7745.
[111] J.B. Zhong, D. Ma, X.Y. He, J.Z. Li, Y.Q. Chen, Sol–gel preparation and photocatalytic performance of $TiO_2/SrAl_2O_4$: Eu^{2+}, Dy^{3+} toward the oxidation of gaseous benzene, J. Sol-Gel Sci. Technol. 52 (2009) 140–145.
[112] J. Zhong, J. Wang, L. Tao, M. Gong, L. Zhimin, Y. Chen, Photocatalytic degradation of gaseous benzene over TiO_2/Sr_2CeO_4: preparation and photocatalytic behavior of TiO_2/Sr_2CeO_4, J. Hazard. Mater. 140 (2007) 200–204.

[113] J. Zhang, F. Pan, W. Hao, Q. Ge, T. Wang, Light-storing photocatalyst, Appl. Phys. Lett. 85 (2004) 5778–5780.
[114] Y.L. Liu, Z.X. Sun, L.H. Huang, W.X. Chen, Study of preparation and photocatalytic properties of new type complex material $CaAl_2O_4$: Eu^{2+}, Dy^{3+} coated by TiO_2, Huaxue Jiancai (in Chinese) 6 (2004).
[115] J.H. Ko, S. Ko, Process for Providing a Titanium Dioxide Layer on a Material that Contains a Light Absorbing Substance and the Product So Formed, Google Patents, 2003.
[116] D. Li, Y. Wang, K. Xu, L. Li, Z. Hu, Persistently luminescent and photocatalytic properties of $ZnGa_2O_4$ phosphors, Mater. Res. Exp. 2 (2015), 046202.
[117] D. Li, Y. Wang, K. Xu, H. Zhao, Z. Hu, Persistent luminescent and photocatalytic properties of $Zn_xGa_2O_{3+x}$ ($0.8 \leq x \leq 1$) phosphors, RSC Adv. 5 (2015) 20972–20975.
[118] C. García, L. Diaz-Torres, P. Salas, M. Guzman, C. Angeles-Chavez, Photoluminescent and photocatalytic properties of bismuth doped strontium aluminates blended with titanium dioxide, Mater. Sci. Semicond. Process. 37 (2015) 105–111.
[119] L. Li, Y.-h. Wang, H. Li, H.-j. Huang, H. Zhao, Suppression of photocatalysis and long-lasting luminescence in $ZnGa_2O_4$ by Cr^{3+} doping, RSC Adv. 5 (2015) 57193–57200.
[120] L. Li, K. Xu, Y. Wang, Z. Hu, H. Zhao, Enhanced persistent luminescence and photocatalytic properties of Ga_2O_3: Cr^{3+} by In^{3+} doping, Opt. Mater. Express 6 (2016) 1122–1130.
[121] Q. Zhou, F. Peng, Y. Ni, J. Kou, C. Lu, Z. Xu, Long afterglow phosphor driven round-the-clock g-C_3N_4 photocatalyst, J. Photochem. Photobiol. A Chem. 328 (2016) 182–188.
[122] H. Wu, W. Peng, Z.-M. Wang, K. Koike, Cerium-doped gehlenite supporting silver/silver chloride for continuous photocatalysis, RSC Adv. 6 (2016) 37995–38003.
[123] H. Li, Y. Wang, L. Li, H. Huang, H. Zhao, Z. Hu, Enhanced photocatalytic activity and persistent luminescence in Zn_2GeO_4: Mn^{2+} by Eu^{3+} doping, Mod. Phys. Lett. B 30 (2016) 1650305.
[124] P. Feng, Y. Wei, Y. Wang, J. Zhang, H. Li, Z. Ci, Long persistent phosphor $CdSiO_3$: Gd^{3+}, Bi^{3+} and potential photocatalytic application of $CdSiO_3$: Gd^{3+}, Bi^{3+}@ TiO_2 in dark, J. Am. Ceram. Soc. 99 (2016) 2368–2375.
[125] P. Feng, J. Zhao, J. Zhang, Y. Wei, Y. Wang, H. Li, Y. Wang, Long persistent photocatalysis of magnesium gallate nanorodes, J. Alloys Compd. 695 (2017) 1884–1890.
[126] V. Havasi, B. Vödrédi, Á. Kukovecz, Photocatalytic performance of Sr4Al14O25: Eu, Dy phosphor assisted ZnO: co^+ Ag nanocomposite under continuous and pulsed illumination, Catal. Today 284 (2017) 107–113.
[127] J.-M. Fan, Z.-H. Zhao, C. Gong, Y.-Q. Xue, S. Yin, Persistent methyl orange degradation ability of MgAl2O4:(Pr3+, Dy3+)/M-TiO2 luminescent photocatalyst, J. Nanosci. Nanotechnol. 18 (2018) 1675–1681.
[128] G. Cui, X. Yang, Y. Zhang, Y. Fan, P. Chen, H. Cui, Y. Liu, X. Shi, Q. Shang, B. Tang, Round-the-clock photocatalytic hydrogen production with high efficiency by a long-afterglow material, Angew. Chem. 131 (2019) 1354–1358.
[129] A. Tuerdi, A. Abdukayum, Dual-functional persistent luminescent nanoparticles with enhanced persistent luminescence and photocatalytic activity, RSC Adv. 9 (2019) 17653–17657.
[130] V. Vaiano, O. Sacco, D. Sannino, Electric energy saving in photocatalytic removal of crystal violet dye through the simultaneous use of long-persistent blue phosphors, nitrogen-doped TiO_2 and UV-light emitting diodes, J. Clean. Prod. 210 (2019) 1015–1021.

Index

Note: Page numbers followed by *f* indicate figures and *t* indicate tables.

Kolibri 360 GmbH
Ferdinand-Jühlke-Straße 7
99095 Erfurt, Deutschland
produktsicherheit@kolibri360.de